Praise for *Designing Machine Learning Systems*

There is so much information one needs to know to be an effective machine learning engineer. It's hard to cut through the chaff to get the most relevant information, but Chip has done that admirably with this book. If you are serious about ML in production, and care about how to design and implement ML systems end to end, this book is essential.

—*Laurence Moroney, AI and ML Lead, Google*

One of the best resources that focuses on the first principles behind designing ML systems for production. A must-read to navigate the ephemeral landscape of tooling and platform options.

—*Goku Mohandas, Founder of Made With ML*

Chip's manual is the book we deserve and the one we need right now. In a blooming but chaotic ecosystem, this principled view on end-to-end ML is both your map and your compass: a must-read for practitioners inside and outside of Big Tech—especially those working at "reasonable scale." This book will also appeal to data leaders looking for best practices on how to deploy, manage, and monitor systems in the wild.

—*Jacopo Tagliabue, Director of AI, Coveo; Adj. Professor of MLSys, NYU*

This is, simply, the very best book you can read about how to build, deploy, and scale machine learning models at a company for maximum impact. Chip is a masterful teacher, and the breadth and depth of her knowledge is unparalleled.

—*Josh Wills, Software Engineer at WeaveGrid and former Director of Data Engineering, Slack*

This is the book I wish I had read when I started as an ML engineer.

—*Shreya Shankar, MLOps PhD Student*

Designing Machine Learning Systems is a welcome addition to the field of applied machine learning. The book provides a detailed guide for people building end-to-end machine learning systems. Chip Huyen writes from her extensive, hands-on experience building real-world machine learning applications.

—Brian Spiering, Data Science Instructor at Metis

Chip is truly a world-class expert on machine learning systems, as well as a brilliant writer. Both are evident in this book, which is a fantastic resource for anyone looking to learn about this topic.

—Andrey Kurenkov, PhD Candidate at the Stanford AI Lab

Chip Huyen has produced an important addition to the canon of machine learning literature—one that is deeply literate in ML fundamentals, but has a much more concrete and practical approach than most. The focus on business requirements alone is uncommon and valuable. This book will resonate with engineers getting started with ML and with others in any part of the organization trying to understand how ML works.

—Todd Underwood, Senior Engineering Director for ML SRE, Google, and Coauthor of Reliable Machine Learning

设计机器学习系统（影印版）

Designing Machine Learning Systems

Chip Huyen 著

Beijing • Boston • Farnham • Sebastopol • Tokyo

O'Reilly Media, Inc.授权东南大学出版社出版

南京 东南大学出版社

图书在版编目(CIP)数据

设计机器学习系统：影印版：英文 / （越）奇普·胡岩（Chip Huyen）著．—南京：东南大学出版社，2022.10

书名原文：Designing Machine Learning Systems

ISBN 978-7-5766-0224-1

Ⅰ.①设… Ⅱ.①奇… Ⅲ.①机器学习-英文 Ⅳ.①TP181

中国版本图书馆 CIP 数据核字(2022)第 152339 号

图字：10-2022-339 号

设计机器学习系统（影印版）

著　　者：Chip Huyen
责任编辑：张　烨　　封面设计：Karen Montgomery，张　健　　责任印制：周荣虎
出版发行：东南大学出版社
社　　址：南京四牌楼 2 号　　邮编：210096　　电话：025-83793330
网　　址：http://www.seupress.com
电子邮件：press@ seupress.com
经　　销：全国各地新华书店
印　　刷：常州市武进第三印刷印刷有限公司
开　　本：787mm×980mm　1/16
印　　张：24.25
字　　数：475 千
版　　次：2022 年 10 月第 1 版
版　　次：2022 年 10 月第 1 次印刷
书　　号：ISBN 978-7-5766-0224-1
定　　价：138.00 元

本社图书若有印装质量问题，请直接与营销部联系。电话（传真）：025-83791830

Table of Contents

Preface

Ever since the first machine learning course I taught at Stanford in 2017, many people have asked me for advice on how to deploy ML models at their organizations. These questions can be generic, such as "What model should I use?" "How often should I retrain my model?" "How can I detect data distribution shifts?" "How do I ensure that the features used during training are consistent with the features used during inference?"

These questions can also be specific, such as "I'm convinced that switching from batch prediction to online prediction will give our model a performance boost, but how do I convince my manager to let me do so?" or "I'm the most senior data scientist at my company and I've recently been tasked with setting up our first machine learning platform; where do I start?"

My short answer to all these questions is always: "It depends." My long answers often involve hours of discussion to understand where the questioner comes from, what they're actually trying to achieve, and the pros and cons of different approaches for their specific use case.

ML systems are both complex and unique. They are complex because they consist of many different components (ML algorithms, data, business logic, evaluation metrics, underlying infrastructure, etc.) and involve many different stakeholders (data scientists, ML engineers, business leaders, users, even society at large). ML systems are unique because they are data dependent, and data varies wildly from one use case to the next.

For example, two companies might be in the same domain (ecommerce) and have the same problem that they want ML to solve (recommender system), but their resulting ML systems can have different model architecture, use different sets of features, be evaluated on different metrics, and bring different returns on investment.

Many blog posts and tutorials on ML production focus on answering one specific question. While the focus helps get the point across, they can create the impression that it's possible to consider each of these questions in isolation. In reality, changes in one component will likely affect other components. Therefore, it's necessary to consider the system as a whole while attempting to make any design decision.

This book takes a holistic approach to ML systems. It takes into account different components of the system and the objectives of different stakeholders involved. The content in this book is illustrated using actual case studies, many of which I've personally worked on, backed by ample references, and reviewed by ML practitioners in both academia and industry. Sections that require in-depth knowledge of a certain topic—e.g., batch processing versus stream processing, infrastructure for storage and compute, and responsible AI—are further reviewed by experts whose work focuses on that one topic. In other words, this book is an attempt to give nuanced answers to the aforementioned questions and more.

When I first wrote the lecture notes that laid the foundation for this book, I thought I wrote them for my students to prepare them for the demands of their future jobs as data scientists and ML engineers. However, I soon realized that I also learned tremendously through the process. The initial drafts I shared with early readers sparked many conversations that tested my assumptions, forced me to consider different perspectives, and introduced me to new problems and new approaches.

I hope that this learning process will continue for me now that the book is in your hand, as you have experiences and perspectives that are unique to you. Please feel free to share with me any feedback you might have for this book, via the MLOps Discord server (*https://discord.gg/Mw77HPrgjF*) that I run (where you can also find other readers of this book), Twitter (*https://twitter.com/chipro*), LinkedIn (*https://www.linkedin.com/in/chiphuyen*), or other channels that you can find on my website (*https://huyenchip.com*).

Who This Book Is For

This book is for anyone who wants to leverage ML to solve real-world problems. ML in this book refers to both deep learning and classical algorithms, with a leaning toward ML systems at scale, such as those seen at medium to large enterprises and fast-growing startups. Systems at a smaller scale tend to be less complex and might benefit less from the comprehensive approach laid out in this book.

Because my background is engineering, the language of this book is geared toward engineers, including ML engineers, data scientists, data engineers, ML platform engineers, and engineering managers. You might be able to relate to one of the following scenarios:

- You have been given a business problem and a lot of raw data. You want to engineer this data and choose the right metrics to solve this problem.
- Your initial models perform well in offline experiments and you want to deploy them.
- You have little feedback on how your models are performing after your models are deployed, and you want to figure out a way to quickly detect, debug, and address any issue your models might run into in production.
- The process of developing, evaluating, deploying, and updating models for your team has been mostly manual, slow, and error-prone. You want to automate and improve this process.
- Each ML use case in your organization has been deployed using its own workflow, and you want to lay down the foundation (e.g., model store, feature store, monitoring tools) that can be shared and reused across use cases.
- You're worried that there might be biases in your ML systems and you want to make your systems responsible!

You can also benefit from the book if you belong to one of the following groups:

- Tool developers who want to identify underserved areas in ML production and figure out how to position your tools in the ecosystem.
- Individuals looking for ML-related roles in the industry.
- Technical and business leaders who are considering adopting ML solutions to improve your products and/or business processes. Readers without strong technical backgrounds might benefit the most from Chapters 1, 2, and 11.

What This Book Is Not

This book is not an introduction to ML. There are many books, courses, and resources available for ML theories, and therefore, this book shies away from these concepts to focus on the practical aspects of ML. To be specific, the book assumes that readers have a basic understanding of the following topics:

- *ML models* such as clustering, logistic regression, decision trees, collaborative filtering, and various neural network architectures including feed-forward, recurrent, convolutional, and transformer
- *ML techniques* such as supervised versus unsupervised, gradient descent, objective/loss function, regularization, generalization, and hyperparameter tuning
- *Metrics* such as accuracy, F1, precision, recall, ROC, mean squared error, and log-likelihood

- *Statistical concepts* such as variance, probability, and normal/long-tail distribution
- *Common ML tasks* such as language modeling, anomaly detection, object classification, and machine translation

You don't have to know these topics inside out—for concepts whose exact definitions can take some effort to remember, e.g., F1 score, we include short notes as references—but you should have a rough sense of what they mean going in.

While this book mentions current tools to illustrate certain concepts and solutions, it's not a tutorial book. Technologies evolve over time. Tools go in and out of style quickly, but fundamental approaches to problem solving should last a bit longer. This book provides a framework for you to evaluate the tool that works best for your use cases. When there's a tool you want to use, it's usually straightforward to find tutorials for it online. As a result, this book has few code snippets and instead focuses on providing a lot of discussion around trade-offs, pros and cons, and concrete examples.

Navigating This Book

The chapters in this book are organized to reflect the problems data scientists might encounter as they progress through the lifecycle of an ML project. The first two chapters lay down the groundwork to set an ML project up for success, starting from the most basic question: does your project need ML? It also covers choosing the objectives for your project and how to frame your problem in a way that makes for simpler solutions. If you're already familiar with these considerations and impatient to get to the technical solutions, feel free to skip the first two chapters.

Chapters 4 to 6 cover the pre-deployment phase of an ML project: from creating the training data and engineering features to developing and evaluating your models in a development environment. This is the phase where expertise in both ML and the problem domain are especially needed.

Chapters 7 to 9 cover the deployment and post-deployment phase of an ML project. We'll learn through a story many readers might be able to relate to that having a model deployed isn't the end of the deployment process. The deployed model will need to be monitored and continually updated to changing environments and business requirements.

Chapters 3 and 10 focus on the infrastructure needed to enable stakeholders from different backgrounds to work together to deliver successful ML systems. Chapter 3 focuses on data systems, whereas Chapter 10 focuses on compute infrastructure and ML platforms. I debated for a long time on how deep to go into data systems and where to introduce it in the book. Data systems, including databases, data formats, data movements, and data processing engines, tend to be sparsely covered in ML coursework, and therefore many data scientists might think of them as low level or irrelevant. After consulting with many of my colleagues, I decided that because ML systems depend on data, covering the basics of data systems early will help us get on the same page to discuss data matters in the rest of the book.

While we cover many technical aspects of an ML system in this book, ML systems are built by people, for people, and can have outsized impact on the life of many. It'd be remiss to write a book on ML production without a chapter on the human side of it, which is the focus of Chapter 11, the last chapter.

Note that "data scientist" is a role that has evolved a lot in the last few years, and there have been many discussions to determine what this role should entail—we'll go into some of these discussions in Chapter 10. In this book, we use "data scientist" as an umbrella term to include anyone who works developing and deploying ML models, including people whose job titles might be ML engineers, data engineers, data analysts, etc.

GitHub Repository and Community

This book is accompanied by a GitHub repository (*https://oreil.ly/designing-machine-learning-systems-code*) that contains:

- A review of basic ML concepts
- A list of references used in this book and other advanced, updated resources
- Code snippets used in this book
- A list of tools you can use for certain problems you might encounter in your workflows

I also run a Discord server on MLOps (*https://discord.gg/Mw77HPrgjF*) where you're encouraged to discuss and ask questions about the book.

Conventions Used in This Book

The following typographical conventions are used in this book:

Italic
: Indicates new terms, URLs, email addresses, filenames, and file extensions.

`Constant width`
: Used for program listings, as well as within paragraphs to refer to program elements such as variable or function names, databases, data types, environment variables, statements, and keywords.

This element signifies a general note.

This element indicates a warning or caution.

Using Code Examples

As mentioned, supplemental material (code examples, exercises, etc.) is available for download at *https://oreil.ly/designing-machine-learning-systems-code*.

If you have a technical question or a problem using the code examples, please send email to *bookquestions@oreilly.com*.

This book is here to help you get your job done. In general, if example code is offered with this book, you may use it in your programs and documentation. You do not need to contact us for permission unless you're reproducing a significant portion of the code. For example, writing a program that uses several chunks of code from this book does not require permission. Selling or distributing examples from O'Reilly books does require permission. Answering a question by citing this book and quoting example code does not require permission. Incorporating a significant amount of example code from this book into your product's documentation does require permission.

We appreciate, but generally do not require, attribution. An attribution usually includes the title, author, publisher, and ISBN. For example: "*Designing Machine Learning Systems* by Chip Huyen (O'Reilly). Copyright 2022 Huyen Thi Khanh Nguyen, 978-1-098-10796-3."

If you feel your use of code examples falls outside fair use or the permission given above, feel free to contact us at *permissions@oreilly.com*.

O'Reilly Online Learning

For more than 40 years, *O'Reilly Media* has provided technology and business training, knowledge, and insight to help companies succeed.

Our unique network of experts and innovators share their knowledge and expertise through books, articles, and our online learning platform. O'Reilly's online learning platform gives you on-demand access to live training courses, in-depth learning paths, interactive coding environments, and a vast collection of text and video from O'Reilly and 200+ other publishers. For more information, visit *https://oreilly.com*.

How to Contact Us

Please address comments and questions concerning this book to the publisher:

O'Reilly Media, Inc.
1005 Gravenstein Highway North
Sebastopol, CA 95472
800-998-9938 (in the United States or Canada)
707-829-0515 (international or local)
707-829-0104 (fax)

We have a web page for this book, where we list errata, examples, and any additional information. You can access this page at *https://oreil.ly/designing-machine-learning-systems*.

Email *bookquestions@oreilly.com* to comment or ask technical questions about this book.

For news and information about our books and courses, visit *https://oreilly.com*.

Find us on LinkedIn: *https://linkedin.com/company/oreilly-media*

Follow us on Twitter: *https://twitter.com/oreillymedia*

Watch us on YouTube: *https://youtube.com/oreillymedia*

Acknowledgments

This book took two years to write, and many more years beforehand to prepare. Looking back, I'm equally amazed and grateful for the enormous amount of help I received in writing this book. I tried my best to include the names of everyone who has helped me here, but due to the inherent faultiness of human memory, I undoubtedly neglected to mention many. If I forgot to include your name, please know that it wasn't because I don't appreciate your contribution and please kindly remind me so that I can rectify as soon as possible!

First and foremost, I'd like to thank the course staff who helped me develop the course and materials this book was based on: Michael Cooper, Xi Yin, Chloe He, Kinbert Chou, Megan Leszczynski, Karan Goel, and Michele Catasta. I'd like to thank my professors, Christopher Ré and Mehran Sahami, without whom the course wouldn't exist in the first place.

I'd like to thank a long list of reviewers who not only gave encouragement but also improved the book by many orders of magnitude: Eugene Yan, Josh Wills, Han-chung Lee, Thomas Dietterich, Irene Tematelewo, Goku Mohandas, Jacopo Tagliabue, Andrey Kurenkov, Zach Nussbaum, Jay Chia, Laurens Geffert, Brian Spiering, Erin Ledell, Rosanne Liu, Chin Ling, Shreya Shankar, and Sara Hooker.

I'd like to thank all the readers who read the early release version of the book and gave me ideas on how to improve the book, including Charles Frye, Xintong Yu, Jordan Zhang, Jonathon Belotti, and Cynthia Yu.

Of course, the book wouldn't have been possible with the team at O'Reilly, especially my development editor, Jill Leonard, and my production editors, Kristen Brown, Sharon Tripp, and Gregory Hyman. I'd like to thank Laurence Moroney, Hannes Hapke, and Rebecca Novack, who helped me get this book from an idea to a proposal.

This book, after all, is an accumulation of invaluable lessons I learned throughout my career to date. I owe these lessons to my extremely competent and patient coworkers and former coworkers at Claypot AI, Primer AI, Netflix, NVIDIA, and Snorkel AI. Every person I've worked with has taught me something new about bringing ML into the world.

A special thanks to my cofounder Zhenzhong Xu for putting out the fires at our startup and allowing me to spend time on this book. Thank you, Luke, for always being so supportive of everything that I want to do, no matter how ambitious it is.

CHAPTER 1

Overview of Machine Learning Systems

In November 2016, Google announced that it had incorporated its multilingual neural machine translation system into Google Translate, marking one of the first success stories of deep artificial neural networks in production at scale.[1] According to Google, with this update, the quality of translation improved more in a single leap than they had seen in the previous 10 years combined.

This success of deep learning renewed the interest in machine learning (ML) at large. Since then, more and more companies have turned toward ML for solutions to their most challenging problems. In just five years, ML has found its way into almost every aspect of our lives: how we access information, how we communicate, how we work, how we find love. The spread of ML has been so rapid that it's already hard to imagine life without it. Yet there are still many more use cases for ML waiting to be explored in fields such as health care, transportation, farming, and even in helping us understand the universe.[2]

Many people, when they hear "machine learning system," think of just the ML algorithms being used such as logistic regression or different types of neural networks. However, the algorithm is only a small part of an ML system in production. The system also includes the business requirements that gave birth to the ML project in the first place, the interface where users and developers interact with your system, the data stack, and the logic for developing, monitoring, and updating your models, as well as the infrastructure that enables the delivery of that logic. Figure 1-1 shows you the different components of an ML system and in which chapters of this book they will be covered.

1 Mike Schuster, Melvin Johnson, and Nikhil Thorat, "Zero-Shot Translation with Google's Multilingual Neural Machine Translation System," *Google AI Blog*, November 22, 2016, *https://oreil.ly/2R1CB*.

2 Larry Hardesty, "A Method to Image Black Holes," *MIT News*, June 6, 2016, *https://oreil.ly/HpL2F*.

The Relationship Between MLOps and ML Systems Design

Ops in MLOps comes from DevOps, short for Developments and Operations. To operationalize something means to bring it into production, which includes deploying, monitoring, and maintaining it. MLOps is a set of tools and best practices for bringing ML into production.

ML systems design takes a system approach to MLOps, which means that it considers an ML system holistically to ensure that all the components and their stakeholders can work together to satisfy the specified objectives and requirements.

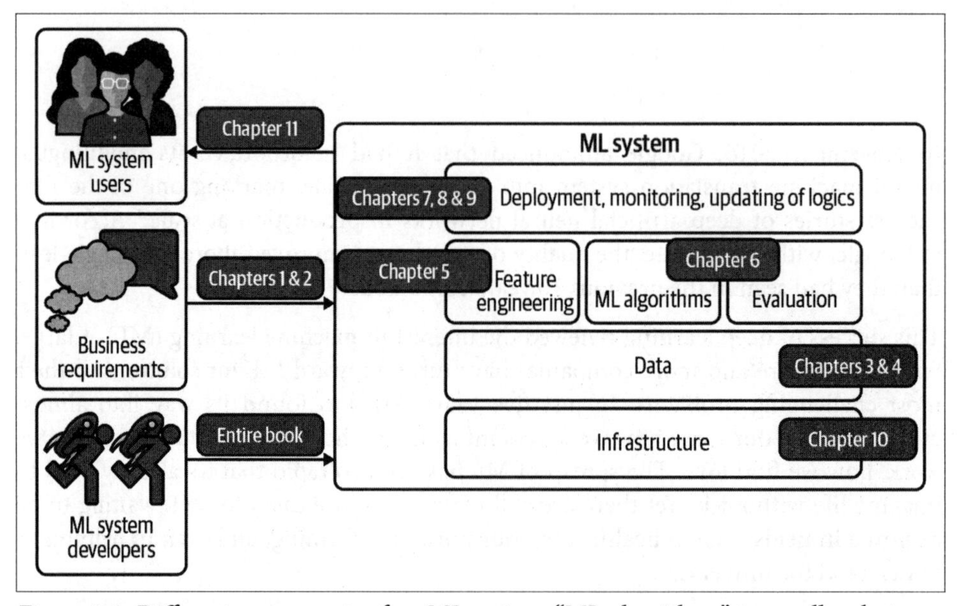

Figure 1-1. Different components of an ML system. "ML algorithms" is usually what people think of when they say machine learning, but it's only a small part of the entire system.

There are many excellent books about various ML algorithms. This book doesn't cover any specific algorithms in detail but rather helps readers understand the entire ML system as a whole. In other words, this book's goal is to provide you with a framework to develop a solution that best works for your problem, regardless of which algorithm you might end up using. Algorithms might become outdated quickly as new algorithms are constantly being developed, but the framework proposed in this book should still work with new algorithms.

The first chapter of the book aims to give you an overview of what it takes to bring an ML model to production. Before discussing how to develop an ML system, it's important to ask a fundamental question of when and when not to use ML. We'll cover some of the popular use cases of ML to illustrate this point.

After the use cases, we'll move on to the challenges of deploying ML systems, and we'll do so by comparing ML in production to ML in research as well as to traditional software. If you've been in the trenches of developing applied ML systems, you might already be familiar with what's written in this chapter. However, if you have only had experience with ML in an academic setting, this chapter will give an honest view of ML in the real world and set your first application up for success.

When to Use Machine Learning

As its adoption in the industry quickly grows, ML has proven to be a powerful tool for a wide range of problems. Despite an incredible amount of excitement and hype generated by people both inside and outside the field, ML is not a magic tool that can solve all problems. Even for problems that ML can solve, ML solutions might not be the optimal solutions. Before starting an ML project, you might want to ask whether ML is necessary or cost-effective.[3]

To understand what ML can do, let's examine what ML solutions generally do:

> Machine learning is an approach to (1) *learn* (2) *complex patterns* from (3) *existing data* and use these patterns to make (4) *predictions* on (5) *unseen data*.

We'll look at each of the italicized keyphrases in the above framing to understand its implications to the problems ML can solve:

1. *Learn: the system has the capacity to learn*

    A relational database isn't an ML system because it doesn't have the capacity to learn. You can explicitly state the relationship between two columns in a relational database, but it's unlikely to have the capacity to figure out the relationship between these two columns by itself.

    For an ML system to learn, there must be something for it to learn from. In most cases, ML systems learn from data. In supervised learning, based on example input and output pairs, ML systems learn how to generate outputs for arbitrary inputs. For example, if you want to build an ML system to learn to predict the rental price for Airbnb listings, you need to provide a dataset where each input is a listing with relevant characteristics (square footage, number of rooms, neighborhood, amenities, rating of that listing, etc.) and the associated output is the rental price of that listing. Once learned, this ML system should be able to predict the price of a new listing given its characteristics.

3 I didn't ask whether ML is sufficient because the answer is always no.

2. Complex patterns: there are patterns to learn, and they are complex

ML solutions are only useful when there are patterns to learn. Sane people don't invest money into building an ML system to predict the next outcome of a fair die because there's no pattern in how these outcomes are generated.[4] However, there are patterns in how stocks are priced, and therefore companies have invested billions of dollars in building ML systems to learn those patterns.

Whether a pattern exists might not be obvious, or if patterns exist, your dataset or ML algorithms might not be sufficient to capture them. For example, there might be a pattern in how Elon Musk's tweets affect cryptocurrency prices. However, you wouldn't know until you've rigorously trained and evaluated your ML models on his tweets. Even if all your models fail to make reasonable predictions of cryptocurrency prices, it doesn't mean there's no pattern.

Consider a website like Airbnb with a lot of house listings; each listing comes with a zip code. If you want to sort listings into the states they are located in, you wouldn't need an ML system. Since the pattern is simple—each zip code corresponds to a known state—you can just use a lookup table.

The relationship between a rental price and all its characteristics follows a much more complex pattern, which would be very challenging to manually specify. ML is a good solution for this. Instead of telling your system how to calculate the price from a list of characteristics, you can provide prices and characteristics, and let your ML system figure out the pattern. The difference between ML solutions and the lookup table solution as well as general traditional software solutions is shown in Figure 1-2. For this reason, ML is also called Software 2.0.[5]

ML has been very successful with tasks with complex patterns such as object detection and speech recognition. What is complex to machines is different from what is complex to humans. Many tasks that are hard for humans to do are easy for machines—for example, raising a number of the power of 10. On the other hand, many tasks that are easy for humans can be hard for machines—for example, deciding whether there's a cat in a picture.

---

4 Patterns are different from distributions. We know the distribution of the outcomes of a fair die, but there are no patterns in the way the outcomes are generated.

5 Andrej Karpathy, "Software 2.0," *Medium*, November 11, 2017, *https://oreil.ly/yHZrE*.

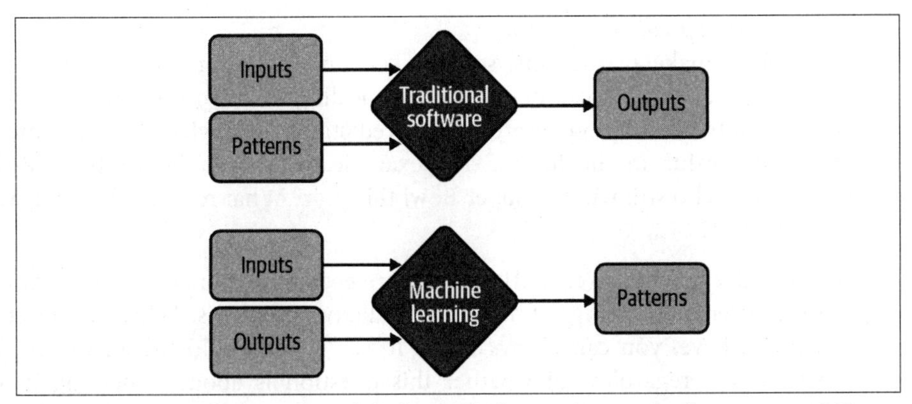

Figure 1-2. Instead of requiring hand-specified patterns to calculate outputs, ML solutions learn patterns from inputs and outputs

3. *Existing data: data is available, or it's possible to collect data*

   Because ML learns from data, there must be data for it to learn from. It's amusing to think about building a model to predict how much tax a person should pay a year, but it's not possible unless you have access to tax and income data of a large population.

   In the zero-shot learning (*https://oreil.ly/7shSg*) (sometimes known as zero-data learning) context, it's possible for an ML system to make good predictions for a task without having been trained on data for that task. However, this ML system was previously trained on data for other tasks, often related to the task in consideration. So even though the system doesn't require data for the task at hand to learn from, it still requires data to learn.

   It's also possible to launch an ML system without data. For example, in the context of continual learning, ML models can be deployed without having been trained on any data, but they will learn from incoming data in production.[6] However, serving insufficiently trained models to users comes with certain risks, such as poor customer experience.

   Without data and without continual learning, many companies follow a "fake-it-til-you make it" approach: launching a product that serves predictions made by humans, instead of ML models, with the hope of using the generated data to train ML models later.

6 We'll go over online learning in Chapter 9.

4. *Predictions: it's a predictive problem*

   ML models make predictions, so they can only solve problems that require predictive answers. ML can be especially appealing when you can benefit from a large quantity of cheap but approximate predictions. In English, "predict" means "estimate a value in the future." For example, what will the weather be like tomorrow? Who will win the Super Bowl this year? What movie will a user want to watch next?

   As predictive machines (e.g., ML models) are becoming more effective, more and more problems are being reframed as predictive problems. Whatever question you might have, you can always frame it as: "What would the answer to this question be?" regardless of whether this question is about something in the future, the present, or even the past.

   Compute-intensive problems are one class of problems that have been very successfully reframed as predictive. Instead of computing the exact outcome of a process, which might be even more computationally costly and time-consuming than ML, you can frame the problem as: "What would the outcome of this process look like?" and approximate it using an ML model. The output will be an approximation of the exact output, but often, it's good enough. You can see a lot of it in graphic renderings, such as image denoising and screen-space shading.[7]

5. *Unseen data: unseen data shares patterns with the training data*

   The patterns your model learns from existing data are only useful if unseen data also share these patterns. A model to predict whether an app will get downloaded on Christmas 2020 won't perform very well if it's trained on data from 2008, when the most popular app on the App Store was Koi Pond. What's Koi Pond? Exactly.

   In technical terms, it means your unseen data and training data should come from similar distributions. You might ask: "If the data is unseen, how do we know what distribution it comes from?" We don't, but we can make assumptions—such as we can assume that users' behaviors tomorrow won't be too different from users' behaviors today—and hope that our assumptions hold. If they don't, we'll have a model that performs poorly, which we might be able to find out with monitoring, as covered in Chapter 8, and test in production, as covered in Chapter 9.

7 Steke Bako, Thijs Vogels, Brian McWilliams, Mark Meyer, Jan Novák, Alex Harvill, Pradeep Sen, Tony Derose, and Fabrice Rousselle, "Kernel-Predicting Convolutional Networks for Denoising Monte Carlo Renderings," *ACM Transactions on Graphics* 36, no. 4 (2017): 97, *https://oreil.ly/EeI3j*; Oliver Nalbach, Elena Arabadzhiyska, Dushyant Mehta, Hans-Peter Seidel, and Tobias Ritschel, "Deep Shading: Convolutional Neural Networks for Screen-Space Shading," *arXiv*, 2016, *https://oreil.ly/dSspz*.

Due to the way most ML algorithms today learn, ML solutions will especially shine if your problem has these additional following characteristics:

6. It's repetitive
: Humans are great at few-shot learning: you can show kids a few pictures of cats and most of them will recognize a cat the next time they see one. Despite exciting progress in few-shot learning research, most ML algorithms still require many examples to learn a pattern. When a task is repetitive, each pattern is repeated multiple times, which makes it easier for machines to learn it.

7. The cost of wrong predictions is cheap
: Unless your ML model's performance is 100% all the time, which is highly unlikely for any meaningful tasks, your model is going to make mistakes. ML is especially suitable when the cost of a wrong prediction is low. For example, one of the biggest use cases of ML today is in recommender systems because with recommender systems, a bad recommendation is usually forgiving—the user just won't click on the recommendation.

  If one prediction mistake can have catastrophic consequences, ML might still be a suitable solution if, on average, the benefits of correct predictions outweigh the cost of wrong predictions. Developing self-driving cars is challenging because an algorithmic mistake can lead to death. However, many companies still want to develop self-driving cars because they have the potential to save many lives once self-driving cars are statistically safer than human drivers.

8. It's at scale
: ML solutions often require nontrivial up-front investment on data, compute, infrastructure, and talent, so it'd make sense if we can use these solutions a lot.

  "At scale" means different things for different tasks, but, in general, it means making a lot of predictions. Examples include sorting through millions of emails a year or predicting which departments thousands of support tickets should be routed to a day.

  A problem might appear to be a singular prediction, but it's actually a series of predictions. For example, a model that predicts who will win a US presidential election seems like it only makes one prediction every four years, but it might actually be making a prediction every hour or even more frequently because that prediction has to be continually updated to incorporate new information.

  Having a problem at scale also means that there's a lot of data for you to collect, which is useful for training ML models.

9. *The patterns are constantly changing*

Cultures change. Tastes change. Technologies change. What's trendy today might be old news tomorrow. Consider the task of email spam classification. Today an indication of a spam email is a Nigerian prince, but tomorrow it might be a distraught Vietnamese writer.

If your problem involves one or more constantly changing patterns, hardcoded solutions such as handwritten rules can become outdated quickly. Figuring how your problem has changed so that you can update your handwritten rules accordingly can be too expensive or impossible. Because ML learns from data, you can update your ML model with new data without having to figure out how the data has changed. It's also possible to set up your system to adapt to the changing data distributions, an approach we'll discuss in the section "Continual Learning" on page 264.

The list of use cases can go on and on, and it'll grow even longer as ML adoption matures in the industry. Even though ML can solve a subset of problems very well, it can't solve and/or shouldn't be used for a lot of problems. Most of today's ML algorithms shouldn't be used under any of the following conditions:

- It's unethical. We'll go over one case study where the use of ML algorithms can be argued as unethical in the section "Case study I: Automated grader's biases" on page 341.
- Simpler solutions do the trick. In Chapter 6, we'll cover the four phases of ML model development where the first phase should be non-ML solutions.
- It's not cost-effective.

However, even if ML can't solve your problem, it might be possible to break your problem into smaller components, and use ML to solve some of them. For example, if you can't build a chatbot to answer all your customers' queries, it might be possible to build an ML model to predict whether a query matches one of the frequently asked questions. If yes, direct the customer to the answer. If not, direct them to customer service.

I'd also want to caution against dismissing a new technology because it's not as cost-effective as the existing technologies at the moment. Most technological advances are incremental. A type of technology might not be efficient now, but it might be over time with more investments. If you wait for the technology to prove its worth to the rest of the industry before jumping in, you might end up years or decades behind your competitors.

Machine Learning Use Cases

ML has found increasing usage in both enterprise and consumer applications. Since the mid-2010s, there has been an explosion of applications that leverage ML to deliver superior or previously impossible services to consumers.

With the explosion of information and services, it would have been very challenging for us to find what we want without the help of ML, manifested in either a *search engine* or a *recommender system*. When you visit a website like Amazon or Netflix, you're recommended items that are predicted to best match your taste. If you don't like any of your recommendations, you might want to search for specific items, and your search results are likely powered by ML.

If you have a smartphone, ML is likely already assisting you in many of your daily activities. Typing on your phone is made easier with *predictive typing*, an ML system that gives you suggestions on what you might want to say next. An ML system might run in your photo editing app to suggest how best to enhance your photos. You might authenticate your phone using your fingerprint or your face, which requires an ML system to predict whether a fingerprint or a face matches yours.

The ML use case that drew me into the field was *machine translation*, automatically translating from one language to another. It has the potential to allow people from different cultures to communicate with each other, erasing the language barrier. My parents don't speak English, but thanks to Google Translate, now they can read my writing and talk to my friends who don't speak Vietnamese.

ML is increasingly present in our homes with smart personal assistants such as Alexa and Google Assistant. Smart security cameras can let you know when your pets leave home or if you have an uninvited guest. A friend of mine was worried about his aging mother living by herself—if she falls, no one is there to help her get up—so he relied on an at-home health monitoring system that predicts whether someone has fallen in the house.

Even though the market for consumer ML applications is booming, the majority of ML use cases are still in the enterprise world. Enterprise ML applications tend to have vastly different requirements and considerations from consumer applications. There are many exceptions, but for most cases, enterprise applications might have stricter accuracy requirements but be more forgiving with latency requirements. For example, improving a speech recognition system's accuracy from 95% to 95.5% might not be noticeable to most consumers, but improving a resource allocation system's efficiency by just 0.1% can help a corporation like Google or General Motors save millions of dollars. At the same time, latency of a second might get a consumer distracted and opening something else, but enterprise users might be more tolerant of high latency. For people interested in building companies out of ML applications, consumer apps might be easier to distribute but much harder to monetize. However, most enterprise use cases aren't obvious unless you've encountered them yourself.

According to Algorithmia's 2020 state of enterprise machine learning survey, ML applications in enterprises are diverse, serving both internal use cases (reducing costs, generating customer insights and intelligence, internal processing automation) and external use cases (improving customer experience, retaining customers, interacting with customers) as shown in Figure 1-3.[8]

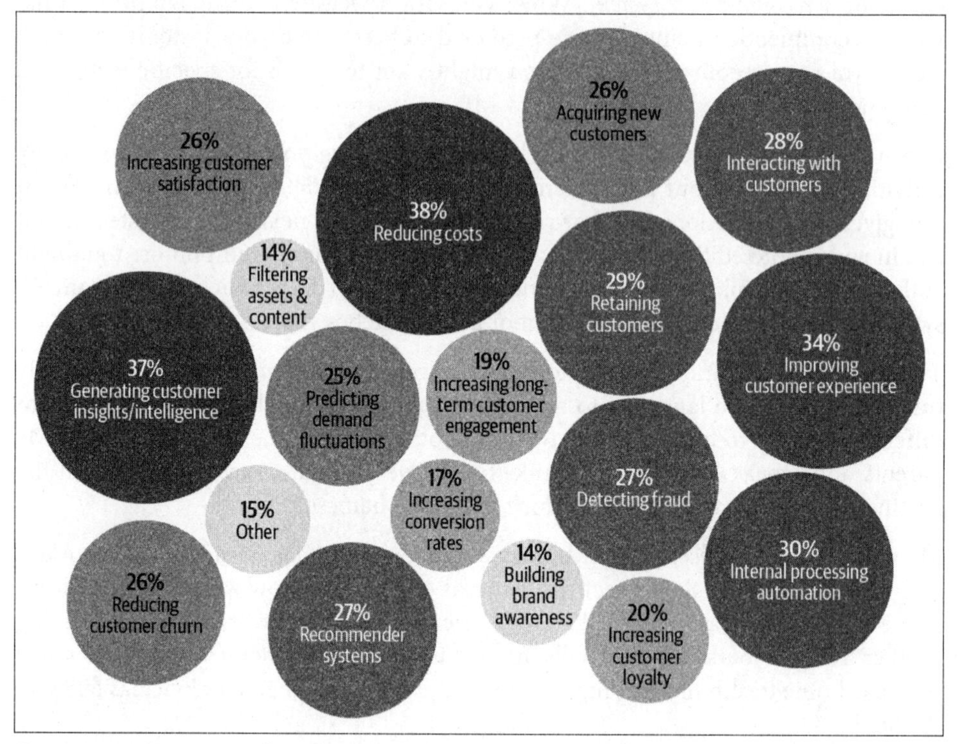

Figure 1-3. 2020 state of enterprise machine learning. Source: Adapted from an image by Algorithmia

8 "2020 State of Enterprise Machine Learning," *Algorithmia*, 2020, *https://oreil.ly/wKMZB*.

Fraud detection is among the oldest applications of ML in the enterprise world. If your product or service involves transactions of any value, it'll be susceptible to fraud. By leveraging ML solutions for anomaly detection, you can have systems that learn from historical fraud transactions and predict whether a future transaction is fraudulent.

Deciding how much to charge for your product or service is probably one of the hardest business decisions; why not let ML do it for you? *Price optimization* is the process of estimating a price at a certain time period to maximize a defined objective function, such as the company's margin, revenue, or growth rate. ML-based pricing optimization is most suitable for cases with a large number of transactions where demand fluctuates and consumers are willing to pay a dynamic price—for example, internet ads, flight tickets, accommodation bookings, ride-sharing, and events.

To run a business, it's important to be able to forecast customer demand so that you can prepare a budget, stock inventory, allocate resources, and update pricing strategy. For example, if you run a grocery store, you want to stock enough so that customers find what they're looking for, but you don't want to overstock, because if you do, your groceries might go bad and you lose money.

Acquiring a new user is expensive. As of 2019, the average cost for an app to acquire a user who'll make an in-app purchase is $86.61.[9] The acquisition cost for Lyft is estimated at $158/rider.[10] This cost is so much higher for enterprise customers. Customer acquisition cost is hailed by investors as a startup killer.[11] Reducing customer acquisition costs by a small amount can result in a large increase in profit. This can be done through better identifying potential customers, showing better-targeted ads, giving discounts at the right time, etc.—all of which are suitable tasks for ML.

After you've spent so much money acquiring a customer, it'd be a shame if they leave. The cost of acquiring a new user is approximated to be 5 to 25 times more expensive than retaining an existing one.[12] *Churn prediction* is predicting when a specific customer is about to stop using your products or services so that you can take appropriate actions to win them back. Churn prediction can be used not only for customers but also for employees.

9 "Average Mobile App User Acquisition Costs Worldwide from September 2018 to August 2019, by User Action and Operating System," *Statista*, 2019, *https://oreil.ly/2pTCH*.

10 Jeff Henriksen, "Valuing Lyft Requires a Deep Look into Unit Economics," *Forbes*, May 17, 2019, *https://oreil.ly/VeSt4*.

11 David Skok, "Startup Killer: The Cost of Customer Acquisition," *For Entrepreneurs*, 2018, *https://oreil.ly/L3tQ7*.

12 Amy Gallo, "The Value of Keeping the Right Customers," *Harvard Business Review*, October 29, 2014, *https://oreil.ly/OlNkl*.

To prevent customers from leaving, it's important to keep them happy by addressing their concerns as soon as they arise. Automated support ticket classification can help with that. Previously, when a customer opened a support ticket or sent an email, it needed to first be processed then passed around to different departments until it arrived at the inbox of someone who could address it. An ML system can analyze the ticket content and predict where it should go, which can shorten the response time and improve customer satisfaction. It can also be used to classify internal IT tickets.

Another popular use case of ML in enterprise is brand monitoring. The brand is a valuable asset of a business.[13] It's important to monitor how the public and your customers perceive your brand. You might want to know when/where/how it's mentioned, both explicitly (e.g., when someone mentions "Google") or implicitly (e.g., when someone says "the search giant"), as well as the sentiment associated with it. If there's suddenly a surge of negative sentiment in your brand mentions, you might want to address it as soon as possible. Sentiment analysis is a typical ML task.

A set of ML use cases that has generated much excitement recently is in health care. There are ML systems that can detect skin cancer and diagnose diabetes. Even though many health-care applications are geared toward consumers, because of their strict requirements with accuracy and privacy, they are usually provided through a health-care provider such as a hospital or used to assist doctors in providing diagnosis.

Understanding Machine Learning Systems

Understanding ML systems will be helpful in designing and developing them. In this section, we'll go over how ML systems are different from both ML in research (or as often taught in school) and traditional software, which motivates the need for this book.

Machine Learning in Research Versus in Production

As ML usage in the industry is still fairly new, most people with ML expertise have gained it through academia: taking courses, doing research, reading academic papers. If that describes your background, it might be a steep learning curve for you to understand the challenges of deploying ML systems in the wild and navigate an overwhelming set of solutions to these challenges. ML in production is very different from ML in research. Table 1-1 shows five of the major differences.

13 Marty Swant, "The World's 20 Most Valuable Brands," *Forbes*, 2020, *https://oreil.ly/4uS5i*.

Table 1-1. Key differences between ML in research and ML in production

	Research	Production
Requirements	State-of-the-art model performance on benchmark datasets	Different stakeholders have different requirements
Computational priority	Fast training, high throughput	Fast inference, low latency
Data	Static[a]	Constantly shifting
Fairness	Often not a focus	Must be considered
Interpretability	Often not a focus	Must be considered

[a] A subfield of research focuses on continual learning: developing models to work with changing data distributions. We'll cover continual learning in Chapter 9.

Different stakeholders and requirements

People involved in a research and leaderboard project often align on one single objective. The most common objective is model performance—develop a model that achieves the state-of-the-art results on benchmark datasets. To edge out a small improvement in performance, researchers often resort to techniques that make models too complex to be useful.

There are many stakeholders involved in bringing an ML system into production. Each stakeholder has their own requirements. Having different, often conflicting, requirements can make it difficult to design, develop, and select an ML model that satisfies all the requirements.

Consider a mobile app that recommends restaurants to users. The app makes money by charging restaurants a 10% service fee on each order. This means that expensive orders give the app more money than cheap orders. The project involves ML engineers, salespeople, product managers, infrastructure engineers, and a manager:

ML engineers
: Want a model that recommends restaurants that users will most likely order from, and they believe they can do so by using a more complex model with more data.

Sales team
: Wants a model that recommends the more expensive restaurants since these restaurants bring in more service fees.

Product team
: Notices that every increase in latency leads to a drop in orders through the service, so they want a model that can return the recommended restaurants in less than 100 milliseconds.

ML platform team

As the traffic grows, this team has been woken up in the middle of the night because of problems with scaling their existing system, so they want to hold off on model updates to prioritize improving the ML platform.

Manager

Wants to maximize the margin, and one way to achieve this might be to let go of the ML team.[14]

"Recommending the restaurants that users are most likely to click on" and "recommending the restaurants that will bring in the most money for the app" are two different objectives, and in the section "Decoupling objectives" on page 41, we'll discuss how to develop an ML system that satisfies different objectives. Spoiler: we'll develop one model for each objective and combine their predictions.

Let's imagine for now that we have two different models. Model A is the model that recommends the restaurants that users are most likely to click on, and model B is the model that recommends the restaurants that will bring in the most money for the app. A and B might be very different models. Which model should be deployed to the users? To make the decision more difficult, neither A nor B satisfies the requirement set forth by the product team: they can't return restaurant recommendations in less than 100 milliseconds.

When developing an ML project, it's important for ML engineers to understand requirements from all stakeholders involved and how strict these requirements are. For example, if being able to return recommendations within 100 milliseconds is a must-have requirement—the company finds that if your model takes over 100 milliseconds to recommend restaurants, 10% of users would lose patience and close the app—then neither model A nor model B will work. However, if it's just a nice-to-have requirement, you might still want to consider model A or model B.

Production having different requirements from research is one of the reasons why successful research projects might not always be used in production. For example, ensembling is a technique popular among the winners of many ML competitions, including the famed $1 million Netflix Prize, and yet it's not widely used in production. Ensembling combines "multiple learning algorithms to obtain better predictive performance than could be obtained from any of the constituent learning algorithms alone."[15] While it can give your ML system a small performance improvement, ensembling tends to make a system too complex to be useful in production, e.g.,

14 It's not unusual for the ML and data science teams to be among the first to go during a company's mass layoff, as has been reported at IBM (*https://oreil.ly/AfUB5*), Uber (*https://oreil.ly/t0QpY*), Airbnb (*https://oreil.ly/q4M4E*). See also Sejuti Das's analysis "How Data Scientists Are Also Susceptible to the Layoffs Amid Crisis," *Analytics India Magazine*, May 21, 2020, *https://oreil.ly/jobmz*.

15 Wikipedia, s.v. "Ensemble learning," *https://oreil.ly/5qkgp*.

slower to make predictions or harder to interpret the results. We'll discuss ensembling further in the section "Ensembles" on page 156.

For many tasks, a small improvement in performance can result in a huge boost in revenue or cost savings. For example, a 0.2% improvement in the click-through rate for a product recommender system can result in millions of dollars increase in revenue for an ecommerce site. However, for many tasks, a small improvement might not be noticeable for users. For the second type of task, if a simple model can do a reasonable job, complex models must perform significantly better to justify the complexity.

Criticism of ML Leaderboards

In recent years, there have been many critics of ML leaderboards, both competitions such as Kaggle and research leaderboards such as ImageNet or GLUE.

An obvious argument is that in these competitions many of the hard steps needed for building ML systems are already done for you.[16]

A less obvious argument is that due to the multiple-hypothesis testing scenario that happens when you have multiple teams testing on the same hold-out test set, a model can do better than the rest just by chance.[17]

The misalignment of interests between research and production has been noticed by researchers. In an EMNLP 2020 paper, Ethayarajh and Jurafsky argued that benchmarks have helped drive advances in natural language processing (NLP) by incentivizing the creation of more accurate models at the expense of other qualities valued by practitioners such as compactness, fairness, and energy efficiency.[18]

Computational priorities

When designing an ML system, people who haven't deployed an ML system often make the mistake of focusing too much on the model development part and not enough on the model deployment and maintenance part.

During the model development process, you might train many different models, and each model does multiple passes over the training data. Each trained model then generates predictions on the validation data once to report the scores. The validation data is usually much smaller than the training data. During model development,

16 Julia Evans, "Machine Learning Isn't Kaggle Competitions," 2014, *https://oreil.ly/p8mZq*.

17 Lauren Oakden-Rayner, "AI Competitions Don't Produce Useful Models," September 19, 2019, *https://oreil.ly/X6RlT*.

18 Kawin Ethayarajh and Dan Jurafsky, "Utility Is in the Eye of the User: A Critique of NLP Leaderboards," EMNLP, 2020, *https://oreil.ly/4Ud8P*.

training is the bottleneck. Once the model has been deployed, however, its job is to generate predictions, so inference is the bottleneck. Research usually prioritizes fast training, whereas production usually prioritizes fast inference.

One corollary of this is that research prioritizes high throughput whereas production prioritizes low latency. In case you need a refresh, latency refers to the time it takes from receiving a query to returning the result. Throughput refers to how many queries are processed within a specific period of time.

Terminology Clash

Some books make the distinction between latency and response time. According to Martin Kleppmann in his book *Designing Data-Intensive Applications*, "The response time is what the client sees: besides the actual time to process the request (the service time), it includes network delays and queueing delays. Latency is the duration that a request is waiting to be handled—during which it is latent, awaiting service."[19]

In this book, to simplify the discussion and to be consistent with the terminology used in the ML community, we use latency to refer to the response time, so the latency of a request measures the time from when the request is sent to the time a response is received.

For example, the average latency of Google Translate is the average time it takes from when a user clicks Translate to when the translation is shown, and the throughput is how many queries it processes and serves a second.

If your system always processes one query at a time, higher latency means lower throughput. If the average latency is 10 ms, which means it takes 10 ms to process a query, the throughput is 100 queries/second. If the average latency is 100 ms, the throughput is 10 queries/second.

However, because most modern distributed systems batch queries to process them together, often concurrently, *higher latency might also mean higher throughput*. If you process 10 queries at a time and it takes 10 ms to run a batch, the average latency is still 10 ms but the throughput is now 10 times higher—1,000 queries/second. If you process 50 queries at a time and it takes 20 ms to run a batch, the average latency now is 20 ms and the throughput is 2,500 queries/second. Both latency and throughput have increased! The difference in latency and throughput trade-off for processing queries one at a time and processing queries in batches is illustrated in Figure 1-4.

19 Martin Kleppmann, *Designing Data-Intensive Applications* (Sebastopol, CA: O'Reilly, 2017).

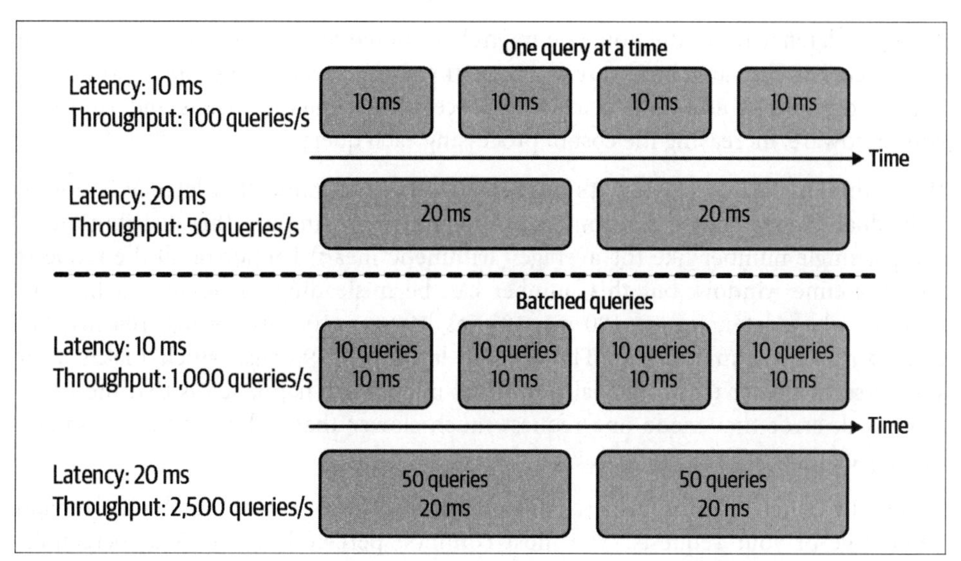

Figure 1-4. When processing queries one at a time, higher latency means lower throughput. When processing queries in batches, however, higher latency might also mean higher throughput.

This is even more complicated if you want to batch online queries. Batching requires your system to wait for enough queries to arrive in a batch before processing them, which further increases latency.

In research, you care more about how many samples you can process in a second (throughput) and less about how long it takes for each sample to be processed (latency). You're willing to increase latency to increase throughput, for example, with aggressive batching.

However, once you deploy your model into the real world, latency matters a lot. In 2017, an Akamai study found that a 100 ms delay can hurt conversion rates by 7%.[20] In 2019, Booking.com found that an increase of about 30% in latency cost about 0.5% in conversion rates—"a relevant cost for our business."[21] In 2016, Google found that more than half of mobile users will leave a page if it takes more than three seconds to load.[22] Users today are even less patient.

20 Akamai Technologies, *Akamai Online Retail Performance Report: Milliseconds Are Critical*, April 19, 2017, *https://oreil.ly/bEtRu*.

21 Lucas Bernardi, Themis Mavridis, and Pablo Estevez, "150 Successful Machine Learning Models: 6 Lessons Learned at Booking.com," KDD '19, August 4–8, 2019, Anchorage, AK, *https://oreil.ly/G5QNA*.

22 "Consumer Insights," Think with Google, *https://oreil.ly/JCp6Z*.

To reduce latency in production, you might have to reduce the number of queries you can process on the same hardware at a time. If your hardware is capable of processing many more queries at a time, using it to process fewer queries means underutilizing your hardware, increasing the cost of processing each query.

When thinking about latency, it's important to keep in mind that latency is not an individual number but a distribution. It's tempting to simplify this distribution by using a single number like the average (arithmetic mean) latency of all the requests within a time window, but this number can be misleading. Imagine you have 10 requests whose latencies are 100 ms, 102 ms, 100 ms, 100 ms, 99 ms, 104 ms, 110 ms, 90 ms, 3,000 ms, 95 ms. The average latency is 390 ms, which makes your system seem slower than it actually is. What might have happened is that there was a network error that made one request much slower than others, and you should investigate that troublesome request.

It's usually better to think in percentiles, as they tell you something about a certain percentage of your requests. The most common percentile is the 50th percentile, abbreviated as p50. It's also known as the median. If the median is 100 ms, half of the requests take longer than 100 ms, and half of the requests take less than 100 ms.

Higher percentiles also help you discover outliers, which might be symptoms of something wrong. Typically, the percentiles you'll want to look at are p90, p95, and p99. The 90th percentile (p90) for the 10 requests above is 3,000 ms, which is an outlier.

Higher percentiles are important to look at because even though they account for a small percentage of your users, sometimes they can be the most important users. For example, on the Amazon website, the customers with the slowest requests are often those who have the most data on their accounts because they have made many purchases—that is, they're the most valuable customers.[23]

It's a common practice to use high percentiles to specify the performance requirements for your system; for example, a product manager might specify that the 90th percentile or 99.9th percentile latency of a system must be below a certain number.

Data

During the research phase, the datasets you work with are often clean and well-formatted, freeing you to focus on developing models. They are static by nature so that the community can use them to benchmark new architectures and techniques. This means that many people might have used and discussed the same datasets, and quirks of the dataset are known. You might even find open source scripts to process and feed the data directly into your models.

23 Kleppmann, *Designing Data-Intensive Applications.*

In production, data, if available, is a lot more messy. It's noisy, possibly unstructured, constantly shifting. It's likely biased, and you likely don't know how it's biased. Labels, if there are any, might be sparse, imbalanced, or incorrect. Changing project or business requirements might require updating some or all of your existing labels. If you work with users' data, you'll also have to worry about privacy and regulatory concerns. We'll discuss a case study where users' data is inadequately handled in the section "Case study II: The danger of "anonymized" data" on page 344.

In research, you mostly work with historical data, e.g., data that already exists and is stored somewhere. In production, most likely you'll also have to work with data that is being constantly generated by users, systems, and third-party data.

Figure 1-5 has been adapted from a great graphic by Andrej Karpathy, director of AI at Tesla, that illustrates the data problems he encountered during his PhD compared to his time at Tesla.

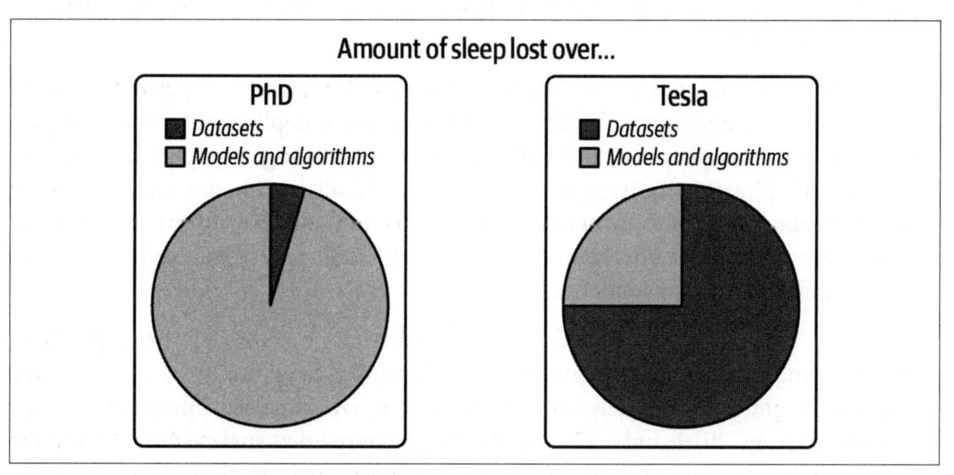

Figure 1-5. Data in research versus data in production. Source: Adapted from an image by Andrej Karpathy[24]

Fairness

During the research phase, a model is not yet used on people, so it's easy for researchers to put off fairness as an afterthought: "Let's try to get state of the art first and worry about fairness when we get to production." When it gets to production, it's too late. If you optimize your models for better accuracy or lower latency, you can show that your models beat state of the art. But, as of writing this book, there's no equivalent state of the art for fairness metrics.

24 Andrej Karpathy, "Building the Software 2.0 Stack," Spark+AI Summit 2018, video, 17:54, *https://oreil.ly/Z21Oz*.

You or someone in your life might already be a victim of biased mathematical algorithms without knowing it. Your loan application might be rejected because the ML algorithm picks on your zip code, which embodies biases about one's socioeconomic background. Your resume might be ranked lower because the ranking system employers use picks on the spelling of your name. Your mortgage might get a higher interest rate because it relies partially on credit scores, which favor the rich and punish the poor. Other examples of ML biases in the real world are in predictive policing algorithms, personality tests administered by potential employers, and college rankings.

In 2019, "Berkeley researchers found that both face-to-face and online lenders rejected a total of 1.3 million creditworthy Black and Latino applicants between 2008 and 2015." When the researchers "used the income and credit scores of the rejected applications but deleted the race identifiers, the mortgage application was accepted."[25] For even more galling examples, I recommend Cathy O'Neil's *Weapons of Math Destruction*.[26]

ML algorithms don't predict the future, but encode the past, thus perpetuating the biases in the data and more. When ML algorithms are deployed at scale, they can discriminate against people at scale. If a human operator might only make sweeping judgments about a few individuals at a time, an ML algorithm can make sweeping judgments about millions in split seconds. This can especially hurt members of minority groups because misclassification on them could only have a minor effect on models' overall performance metrics.

If an algorithm can already make correct predictions on 98% of the population, and improving the predictions on the other 2% would incur multiples of cost, some companies might, unfortunately, choose not to do it. During a McKinsey & Company research study in 2019, only 13% of the large companies surveyed said they are taking steps to mitigate risks to equity and fairness, such as algorithmic bias and discrimination.[27] However, this is changing rapidly. We'll cover fairness and other aspects of responsible AI in Chapter 11.

Interpretability

In early 2020, the Turing Award winner Professor Geoffrey Hinton proposed a heatedly debated question about the importance of interpretability in ML systems. "Suppose you have cancer and you have to choose between a black box AI surgeon

25 Khristopher J. Brooks, "Disparity in Home Lending Costs Minorities Millions, Researchers Find," *CBS News*, November 15, 2019, *https://oreil.ly/UiHUB*.

26 Cathy O'Neil, *Weapons of Math Destruction* (New York: Crown Books, 2016).

27 Stanford University Human-Centered Artificial Intelligence (HAI), *The 2019 AI Index Report*, 2019, *https://oreil.ly/xs8mG*.

that cannot explain how it works but has a 90% cure rate and a human surgeon with an 80% cure rate. Do you want the AI surgeon to be illegal?"[28]

A couple of weeks later, when I asked this question to a group of 30 technology executives at public nontech companies, only half of them would want the highly effective but unable-to-explain AI surgeon to operate on them. The other half wanted the human surgeon.

While most of us are comfortable with using a microwave without understanding how it works, many don't feel the same way about AI yet, especially if that AI makes important decisions about their lives.

Since most ML research is still evaluated on a single objective, model performance, researchers aren't incentivized to work on model interpretability. However, interpretability isn't just optional for most ML use cases in the industry, but a requirement.

First, interpretability is important for users, both business leaders and end users, to understand why a decision is made so that they can trust a model and detect potential biases mentioned previously.[29] Second, it's important for developers to be able to debug and improve a model.

Just because interpretability is a requirement doesn't mean everyone is doing it. As of 2019, only 19% of large companies are working to improve the explainability of their algorithms.[30]

Discussion

Some might argue that it's OK to know only the academic side of ML because there are plenty of jobs in research. The first part—it's OK to know only the academic side of ML—is true. The second part is false.

While it's important to pursue pure research, most companies can't afford it unless it leads to short-term business applications. This is especially true now that the research community took the "bigger, better" approach. Oftentimes, new models require a massive amount of data and tens of millions of dollars in compute alone.

As ML research and off-the-shelf models become more accessible, more people and organizations would want to find applications for them, which increases the demand for ML in production.

The vast majority of ML-related jobs will be, and already are, in productionizing ML.

---

28 Tweet by Geoffrey Hinton (@geoffreyhinton), February 20, 2020, *https://oreil.ly/KdfD8*.

29 For certain use cases in certain countries, users have a "right to explanation": a right to be given an explanation for an output of the algorithm.

30 Stanford HAI, *The 2019 AI Index Report*.

Machine Learning Systems Versus Traditional Software

Since ML is part of software engineering (SWE), and software has been successfully used in production for more than half a century, some might wonder why we don't just take tried-and-true best practices in software engineering and apply them to ML.

That's an excellent idea. In fact, ML production would be a much better place if ML experts were better software engineers. Many traditional SWE tools can be used to develop and deploy ML applications.

However, many challenges are unique to ML applications and require their own tools. In SWE, there's an underlying assumption that code and data are separated. In fact, in SWE, we want to keep things as modular and separate as possible (see the Wikipedia page on separation of concerns (*https://oreil.ly/kH67y*)).

On the contrary, ML systems are part code, part data, and part artifacts created from the two. The trend in the last decade shows that applications developed with the most/best data win. Instead of focusing on improving ML algorithms, most companies will focus on improving their data. Because data can change quickly, ML applications need to be adaptive to the changing environment, which might require faster development and deployment cycles.

In traditional SWE, you only need to focus on testing and versioning your code. With ML, we have to test and version our data too, and that's the hard part. How to version large datasets? How to know if a data sample is good or bad for your system? Not all data samples are equal—some are more valuable to your model than others. For example, if your model has already trained on one million scans of normal lungs and only one thousand scans of cancerous lungs, a scan of a cancerous lung is much more valuable than a scan of a normal lung. Indiscriminately accepting all available data might hurt your model's performance and even make it susceptible to data poisoning attacks.[31]

The size of ML models is another challenge. As of 2022, it's common for ML models to have hundreds of millions, if not billions, of parameters, which requires gigabytes of random-access memory (RAM) to load them into memory. A few years from now, a billion parameters might seem quaint—like, "Can you believe the computer that sent men to the moon only had 32 MB of RAM?"

However, for now, getting these large models into production, especially on edge devices,[32] is a massive engineering challenge. Then there is the question of how to get these models to run fast enough to be useful. An autocompletion model is useless if

---

31 Xinyun Chen, Chang Liu, Bo Li, Kimberly Lu, and Dawn Song, "Targeted Backdoor Attacks on Deep Learning Systems Using Data Poisoning," *arXiv*, December 15, 2017, *https://oreil.ly/OkAjb*.

32 We'll cover edge devices in Chapter 7.

the time it takes to suggest the next character is longer than the time it takes for you to type.

Monitoring and debugging these models in production is also nontrivial. As ML models get more complex, coupled with the lack of visibility into their work, it's hard to figure out what went wrong or be alerted quickly enough when things go wrong.

The good news is that these engineering challenges are being tackled at a breakneck pace. Back in 2018, when the Bidirectional Encoder Representations from Transformers (BERT) paper first came out, people were talking about how BERT was too big, too complex, and too slow to be practical. The pretrained large BERT model has 340 million parameters and is 1.35 GB.[33] Fast-forward two years later, BERT and its variants were already used in almost every English search on Google.[34]

Summary

This opening chapter aimed to give readers an understanding of what it takes to bring ML into the real world. We started with a tour of the wide range of use cases of ML in production today. While most people are familiar with ML in consumer-facing applications, the majority of ML use cases are for enterprise. We also discussed when ML solutions would be appropriate. Even though ML can solve many problems very well, it can't solve all the problems and it's certainly not appropriate for all the problems. However, for problems that ML can't solve, it's possible that ML can be one part of the solution.

This chapter also highlighted the differences between ML in research and ML in production. The differences include the stakeholder involvement, computational priority, the properties of data used, the gravity of fairness issues, and the requirements for interpretability. This section is the most helpful to those coming to ML production from academia. We also discussed how ML systems differ from traditional software systems, which motivated the need for this book.

ML systems are complex, consisting of many different components. Data scientists and ML engineers working with ML systems in production will likely find that focusing only on the ML algorithms part is far from enough. It's important to know about other aspects of the system, including the data stack, deployment, monitoring, maintenance, infrastructure, etc. This book takes a system approach to developing ML systems, which means that we'll consider all components of a system holistically instead of just looking at ML algorithms. We'll provide detail on what this holistic approach means in the next chapter.

---

33 Jacob Devlin, Ming-Wei Chang, Kenton Lee, and Kristina Toutanova, "BERT: Pre-training of Deep Bidirectional Transformers for Language Understanding," *arXiv*, October 11, 2018, *https://oreil.ly/TG3ZW*.

34 Google Search On, 2020, *https://oreil.ly/M7YjM*.

CHAPTER 2

Introduction to Machine Learning Systems Design

Now that we've walked through an overview of ML systems in the real world, we can get to the fun part of actually designing an ML system. To reiterate from the first chapter, ML systems design takes a system approach to MLOps, which means that we'll consider an ML system holistically to ensure that all the components—the business requirements, the data stack, infrastructure, deployment, monitoring, etc.—and their stakeholders can work together to satisfy the specified objectives and requirements.

We'll start the chapter with a discussion on objectives. Before we develop an ML system, we must understand why this system is needed. If this system is built for a business, it must be driven by business objectives, which will need to be translated into ML objectives to guide the development of ML models.

Once everyone is on board with the objectives for our ML system, we'll need to set out some requirements to guide the development of this system. In this book, we'll consider the four requirements: reliability, scalability, maintainability, and adaptability. We will then introduce the iterative process for designing systems to meet those requirements.

You might wonder: with all these objectives, requirements, and processes in place, can I finally start building my ML model yet? Not so soon! Before using ML algorithms to solve your problem, you first need to frame your problem into a task that ML can solve. We'll continue this chapter with how to frame your ML problems. The difficulty of your job can change significantly depending on how you frame your problem.

Because ML is a data-driven approach, a book on ML systems design will be amiss if it fails to discuss the importance of data in ML systems. The last part of this chapter touches on a debate that has consumed much of the ML literature in recent years: which is more important—data or intelligent algorithms?

Let's get started!

Business and ML Objectives

We first need to consider the objectives of the proposed ML projects. When working on an ML project, data scientists tend to care about the ML objectives: the metrics they can measure about the performance of their ML models such as accuracy, F1 score, inference latency, etc. They get excited about improving their model's accuracy from 94% to 94.2% and might spend a ton of resources—data, compute, and engineering time—to achieve that.

But the truth is: most companies don't care about the fancy ML metrics. They don't care about increasing a model's accuracy from 94% to 94.2% unless it moves some business metrics. A pattern I see in many short-lived ML projects is that the data scientists become too focused on hacking ML metrics without paying attention to business metrics. Their managers, however, only care about business metrics and, after failing to see how an ML project can help push their business metrics, kill the projects prematurely (and possibly let go of the data science team involved).[1]

So what metrics do companies care about? While most companies want to convince you otherwise, the sole purpose of businesses, according to the Nobel-winning economist Milton Friedman, is to maximize profits for shareholders.[2]

The ultimate goal of any project within a business is, therefore, to increase profits, either directly or indirectly: directly such as increasing sales (conversion rates) and cutting costs; indirectly such as higher customer satisfaction and increasing time spent on a website.

For an ML project to succeed within a business organization, it's crucial to tie the performance of an ML system to the overall business performance. What business performance metrics is the new ML system supposed to influence, e.g., the amount of ads revenue, the number of monthly active users?

---

1 Eugene Yan has a great post (*https://oreil.ly/thQCV*) on how data scientists can understand the business intent and context of the projects they work on.

2 Milton Friedman, "A Friedman Doctrine—The Social Responsibility of Business Is to Increase Its Profits," *New York Times Magazine*, September 13, 1970, *https://oreil.ly/Fmbem*.

Imagine that you work for an ecommerce site that cares about purchase-through rate and you want to move your recommender system from batch prediction to online prediction.[3] You might reason that online prediction will enable recommendations more relevant to users right now, which can lead to a higher purchase-through rate. You can even do an experiment to show that online prediction can improve your recommender system's predictive accuracy by *X*% and, historically on your site, each percent increase in the recommender system's predictive accuracy led to a certain increase in purchase-through rate.

One of the reasons why predicting ad click-through rates and fraud detection are among the most popular use cases for ML today is that it's easy to map ML models' performance to business metrics: every increase in click-through rate results in actual ad revenue, and every fraudulent transaction stopped results in actual money saved.

Many companies create their own metrics to map business metrics to ML metrics. For example, Netflix measures the performance of their recommender system using *take-rate*: the number of quality plays divided by the number of recommendations a user sees.[4] The higher the take-rate, the better the recommender system. Netflix also put a recommender system's take-rate in the context of their other business metrics like total streaming hours and subscription cancellation rate. They found that a higher take-rate also results in higher total streaming hours and lower subscription cancellation rates.[5]

The effect of an ML project on business objectives can be hard to reason about. For example, an ML model that gives customers more personalized solutions can make them happier, which makes them spend more money on your services. The same ML model can also solve their problems faster, which makes them spend less money on your services.

To gain a definite answer on the question of how ML metrics influence business metrics, experiments are often needed. Many companies do that with experiments like A/B testing and choose the model that leads to better business metrics, regardless of whether this model has better ML metrics.

---

3 We'll cover batch prediction and online prediction in Chapter 7.

4 Ashok Chandrashekar, Fernando Amat, Justin Basilico, and Tony Jebara, "Artwork Personalization at Netflix," *Netflix Technology Blog*, December 7, 2017, *https://oreil.ly/UEDmw*.

5 Carlos A. Gomez-Uribe and Neil Hunt, "The Netflix Recommender System: Algorithms, Business Value, and Innovation," *ACM Transactions on Management Information Systems* 6, no. 4 (January 2016): 13, *https://oreil.ly/JkEPB*.

Yet, even rigorous experiments might not be sufficient to understand the relationship between an ML model's outputs and business metrics. Imagine you work for a cybersecurity company that detects and stops security threats, and ML is just a component in their complex process. An ML model is used to detect anomalies in the traffic pattern. These anomalies then go through a logic set (e.g., a series of if-else statements) that categorizes whether they constitute potential threats. These potential threats are then reviewed by security experts to determine whether they are actual threats. Actual threats will then go through another, different process aimed at stopping them. When this process fails to stop a threat, it might be impossible to figure out whether the ML component has anything to do with it.

Many companies like to say that they use ML in their systems because "being AI-powered" alone already helps them attract customers, regardless of whether the AI part actually does anything useful.[6]

When evaluating ML solutions through the business lens, it's important to be realistic about the expected returns. Due to all the hype surrounding ML, generated both by the media and by practitioners with a vested interest in ML adoption, some companies might have the notion that ML can magically transform their businesses overnight.

Magically: possible. Overnight: no.

There are many companies that have seen payoffs from ML. For example, ML has helped Google search better, sell more ads at higher prices, improve translation quality, and build better Android applications. But this gain hardly happened overnight. Google has been investing in ML for decades.

Returns on investment in ML depend a lot on the maturity stage of adoption. The longer you've adopted ML, the more efficient your pipeline will run, the faster your development cycle will be, the less engineering time you'll need, and the lower your cloud bills will be, which all lead to higher returns. According to a 2020 survey by Algorithmia, among companies that are more sophisticated in their ML adoption (having had models in production for over five years), almost 75% can deploy a model in under 30 days. Among those just getting started with their ML pipeline, 60% take over 30 days to deploy a model (see Figure 2-1).[7]

6 Parmy Olson, "Nearly Half of All 'AI Startups' Are Cashing In on Hype," *Forbes*, March 4, 2019, *https://oreil.ly/w5kOr*.

7 "2020 State of Enterprise Machine Learning," Algorithmia, 2020, *https://oreil.ly/FlIV1*.

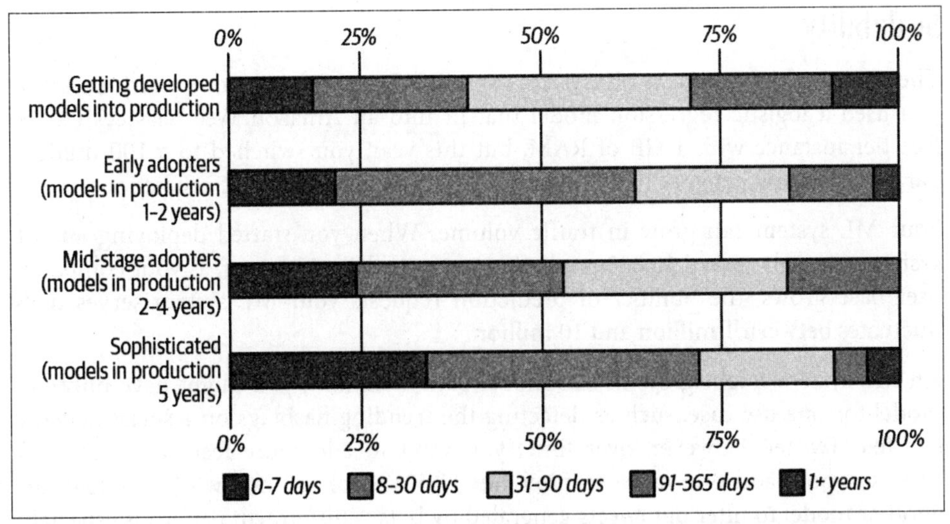

Figure 2-1. How long it takes for a company to bring a model to production is proportional to how long it has used ML. Source: Adapted from an image by Algorithmia

Requirements for ML Systems

We can't say that we've successfully built an ML system without knowing what requirements the system has to satisfy. The specified requirements for an ML system vary from use case to use case. However, most systems should have these four characteristics: reliability, scalability, maintainability, and adaptability. We'll walk through each of these concepts in detail. Let's take a closer look at reliability first.

Reliability

The system should continue to perform the correct function at the desired level of performance even in the face of adversity (hardware or software faults, and even human error).

"Correctness" might be difficult to determine for ML systems. For example, your system might call the predict function—e.g., `model.predict()`—correctly, but the predictions are wrong. How do we know if a prediction is wrong if we don't have ground truth labels to compare it with?

With traditional software systems, you often get a warning, such as a system crash or runtime error or 404. However, ML systems can fail silently. End users don't even know that the system has failed and might have kept on using it as if it were working. For example, if you use Google Translate to translate a sentence into a language you don't know, it might be very hard for you to tell even if the translation is wrong. We'll discuss how ML systems fail in production in Chapter 8.

Scalability

There are multiple ways an ML system can grow. It can grow in complexity. Last year you used a logistic regression model that fit into an Amazon Web Services (AWS) free tier instance with 1 GB of RAM, but this year, you switched to a 100-million-parameter neural network that requires 16 GB of RAM to generate predictions.

Your ML system can grow in traffic volume. When you started deploying an ML system, you only served 10,000 prediction requests daily. However, as your company's user base grows, the number of prediction requests your ML system serves daily fluctuates between 1 million and 10 million.

An ML system might grow in ML model count. Initially, you might have only one model for one use case, such as detecting the trending hashtags on a social network site like Twitter. However, over time, you want to add more features to this use case, so you'll add one more to filter out NSFW (not safe for work) content and another model to filter out tweets generated by bots. This growth pattern is especially common in ML systems that target enterprise use cases. Initially, a startup might serve only one enterprise customer, which means this startup only has one model. However, as this startup gains more customers, they might have one model for each customer. A startup I worked with had 8,000 models in production for their 8,000 enterprise customers.

Whichever way your system grows, there should be reasonable ways of dealing with that growth. When talking about scalability most people think of resource scaling, which consists of up-scaling (expanding the resources to handle growth) and down-scaling (reducing the resources when not needed).[8]

For example, at peak, your system might require 100 GPUs (graphics processing units). However, most of the time, it needs only 10 GPUs. Keeping 100 GPUs up all the time can be costly, so your system should be able to scale down to 10 GPUs.

An indispensable feature in many cloud services is autoscaling: automatically scaling up and down the number of machines depending on usage. This feature can be tricky to implement. Even Amazon fell victim to this when their autoscaling feature failed on Prime Day, causing their system to crash. An hour of downtime was estimated to cost Amazon between $72 million and $99 million.[9]

---

8 Up-scaling and down-scaling are two aspects of "scaling out," which is different from "scaling up." Scaling out is adding more equivalently functional components in parallel to spread out a load. Scaling up is making a component larger or faster to handle a greater load (Leah Schoeb, "Cloud Scalability: Scale Up vs Scale Out," *Turbonomic Blog*, March 15, 2018, *https://oreil.ly/CFPtb*).

9 Sean Wolfe, "Amazon's One Hour of Downtime on Prime Day May Have Cost It up to $100 Million in Lost Sales," *Business Insider*, July 19, 2018, *https://oreil.ly/VBezI*.

However, handling growth isn't just resource scaling, but also artifact management. Managing one hundred models is very different from managing one model. With one model, you can, perhaps, manually monitor this model's performance and manually update the model with new data. Since there's only one model, you can just have a file that helps you reproduce this model whenever needed. However, with one hundred models, both the monitoring and retraining aspect will need to be automated. You'll need a way to manage the code generation so that you can adequately reproduce a model when you need to.

Because scalability is such an important topic throughout the ML project workflow, we'll discuss it in different parts of the book. Specifically, we'll touch on the resource scaling aspect in the section "Distributed Training" on page 168, the section "Model optimization" on page 216, and the section "Resource Management" on page 311. We'll discuss the artifact management aspect in the section "Experiment Tracking and Versioning" on page 162 and the section "Development Environment" on page 302.

Maintainability

There are many people who will work on an ML system. They are ML engineers, DevOps engineers, and subject matter experts (SMEs). They might come from very different backgrounds, with very different programming languages and tools, and might own different parts of the process.

It's important to structure your workloads and set up your infrastructure in such a way that different contributors can work using tools that they are comfortable with, instead of one group of contributors forcing their tools onto other groups. Code should be documented. Code, data, and artifacts should be versioned. Models should be sufficiently reproducible so that even when the original authors are not around, other contributors can have sufficient contexts to build on their work. When a problem occurs, different contributors should be able to work together to identify the problem and implement a solution without finger-pointing.

We'll go more into this in the section "Team Structure" on page 334.

Adaptability

To adapt to shifting data distributions and business requirements, the system should have some capacity for both discovering aspects for performance improvement and allowing updates without service interruption.

Because ML systems are part code, part data, and data can change quickly, ML systems need to be able to evolve quickly. This is tightly linked to maintainability. We'll discuss changing data distributions in the section "Data Distribution Shifts" on page 237, and how to continually update your model with new data in the section "Continual Learning" on page 264.

Iterative Process

Developing an ML system is an iterative and, in most cases, never-ending process.[10] Once a system is put into production, it'll need to be continually monitored and updated.

Before deploying my first ML system, I thought the process would be linear and straightforward. I thought all I had to do was to collect data, train a model, deploy that model, and be done. However, I soon realized that the process looks more like a cycle with a lot of back and forth between different steps.

For example, here is one workflow that you might encounter when building an ML model to predict whether an ad should be shown when users enter a search query:[11]

1. Choose a metric to optimize. For example, you might want to optimize for impressions—the number of times an ad is shown.
2. Collect data and obtain labels.
3. Engineer features.
4. Train models.
5. During error analysis, you realize that errors are caused by the wrong labels, so you relabel the data.
6. Train the model again.
7. During error analysis, you realize that your model always predicts that an ad shouldn't be shown, and the reason is because 99.99% of the data you have have NEGATIVE labels (ads that shouldn't be shown). So you have to collect more data of ads that should be shown.

10 Which, as an early reviewer pointed out, is a property of traditional software.

11 Praying and crying not featured, but present through the entire process.

8. Train the model again.
9. The model performs well on your existing test data, which is by now two months old. However, it performs poorly on the data from yesterday. Your model is now stale, so you need to update it on more recent data.
10. Train the model again.
11. Deploy the model.
12. The model seems to be performing well, but then the businesspeople come knocking on your door asking why the revenue is decreasing. It turns out the ads are being shown, but few people click on them. So you want to change your model to optimize for ad click-through rate instead.
13. Go to step 1.

Figure 2-2 shows an oversimplified representation of what the iterative process for developing ML systems in production looks like from the perspective of a data scientist or an ML engineer. This process looks different from the perspective of an ML platform engineer or a DevOps engineer, as they might not have as much context into model development and might spend a lot more time on setting up infrastructure.

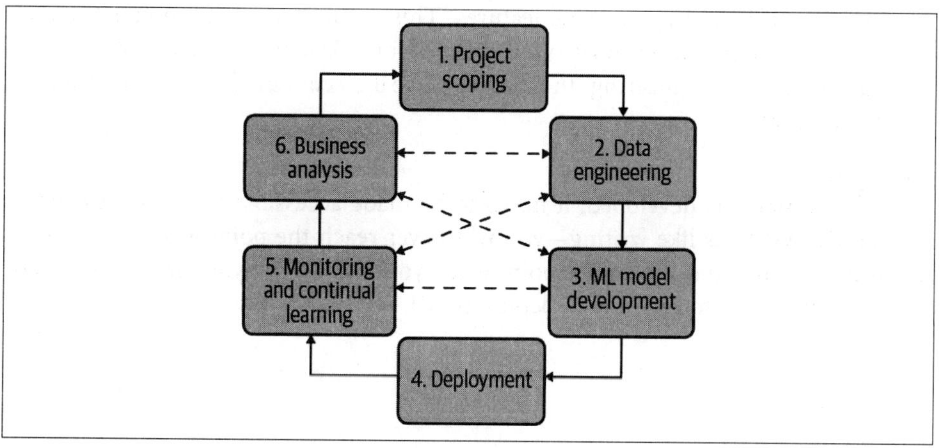

Figure 2-2. The process of developing an ML system looks more like a cycle with a lot of back and forth between steps

Later chapters will dive deeper into what each of these steps requires in practice. Here, let's take a brief look at what they mean:

Step 1. Project scoping
: A project starts with scoping the project, laying out goals, objectives, and constraints. Stakeholders should be identified and involved. Resources should be estimated and allocated. We already discussed different stakeholders and some of the foci for ML projects in production in Chapter 1. We also already discussed how to scope an ML project in the context of a business earlier in this chapter. We'll discuss how to organize teams to ensure the success of an ML project in Chapter 11.

Step 2. Data engineering
: A vast majority of ML models today learn from data, so developing ML models starts with engineering data. In Chapter 3, we'll discuss the fundamentals of data engineering, which covers handling data from different sources and formats. With access to raw data, we'll want to curate training data out of it by sampling and generating labels, which is discussed in Chapter 4.

Step 3. ML model development
: With the initial set of training data, we'll need to extract features and develop initial models leveraging these features. This is the stage that requires the most ML knowledge and is most often covered in ML courses. In Chapter 5, we'll discuss feature engineering. In Chapter 6, we'll discuss model selection, training, and evaluation.

Step 4. Deployment
: After a model is developed, it needs to be made accessible to users. Developing an ML system is like writing—you will never reach the point when your system is done. But you do reach the point when you have to put your system out there. We'll discuss different ways to deploy an ML model in Chapter 7.

Step 5. Monitoring and continual learning

Once in production, models need to be monitored for performance decay and maintained to be adaptive to changing environments and changing requirements. This step will be discussed in Chapters 8 and 9.

Step 6. Business analysis

Model performance needs to be evaluated against business goals and analyzed to generate business insights. These insights can then be used to eliminate unproductive projects or scope out new projects. This step is closely related to the first step.

Framing ML Problems

Imagine you're an ML engineering tech lead at a bank that targets millennial users. One day, your boss hears about a rival bank that uses ML to speed up their customer service support that supposedly helps the rival bank process their customer requests two times faster. He orders your team to look into using ML to speed up your customer service support too.

Slow customer support is a problem, but it's not an ML problem. An ML problem is defined by inputs, outputs, and the objective function that guides the learning process—none of these three components are obvious from your boss's request. It's your job, as a seasoned ML engineer, to use your knowledge of what problems ML can solve to frame this request as an ML problem.

Upon investigation, you discover that the bottleneck in responding to customer requests lies in routing customer requests to the right department among four departments: accounting, inventory, HR (human resources), and IT. You can alleviate this bottleneck by developing an ML model to predict which of these four departments a request should go to. This makes it a classification problem. The input is the customer request. The output is the department the request should go to. The objective function is to minimize the difference between the predicted department and the actual department.

We'll discuss extensively how to extract features from raw data to input into your ML model in Chapter 5. In this section, we'll focus on two aspects: the output of your model and the objective function that guides the learning process.

Types of ML Tasks

The output of your model dictates the task type of your ML problem. The most general types of ML tasks are classification and regression. Within classification, there are more subtypes, as shown in Figure 2-3. We'll go over each of these task types.

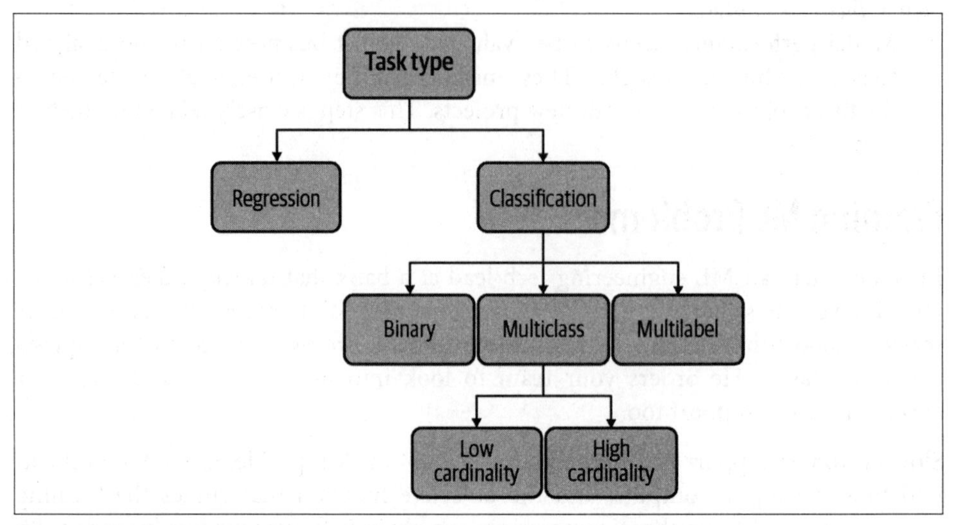

Figure 2-3. Common task types in ML

Classification versus regression

Classification models classify inputs into different categories. For example, you want to classify each email to be either spam or not spam. Regression models output a continuous value. An example is a house prediction model that outputs the price of a given house.

A regression model can easily be framed as a classification model and vice versa. For example, house prediction can become a classification task if we quantize the house prices into buckets such as under $100,000, $100,000–$200,000, $200,000–$500,000, and so forth and predict the bucket the house should be in.

The email classification model can become a regression model if we make it output values between 0 and 1, and decide on a threshold to determine which values should be SPAM (for example, if the value is above 0.5, the email is spam), as shown in Figure 2-4.

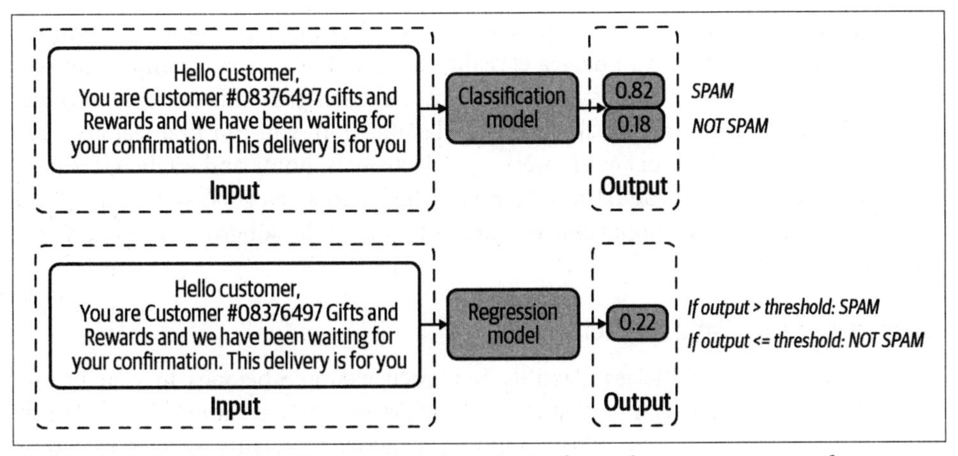

Figure 2-4. The email classification task can also be framed as a regression task

Binary versus multiclass classification

Within classification problems, the fewer classes there are to classify, the simpler the problem is. The simplest is *binary classification*, where there are only two possible classes. Examples of binary classification include classifying whether a comment is toxic, whether a lung scan shows signs of cancer, whether a transaction is fraudulent. It's unclear whether this type of problem is common in the industry because they are common in nature or simply because ML practitioners are most comfortable handling them.

When there are more than two classes, the problem becomes *multiclass classification*. Dealing with binary classification problems is much easier than dealing with multiclass problems. For example, calculating F1 and visualizing confusion matrices are a lot more intuitive when there are only two classes.

When the number of classes is high, such as disease diagnosis where the number of diseases can go up to thousands or product classifications where the number of products can go up to tens of thousands, we say the classification task has *high cardinality*. High cardinality problems can be very challenging. The first challenge is in data collection. In my experience, ML models typically need at least 100 examples for each class to learn to classify that class. So if you have 1,000 classes, you already need at least 100,000 examples. The data collection can be especially difficult for rare classes. When you have thousands of classes, it's likely that some of them are rare.

When the number of classes is large, hierarchical classification might be useful. In hierarchical classification, you have a classifier to first classify each example into one of the large groups. Then you have another classifier to classify this example into one of the subgroups. For example, for product classification, you can first classify each product into one of the four main categories: electronics, home and kitchen, fashion, or pet supplies. After a product has been classified into a category, say fashion, you can use another classifier to put this product into one of the subgroups: shoes, shirts, jeans, or accessories.

Multiclass versus multilabel classification

In both binary and multiclass classification, each example belongs to exactly one class. When an example can belong to multiple classes, we have a *multilabel classification* problem. For example, when building a model to classify articles into four topics—tech, entertainment, finance, and politics—an article can be in both tech and finance.

There are two major approaches to multilabel classification problems. The first is to treat it as you would a multiclass classification. In multiclass classification, if there are four possible classes [tech, entertainment, finance, politics] and the label for an example is entertainment, you represent this label with the vector [0, 1, 0, 0]. In multilabel classification, if an example has both labels entertainment and finance, its label will be represented as [0, 1, 1, 0].

The second approach is to turn it into a set of binary classification problems. For the article classification problem, you can have four models corresponding to four topics, each model outputting whether an article is in that topic or not.

Out of all task types, multilabel classification is usually the one that I've seen companies having the most problems with. Multilabel means that the number of classes an example can have varies from example to example. First, this makes it difficult for label annotation since it increases the label multiplicity problem that we discuss in Chapter 4. For example, an annotator might believe an example belongs to two classes while another annotator might believe the same example to belong in only one class, and it might be difficult resolving their disagreements.

Second, this varying number of classes makes it hard to extract predictions from raw probability. Consider the same task of classifying articles into four topics. Imagine that, given an article, your model outputs this raw probability distribution: [0.45, 0.2, 0.02, 0.33]. In the multiclass setting, when you know that an example can belong to only one category, you simply pick the category with the highest probability, which is 0.45 in this case. In the multilabel setting, because you don't know how many categories an example can belong to, you might pick the two highest probability categories (corresponding to 0.45 and 0.33) or three highest probability categories (corresponding to 0.45, 0.2, and 0.33).

Multiple ways to frame a problem

Changing the way you frame your problem might make your problem significantly harder or easier. Consider the task of predicting what app a phone user wants to use next. A naive setup would be to frame this as a multiclass classification task—use the user's and environment's features (user demographic information, time, location, previous apps used) as input, and output a probability distribution for every single app on the user's phone. Let *N* be the number of apps you want to consider recommending to a user. In this framing, for a given user at a given time, there is only one prediction to make, and the prediction is a vector of the size *N*. This setup is visualized in Figure 2-5.

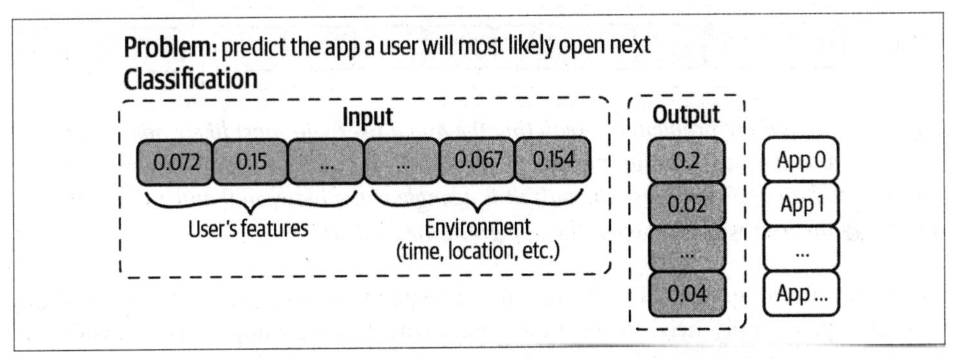

Figure 2-5. Given the problem of predicting the app a user will most likely open next, you can frame it as a classification problem. The input is the user's features and environment's features. The output is a distribution over all apps on the phone.

This is a bad approach because whenever a new app is added, you might have to retrain your model from scratch, or at least retrain all the components of your model whose number of parameters depends on *N*. A better approach is to frame this as a regression task. The input is the user's, the environment's, and the app's features. The output is a single value between 0 and 1; the higher the value, the more likely the user will open the app given the context. In this framing, for a given user at a given time, there are *N* predictions to make, one for each app, but each prediction is just a number. This improved setup is visualized in Figure 2-6.

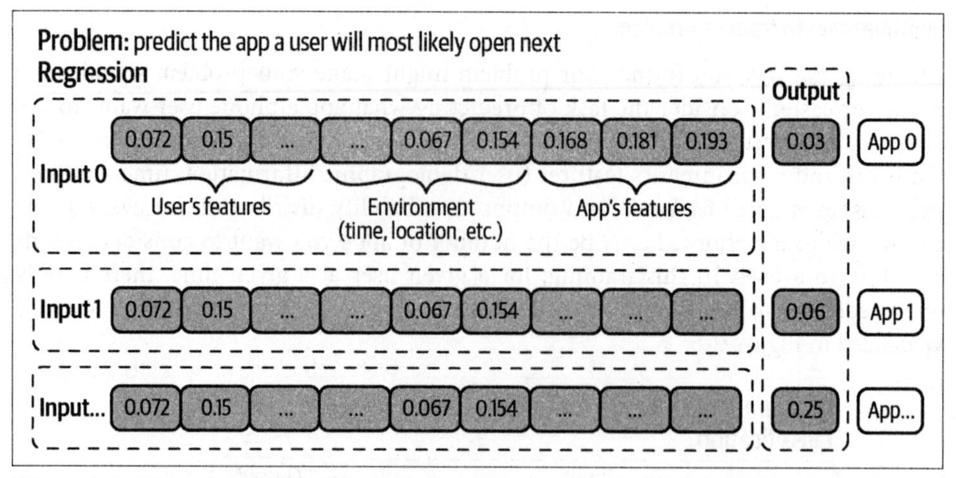

Figure 2-6. Given the problem of predicting the app a user will most likely open next, you can frame it as a regression problem. The input is the user's features, environment's features, and an app's features. The output is a single value between 0 and 1 denoting how likely the user will be to open the app given the context.

In this new framing, whenever there's a new app you want to consider recommending to a user, you simply need to use new inputs with this new app's feature instead of having to retrain your model or part of your model from scratch.

Objective Functions

To learn, an ML model needs an objective function to guide the learning process.[12] An objective function is also called a loss function, because the objective of the learning process is usually to minimize (or optimize) the loss caused by wrong predictions. For supervised ML, this loss can be computed by comparing the model's outputs with the ground truth labels using a measurement like root mean squared error (RMSE) or cross entropy.

To illustrate this point, let's again go back to the previous task of classifying articles into four topics [tech, entertainment, finance, politics]. Consider an article that belongs to the politics class, e.g., its ground truth label is [0, 0, 0, 1]. Imagine that, given this article, your model outputs this raw probability distribution: [0.45, 0.2, 0.02, 0.33]. The cross entropy loss of this model, given this example, is the cross entropy of [0.45, 0.2, 0.02, 0.33] relative to [0, 0, 0, 1]. In Python, you can calculate cross entropy with the following code:

12 Note that objective functions are mathematical functions, which are different from the business and ML objectives we discussed earlier in this chapter.

```
import numpy as np

def cross_entropy(p, q):
return -sum([p[i] * np.log(q[i]) for i in range(len(p))])

p = [0, 0, 0, 1]
q = [0.45, 0.2, 0.02, 0.33]
cross_entropy(p, q)
```

Choosing an objective function is usually straightforward, though not because objective functions are easy. Coming up with meaningful objective functions requires algebra knowledge, so most ML engineers just use common loss functions like RMSE or MAE (mean absolute error) for regression, logistic loss (also log loss) for binary classification, and cross entropy for multiclass classification.

Decoupling objectives

Framing ML problems can be tricky when you want to minimize multiple objective functions. Imagine you're building a system to rank items on users' newsfeeds. Your original goal is to maximize users' engagement. You want to achieve this goal through the following three objectives:

- Filter out spam
- Filter out NSFW content
- Rank posts by engagement: how likely users will click on it

However, you quickly learned that optimizing for users' engagement alone can lead to questionable ethical concerns. Because extreme posts tend to get more engagements, your algorithm learned to prioritize extreme content.[13] You want to create a more wholesome newsfeed. So you have a new goal: maximize users' engagement while minimizing the spread of extreme views and misinformation. To obtain this goal, you add two new objectives to your original plan:

- Filter out spam
- Filter out NSFW content
- Filter out misinformation
- Rank posts by quality
- Rank posts by engagement: how likely users will click on it

13 Joe Kukura, "Facebook Employee Raises Powered by 'Really Dangerous' Algorithm That Favors Angry Posts," *SFist*, September 24, 2019, *https://oreil.ly/PXtGi*; Kevin Roose, "The Making of a YouTube Radical," *New York Times*, June 8, 2019, *https://oreil.ly/KYqzF*.

Now two objectives are in conflict with each other. If a post is engaging but it's of questionable quality, should that post rank high or low?

An objective is represented by an objective function. To rank posts by quality, you first need to predict posts' quality, and you want posts' predicted quality to be as close to their actual quality as possible. Essentially, you want to minimize *quality_loss*: the difference between each post's predicted quality and its true quality.[14]

Similarly, to rank posts by engagement, you first need to predict the number of clicks each post will get. You want to minimize *engagement_loss*: the difference between each post's predicted clicks and its actual number of clicks.

One approach is to combine these two losses into one loss and train one model to minimize that loss:

$$loss = \alpha\ quality\_loss + \beta\ engagement\_loss$$

You can randomly test out different values of α and β to find the values that work best. If you want to be more systematic about tuning these values, you can check out Pareto optimization, "an area of multiple criteria decision making that is concerned with mathematical optimization problems involving more than one objective function to be optimized simultaneously."[15]

A problem with this approach is that each time you tune α and β—for example, if the quality of your users' newsfeeds goes up but users' engagement goes down, you might want to decrease α and increase β—you'll have to retrain your model.

Another approach is to train two different models, each optimizing one loss. So you have two models:

quality_model
: Minimizes *quality_loss* and outputs the predicted quality of each post

engagement_model
: Minimizes *engagement_loss* and outputs the predicted number of clicks of each post

14 For simplicity, let's pretend for now that we know how to measure a post's quality.

15 Wikipedia, s.v. "Pareto optimization," *https://oreil.ly/NdApy*. While you're at it, you might also want to read Jin and Sendhoff's great paper on applying Pareto optimization for ML, in which the authors claimed that "machine learning is inherently a multiobjective task" (Yaochu Jin and Bernhard Sendhoff, "Pareto-Based Multiobjective Machine Learning: An Overview and Case Studies," *IEEE Transactions on Systems, Man, and Cybernetics—Part C: Applications and Reviews* 38, no. 3 [May 2008], *https://oreil.ly/f1aKk*).

You can combine the models' outputs and rank posts by their combined scores:

$$\alpha\ quality\_score + \beta\ engagement\_score$$

Now you can tweak α and β without retraining your models!

In general, when there are multiple objectives, it's a good idea to decouple them first because it makes model development and maintenance easier. First, it's easier to tweak your system without retraining models, as previously explained. Second, it's easier for maintenance since different objectives might need different maintenance schedules. Spamming techniques evolve much faster than the way post quality is perceived, so spam filtering systems need updates at a much higher frequency than quality-ranking systems.

Mind Versus Data

Progress in the last decade shows that the success of an ML system depends largely on the data it was trained on. Instead of focusing on improving ML algorithms, most companies focus on managing and improving their data.[16]

Despite the success of models using massive amounts of data, many are skeptical of the emphasis on data as the way forward. In the last five years, at every academic conference I attended, there were always some public debates on the power of mind versus data. *Mind* might be disguised as inductive biases or intelligent architectural designs. *Data* might be grouped together with computation since more data tends to require more computation.

In theory, you can both pursue architectural designs and leverage large data and computation, but spending time on one often takes time away from another.[17]

In the mind-over-data camp, there's Dr. Judea Pearl, a Turing Award winner best known for his work on causal inference and Bayesian networks. The introduction to his book *The Book of Why* is entitled "Mind over Data," in which he emphasizes: "Data is profoundly dumb." In one of his more controversial posts on Twitter in 2020, he expressed his strong opinion against ML approaches that rely heavily on data and warned that data-centric ML people might be out of a job in three to five years: "ML will not be the same in 3–5 years, and ML folks who continue to follow the current data-centric paradigm will find themselves outdated, if not jobless. Take note."[18]

---

16 Anand Rajaraman, "More Data Usually Beats Better Algorithms," *Datawocky*, March 24, 2008, *https://oreil.ly/wNwhV*.

17 Rich Sutton, "The Bitter Lesson," March 13, 2019, *https://oreil.ly/RhOp9*.

18 Tweet by Dr. Judea Pearl (@yudapearl), September 27, 2020, *https://oreil.ly/wFbHb*.

There's also a milder opinion from Professor Christopher Manning, director of the Stanford Artificial Intelligence Laboratory, who argued that huge computation and a massive amount of data with a simple learning algorithm create incredibly bad learners. The structure allows us to design systems that can learn more from less data.[19]

Many people in ML today are in the data-over-mind camp. Professor Richard Sutton, a professor of computing science at the University of Alberta and a distinguished research scientist at DeepMind, wrote a great blog post in which he claimed that researchers who chose to pursue intelligent designs over methods that leverage computation will eventually learn a bitter lesson: "The biggest lesson that can be read from 70 years of AI research is that general methods that leverage computation are ultimately the most effective, and by a large margin.... Seeking an improvement that makes a difference in the shorter term, researchers seek to leverage their human knowledge of the domain, but the only thing that matters in the long run is the leveraging of computation."[20]

When asked how Google Search was doing so well, Peter Norvig, Google's director of search quality, emphasized the importance of having a large amount of data over intelligent algorithms in their success: "We don't have better algorithms. We just have more data."[21]

Dr. Monica Rogati, former VP of data at Jawbone, argued that data lies at the foundation of data science, as shown in Figure 2-7. If you want to use data science, a discipline of which ML is a part of, to improve your products or processes, you need to start with building out your data, both in terms of quality and quantity. Without data, there's no data science.

The debate isn't about whether finite data is necessary, but whether it's sufficient. The term *finite* here is important, because if we had infinite data, it might be possible for us to look up the answer. Having a lot of data is different from having infinite data.

19 "Deep Learning and Innate Priors" (Chris Manning versus Yann LeCun debate), February 2, 2018, video, 1:02:55, *https://oreil.ly/b3hb1*.

20 Sutton, "The Bitter Lesson."

21 Alon Halevy, Peter Norvig, and Fernando Pereira, "The Unreasonable Effectiveness of Data," *IEEE Computer Society*, March/April 2009, *https://oreil.ly/WkN6p*.

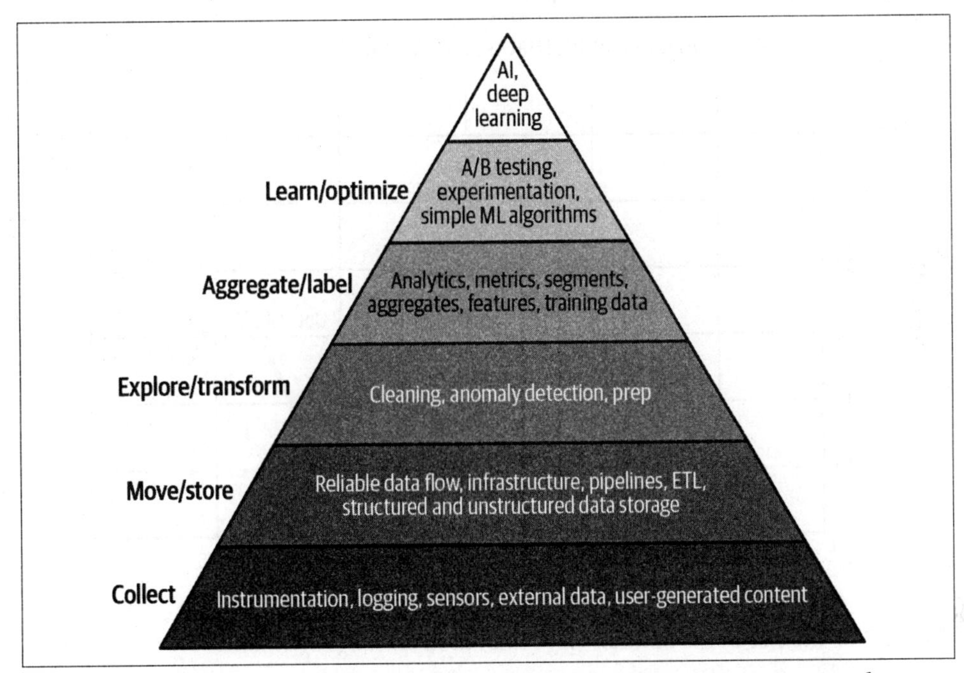

Figure 2-7. The data science hierarchy of needs. Source: Adapted from an image by Monica Rogati[22]

Regardless of which camp will prove to be right eventually, no one can deny that data is essential, for now. Both the research and industry trends in the recent decades show the success of ML relies more and more on the quality and quantity of data. Models are getting bigger and using more data. Back in 2013, people were getting excited when the One Billion Word Benchmark for Language Modeling was released, which contains 0.8 billion tokens.[23] Six years later, OpenAI's GPT-2 used a dataset of 10 billion tokens. And another year later, GPT-3 used 500 billion tokens. The growth rate of the sizes of datasets is shown in Figure 2-8.

22 Monica Rogati, "The AI Hierarchy of Needs," *Hackernoon Newsletter*, June 12, 2017, *https://oreil.ly/3nxJ8*.

23 Ciprian Chelba, Tomas Mikolov, Mike Schuster, Qi Ge, Thorsten Brants, Phillipp Koehn, and Tony Robinson, "One Billion Word Benchmark for Measuring Progress in Statistical Language Modeling," *arXiv*, December 11, 2013, *https://oreil.ly/1AdO6*.

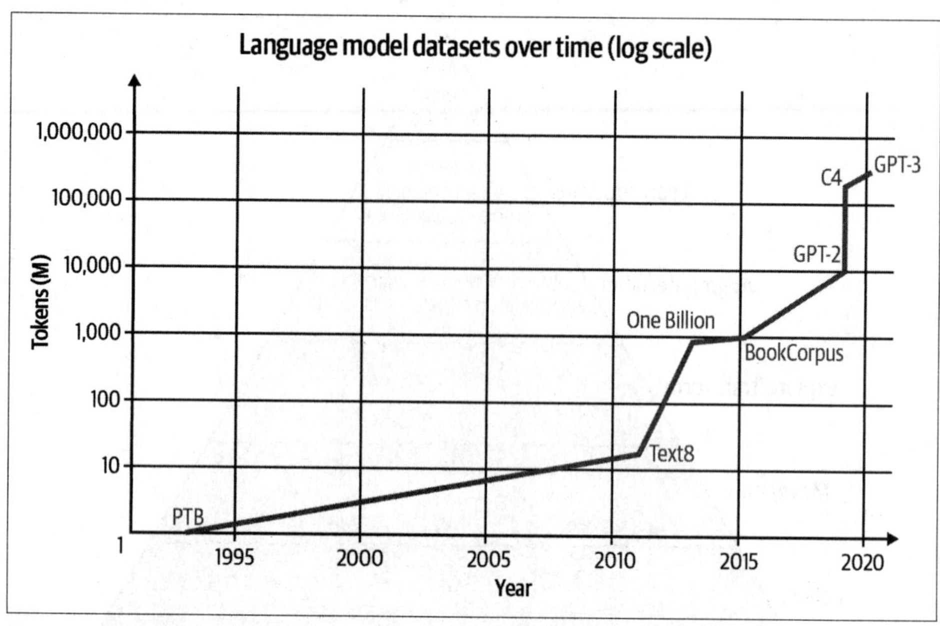

Figure 2-8. The size of the datasets (log scale) used for language models over time

Even though much of the progress in deep learning in the last decade was fueled by an increasingly large amount of data, more data doesn't always lead to better performance for your model. More data at lower quality, such as data that is outdated or data with incorrect labels, might even hurt your model's performance.

Summary

I hope that this chapter has given you an introduction to ML systems design and the considerations we need to take into account when designing an ML system.

Every project must start with why this project needs to happen, and ML projects are no exception. We started the chapter with an assumption that most businesses don't care about ML metrics unless they can move business metrics. Therefore, if an ML system is built for a business, it must be motivated by business objectives, which need to be translated into ML objectives to guide the development of ML models.

Before building an ML system, we need to understand the requirements that the system needs to meet to be considered a good system. The exact requirements vary from use case to use case, and in this chapter, we focused on the four most general requirements: reliability, scalability, maintainability, and adaptability. Techniques to satisfy each of these requirements will be covered throughout the book.

Building an ML system isn't a one-off task but an iterative process. In this chapter, we discussed the iterative process to develop an ML system that met those preceding requirements.

We ended the chapter on a philosophical discussion of the role of data in ML systems. There are still many people who believe that having intelligent algorithms will eventually trump having a large amount of data. However, the success of systems including AlexNet, BERT, and GPT showed that the progress of ML in the last decade relies on having access to a large amount of data.[24] Regardless of whether data can overpower intelligent design, no one can deny the importance of data in ML. A nontrivial part of this book will be devoted to shedding light on various data questions.

Complex ML systems are made up of simpler building blocks. Now that we've covered the high-level overview of an ML system in production, we'll zoom in to its building blocks in the following chapters, starting with the fundamentals of data engineering in the next chapter. If any of the challenges mentioned in this chapter seem abstract to you, I hope that specific examples in the following chapters will make them more concrete.

24 Alex Krizhevsky, Ilya Sutskever, and Geoffrey E Hinton, "ImageNet Classification with Deep Convolutional Neural Networks," in *Advances in Neural Information Processing Systems*, vol. 25, ed. F. Pereira, C.J. Burges, L. Bottou, and K.Q. Weinberger (Curran Associates, 2012), *https://oreil.ly/MFYp9*; Jacob Devlin, Ming-Wei Chang, Kenton Lee, and Kristina Toutanova, "BERT: Pre-training of Deep Bidirectional Transformers for Language Understanding," *arXiv*, 2019, *https://oreil.ly/TN8fN*; "Better Language Models and Their Implications," OpenAI blog, February 14, 2019, *https://oreil.ly/SGV7g*.

CHAPTER 3

Data Engineering Fundamentals

The rise of ML in recent years is tightly coupled with the rise of big data. Large data systems, even without ML, are complex. If you haven't spent years and years working with them, it's easy to get lost in acronyms. There are many challenges and possible solutions that these systems generate. Industry standards, if there are any, evolve quickly as new tools come out and the needs of the industry expand, creating a dynamic and ever-changing environment. If you look into the data stack for different tech companies, it might seem like each is doing its own thing.

In this chapter, we'll cover the basics of data engineering that will, hopefully, give you a steady piece of land to stand on as you explore the landscape for your own needs. We'll start with different sources of data that you might work with in a typical ML project. We'll continue to discuss the formats in which data can be stored. Storing data is only interesting if you intend on retrieving that data later. To retrieve stored data, it's important to know not only how it's formatted but also how it's structured. Data models define how the data stored in a particular data format is structured.

If data models describe the data in the real world, databases specify how the data should be stored on machines. We'll continue to discuss data storage engines, also known as databases, for the two major types of processing: transactional and analytical.

When working with data in production, you usually work with data across multiple processes and services. For example, you might have a feature engineering service that computes features from raw data, and a prediction service to generate predictions based on computed features. This means that you'll have to pass computed features from the feature engineering service to the prediction service. In the following section of the chapter, we'll discuss different modes of data passing across processes.

During the discussion of different modes of data passing, we'll learn about two distinct types of data: historical data in data storage engines, and streaming data in real-time transports. These two different types of data require different processing paradigms, which we'll discuss in the section "Batch Processing Versus Stream Processing" on page 78.

Knowing how to collect, process, store, retrieve, and process an increasingly growing amount of data is essential to people who want to build ML systems in production. If you're already familiar with data systems, you might want to move directly to Chapter 4 to learn more about how to sample and generate labels to create training data. If you want to learn more about data engineering from a systems perspective, I recommend Martin Kleppmann's excellent book *Designing Data-Intensive Applications* (O'Reilly, 2017).

Data Sources

An ML system can work with data from many different sources. They have different characteristics, can be used for different purposes, and require different processing methods. Understanding the sources your data comes from can help you use your data more efficiently. This section aims to give a quick overview of different data sources to those unfamiliar with data in production. If you've already worked with ML in production for a while, feel free to skip this section.

One source is *user input data*, data explicitly input by users. User input can be text, images, videos, uploaded files, etc. If it's even remotely possible for users to input wrong data, they are going to do it. As a result, user input data can be easily malformatted. Text might be too long or too short. Where numerical values are expected, users might accidentally enter text. If you let users upload files, they might upload files in the wrong formats. User input data requires more heavy-duty checking and processing.

On top of that, users also have little patience. In most cases, when we input data, we expect to get results back immediately. Therefore, user input data tends to require fast processing.

Another source is *system-generated data*. This is the data generated by different components of your systems, which include various types of logs and system outputs such as model predictions.

Logs can record the state and significant events of the system, such as memory usage, number of instances, services called, packages used, etc. They can record the results of different jobs, including large batch jobs for data processing and model training. These types of logs provide visibility into how the system is doing. The main purpose of this visibility is for debugging and potentially improving the application. Most of the time, you don't have to look at these types of logs, but they are essential when something is on fire.

Because logs are system generated, they are much less likely to be malformatted the way user input data is. Overall, logs don't need to be processed as soon as they arrive, the way you would want to process user input data. For many use cases, it's acceptable to process logs periodically, such as hourly or even daily. However, you might still want to process your logs fast to be able to detect and be notified whenever something interesting happens.[1]

Because debugging ML systems is hard, it's a common practice to log everything you can. This means that your volume of logs can grow very, very quickly. This leads to two problems. The first is that it can be hard to know where to look because signals are lost in the noise. There have been many services that process and analyze logs, such as Logstash, Datadog, Logz.io, etc. Many of them use ML models to help you process and make sense of your massive number of logs.

The second problem is how to store a rapidly growing number of logs. Luckily, in most cases, you only have to store logs for as long as they are useful and can discard them when they are no longer relevant for you to debug your current system. If you don't have to access your logs frequently, they can also be stored in low-access storage that costs much less than higher-frequency-access storage.[2]

The system also generates data to record users' behaviors, such as clicking, choosing a suggestion, scrolling, zooming, ignoring a pop-up, or spending an unusual amount of time on certain pages. Even though this is system-generated data, it's still considered part of user data and might be subject to privacy regulations.[3]

---

1 "Interesting" in production usually means catastrophic, such as a crash or when your cloud bill hits an astronomical amount.

2 As of November 2021, AWS S3 Standard, the storage option that allows you to access your data with the latency of milliseconds, costs about five times more per GB than S3 Glacier, the storage option that allows you to retrieve your data with a latency from between 1 minute to 12 hours.

3 An ML engineer once mentioned to me that his team only used users' historical product browsing and purchases to make recommendations on what they might like to see next. I responded: "So you don't use personal data at all?" He looked at me, confused. "If you meant demographic data like users' age, location, then no, we don't. But I'd say that a person's browsing and purchasing activities are extremely personal."

There are also *internal databases*, generated by various services and enterprise applications in a company. These databases manage their assets such as inventory, customer relationship, users, and more. This kind of data can be used by ML models directly or by various components of an ML system. For example, when users enter a search query on Amazon, one or more ML models process that query to detect its intention—if someone types in "frozen," are they looking for frozen foods or Disney's *Frozen* franchise?—then Amazon needs to check its internal databases for the availability of these products before ranking them and showing them to users.

Then there's the wonderfully weird world of *third-party data*. First-party data is the data that your company already collects about your users or customers. Second-party data is the data collected by another company on their own customers that they make available to you, though you'll probably have to pay for it. Third-party data companies collect data on the public who aren't their direct customers.

The rise of the internet and smartphones has made it much easier for all types of data to be collected. It used to be especially easy with smartphones since each phone used to have a unique advertiser ID—iPhones with Apple's Identifier for Advertisers (IDFA) and Android phones with their Android Advertising ID (AAID)—which acted as a unique ID to aggregate all activities on a phone. Data from apps, websites, check-in services, etc. are collected and (hopefully) anonymized to generate activity history for each person.

Data of all kinds can be bought, such as social media activities, purchase history, web browsing habits, car rentals, and political leaning for different demographic groups getting as granular as men, age 25–34, working in tech, living in the Bay Area. From this data, you can infer information such as people who like brand A also like brand B. This data can be especially helpful for systems such as recommender systems to generate results relevant to users' interests. Third-party data is usually sold after being cleaned and processed by vendors.

However, as users demand more data privacy, companies have been taking steps to curb the usage of advertiser IDs. In early 2021, Apple made their IDFA opt-in. This change has reduced significantly the amount of third-party data available on iPhones, forcing many companies to focus more on first-party data.[4] To fight back this change, advertisers have been investing in workarounds. For example, the China Advertising Association, a state-supported trade association for China's advertising industry, invested in a device fingerprinting system called CAID that allowed apps like TikTok and Tencent to keep tracking iPhone users.[5]

---

4 John Koetsier, "Apple Just Crippled IDFA, Sending an $80 Billion Industry Into Upheaval," *Forbes*, June 24, 2020, *https://oreil.ly/rqPX9*.

5 Patrick McGee and Yuan Yang, "TikTok Wants to Keep Tracking iPhone Users with State-Backed Workaround," *Ars Technica*, March 16, 2021, *https://oreil.ly/54pkg*.

Data Formats

Once you have data, you might want to store it (or "persist" it, in technical terms). Since your data comes from multiple sources with different access patterns,[6] storing your data isn't always straightforward and, for some cases, can be costly. It's important to think about how the data will be used in the future so that the format you use will make sense. Here are some of the questions you might want to consider:

- How do I store multimodal data, e.g., a sample that might contain both images and texts?
- Where do I store my data so that it's cheap and still fast to access?
- How do I store complex models so that they can be loaded and run correctly on different hardware?

The process of converting a data structure or object state into a format that can be stored or transmitted and reconstructed later is *data serialization*. There are many, many data serialization formats. When considering a format to work with, you might want to consider different characteristics such as human readability, access patterns, and whether it's based on text or binary, which influences the size of its files. Table 3-1 consists of just a few of the common formats that you might encounter in your work. For a more comprehensive list, check out the wonderful Wikipedia page "Comparison of Data-Serialization Formats" (*https://oreil.ly/sgceY*).

Table 3-1. Common data formats and where they are used

Format	Binary/Text	Human-readable	Example use cases
JSON	Text	Yes	Everywhere
CSV	*Text*	*Yes*	*Everywhere*
Parquet	*Binary*	*No*	*Hadoop, Amazon Redshift*
Avro	Binary primary	No	Hadoop
Protobuf	Binary primary	No	Google, TensorFlow (TFRecord)
Pickle	Binary	No	Python, PyTorch serialization

We'll go over a few of these formats, starting with JSON. We'll also go over the two formats that are common and represent two distinct paradigms: CSV and Parquet.

6 "Access pattern" means the pattern in which a system or program reads or writes data.

JSON

JSON, JavaScript Object Notation, is everywhere. Even though it was derived from JavaScript, it's language-independent—most modern programming languages can generate and parse JSON. It's human-readable. Its key-value pair paradigm is simple but powerful, capable of handling data of different levels of structuredness. For example, your data can be stored in a structured format like the following:

```
{
  "firstName": "Boatie",
  "lastName": "McBoatFace",
  "isVibing": true,
  "age": 12,
  "address": {
    "streetAddress": "12 Ocean Drive",
    "city": "Port Royal",
    "postalCode": "10021-3100"
  }
}
```

The same data can also be stored in an unstructured blob of text like the following:

```
{
  "text": "Boatie McBoatFace, aged 12, is vibing, at 12 Ocean Drive, Port Royal,
          10021-3100"
}
```

Because JSON is ubiquitous, the pain it causes can also be felt everywhere. Once you've committed the data in your JSON files to a schema, it's pretty painful to retrospectively go back to change the schema. JSON files are text files, which means they take up a lot of space, as we'll see in the section "Text Versus Binary Format" on page 57.

Row-Major Versus Column-Major Format

The two formats that are common and represent two distinct paradigms are CSV and Parquet. CSV (comma-separated values) is row-major, which means consecutive elements in a row are stored next to each other in memory. Parquet is column-major, which means consecutive elements in a column are stored next to each other.

Because modern computers process sequential data more efficiently than nonsequential data, if a table is row-major, accessing its rows will be faster than accessing its columns in expectation. This means that for row-major formats, accessing data by rows is expected to be faster than accessing data by columns.

Imagine we have a dataset of 1,000 examples, and each example has 10 features. If we consider each example as a row and each feature as a column, as is often the case in ML, then the row-major formats like CSV are better for accessing examples, e.g., accessing all the examples collected today. Column-major formats like Parquet are better for accessing features, e.g., accessing the timestamps of all your examples. See Figure 3-1.

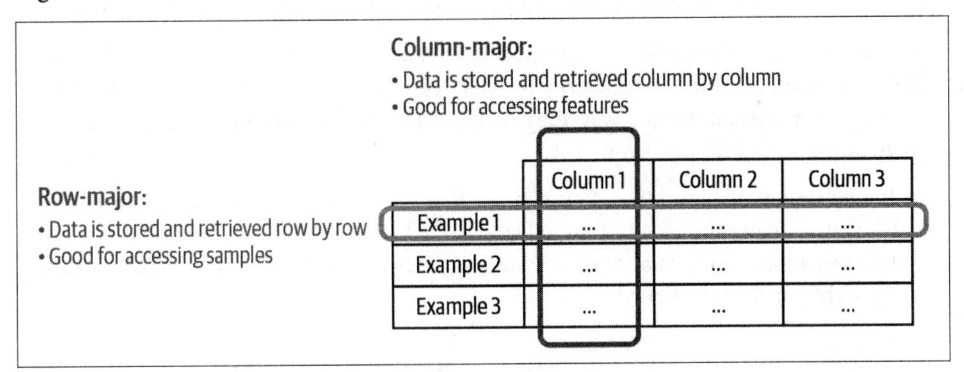

Figure 3-1. Row-major versus column-major formats

Column-major formats allow flexible column-based reads, especially if your data is large with thousands, if not millions, of features. Consider if you have data about ride-sharing transactions that has 1,000 features but you only want 4 features: time, location, distance, price. With column-major formats, you can read the four columns corresponding to these four features directly. However, with row-major formats, if you don't know the sizes of the rows, you will have to read in all columns then filter down to these four columns. Even if you know the sizes of the rows, it can still be slow as you'll have to jump around the memory, unable to take advantage of caching.

Row-major formats allow faster data writes. Consider the situation when you have to keep adding new individual examples to your data. For each individual example, it'd be much faster to write it to a file where your data is already in a row-major format.

Overall, row-major formats are better when you have to do a lot of writes, whereas column-major ones are better when you have to do a lot of column-based reads.

NumPy Versus pandas

One subtle point that a lot of people don't pay attention to, which leads to misuses of pandas, is that this library is built around the columnar format.

pandas is built around DataFrame, a concept inspired by R's Data Frame, which is column-major. A DataFrame is a two-dimensional table with rows and columns.

In NumPy, the major order can be specified. When an `ndarray` is created, it's row-major by default if you don't specify the order. People coming to pandas from NumPy tend to treat DataFrame the way they would `ndarray`, e.g., trying to access data by rows, and find DataFrame slow.

In the left panel of Figure 3-2, you can see that accessing a DataFrame by row is so much slower than accessing the same DataFrame by column. If you convert this same DataFrame to a NumPy `ndarray`, accessing a row becomes much faster, as you can see in the right panel of the figure.[7]

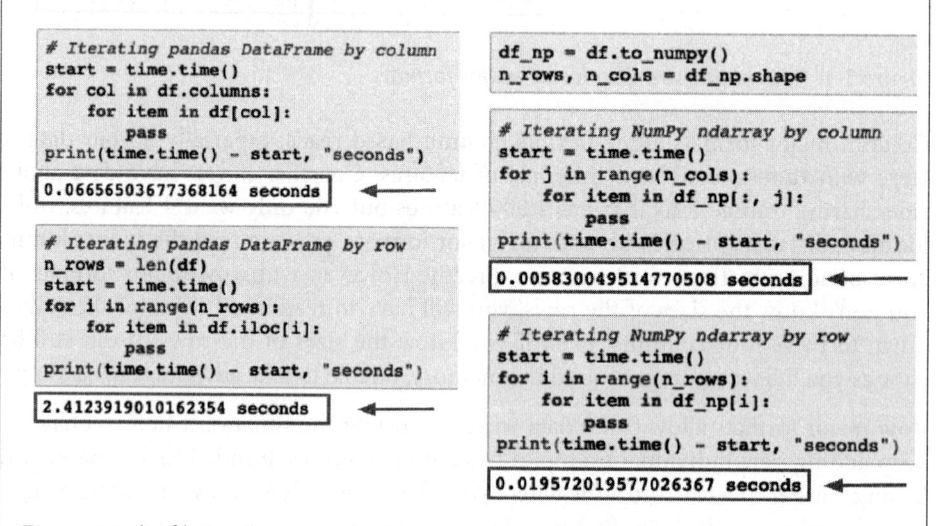

Figure 3-2. (Left) Iterating a pandas DataFrame by column takes 0.07 seconds but iterating the same DataFrame by row takes 2.41 seconds. (Right) When you convert the same DataFrame into a NumPy `ndarray`, accessing its rows becomes much faster.

[7] For more pandas quirks, check out my Just pandas Things (*https://oreil.ly/sFkJX*) GitHub repository.

I use CSV as an example of the row-major format because it's popular and generally recognizable by everyone I've talked to in tech. However, some of the early reviewers of this book pointed out that they believe CSV to be a horrible data format. It serializes nontext characters poorly. For example, when you write float values to a CSV file, some precision might be lost—0.12345678901232323 could be arbitrarily rounded up as "0.12345678901"—as complained about in a Stack Overflow thread (*https://oreil.ly/HjTMM*) and Microsoft Community thread (*https://oreil.ly/cbvQu*). People on Hacker News (*https://oreil.ly/ziCmo*) have passionately argued against using CSV.

Text Versus Binary Format

CSV and JSON are text files, whereas Parquet files are binary files. Text files are files that are in plain text, which usually means they are human-readable. Binary files are the catchall that refers to all nontext files. As the name suggests, binary files are typically files that contain only 0s and 1s, and are meant to be read or used by programs that know how to interpret the raw bytes. A program has to know exactly how the data inside the binary file is laid out to make use of the file. If you open text files in your text editor (e.g., VS Code, Notepad), you'll be able to read the texts in them. If you open a binary file in your text editor, you'll see blocks of numbers, likely in hexadecimal values, for corresponding bytes of the file.

Binary files are more compact. Here's a simple example to show how binary files can save space compared to text files. Consider that you want to store the number `1000000`. If you store it in a text file, it'll require 7 characters, and if each character is 1 byte, it'll require 7 bytes. If you store it in a binary file as int32, it'll take only 32 bits or 4 bytes.

As an illustration, I use *interviews.csv*, which is a CSV file (text format) of 17,654 rows and 10 columns. When I converted it to a binary format (Parquet), the file size went from 14 MB to 6 MB, as shown in Figure 3-3.

AWS recommends using the Parquet format because "the Parquet format is up to 2x faster to unload and consumes up to 6x less storage in Amazon S3, compared to text formats."[8]

8 "Announcing Amazon Redshift Data Lake Export: Share Data in Apache Parquet Format," Amazon AWS, December 3, 2019, *https://oreil.ly/ilDb6*.

```
In [2]: df = pd.read_csv("data/interviews.csv")
        df.info()

        <class 'pandas.core.frame.DataFrame'>
        RangeIndex: 17654 entries, 0 to 17653
        Data columns (total 10 columns):
         #   Column      Non-Null Count  Dtype
        ---  ------      --------------  -----
         0   Company     17654 non-null  object
         1   Title       17654 non-null  object
         2   Job         17654 non-null  object
         3   Level       17654 non-null  object
         4   Date        17652 non-null  object
         5   Upvotes     17654 non-null  int64
         6   Offer       17654 non-null  object
         7   Experience  16365 non-null  float64
         8   Difficulty  16376 non-null  object
         9   Review      17654 non-null  object
        dtypes: float64(1), int64(1), object(8)
        memory usage: 1.3+ MB

In [3]: Path("data/interviews.csv").stat().st_size

Out[3]: 14200063

In [4]: df.to_parquet("data/interviews.parquet")
        Path("data/interviews.parquet").stat().st_size

Out[4]: 6211862
```

Figure 3-3. When stored in CSV format, my interview file is 14 MB. But when stored in Parquet, the same file is 6 MB.

Data Models

Data models describe how data is represented. Consider cars in the real world. In a database, a car can be described using its make, its model, its year, its color, and its price. These attributes make up a data model for cars. Alternatively, you can also describe a car using its owner, its license plate, and its history of registered addresses. This is another data model for cars.

How you choose to represent data not only affects the way your systems are built, but also the problems your systems can solve. For example, the way you represent cars in the first data model makes it easier for people looking to buy cars, whereas the second data model makes it easier for police officers to track down criminals.

In this section, we'll study two types of models that seem opposite to each other but are actually converging: relational models and NoSQL models. We'll go over examples to show the types of problems each model is suited for.

Relational Model

Relational models are among the most persistent ideas in computer science. Invented by Edgar F. Codd in 1970,[9] the relational model is still going strong today, even getting more popular. The idea is simple but powerful. In this model, data is organized into relations; each relation is a set of tuples. A table is an accepted visual representation of a relation, and each row of a table makes up a tuple,[10] as shown in Figure 3-4. Relations are unordered. You can shuffle the order of the rows or the order of the columns in a relation and it's still the same relation. Data following the relational model is usually stored in file formats like CSV or Parquet.

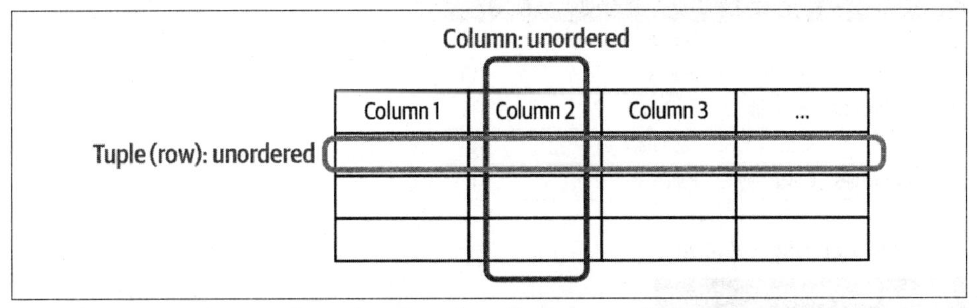

Figure 3-4. In a relation, the order of neither the rows nor the columns matters

It's often desirable for relations to be normalized. Data normalization can follow normal forms such as the first normal form (1NF), second normal form (2NF), etc., and readers interested can read more about it on Wikipedia (*https://oreil.ly/EbrCk*). In this book, we'll go through an example to show how normalization works and how it can reduce data redundancy and improve data integrity.

Consider the relation Book shown in Table 3-2. There are a lot of duplicates in this data. For example, rows 1 and 2 are nearly identical, except for format and price. If the publisher information changes—for example, its name changes from "Banana

9 Edgar F. Codd, "A Relational Model of Data for Large Shared Data Banks," *Communications of the ACM* 13, no. 6 (June 1970): 377–87.

10 For detail-oriented readers, not all tables are relations.

Press" to "Pineapple Press"—or its country changes, we'll have to update rows 1, 2, and 4. If we separate publisher information into its own table, as shown in Tables 3-3 and 3-4, when a publisher's information changes, we only have to update the Publisher relation.[11] This practice allows us to standardize spelling of the same value across different columns. It also makes it easier to make changes to these values, either because these values change or when you want to translate them into different languages.

Table 3-2. Initial Book relation

Title	Author	Format	Publisher	Country	Price
Harry Potter	J.K. Rowling	Paperback	Banana Press	UK	$20
Harry Potter	J.K. Rowling	E-book	Banana Press	UK	$10
Sherlock Holmes	Conan Doyle	Paperback	Guava Press	US	$30
The Hobbit	J.R.R. Tolkien	Paperback	Banana Press	UK	$30
Sherlock Holmes	Conan Doyle	Paperback	Guava Press	US	$15

Table 3-3. Updated Book relation

Title	Author	Format	Publisher ID	Price
Harry Potter	J.K. Rowling	Paperback	1	$20
Harry Potter	J.K. Rowling	E-book	1	$10
Sherlock Holmes	Conan Doyle	Paperback	2	$30
The Hobbit	J.R.R. Tolkien	Paperback	1	$30
Sherlock Holmes	Conan Doyle	Paperback	2	$15

Table 3-4. Publisher relation

Publisher ID	Publisher	Country
1	Banana Press	UK
2	Guava Press	US

One major downside of normalization is that your data is now spread across multiple relations. You can join the data from different relations back together, but joining can be expensive for large tables.

Databases built around the relational data model are relational databases. Once you've put data in your databases, you'll want a way to retrieve it. The language that you can use to specify the data that you want from a database is called a *query language*. The most popular query language for relational databases today is SQL. Even though inspired by the relational model, the data model behind SQL has deviated from the original relational model (*https://oreil.ly/g4waq*). For example, SQL

11 You can further normalize the Book relation, such as separating format into a separate relation.

tables can contain row duplicates, whereas true relations can't contain duplicates. However, this subtle difference has been safely ignored by most people.

The most important thing to note about SQL is that it's a declarative language, as opposed to Python, which is an imperative language. In the imperative paradigm, you specify the steps needed for an action and the computer executes these steps to return the outputs. In the declarative paradigm, you specify the outputs you want, and the computer figures out the steps needed to get you the queried outputs.

With an SQL database, you specify the pattern of data you want—the tables you want the data from, the conditions the results must meet, the basic data transformations such as join, sort, group, aggregate, etc.—but not how to retrieve the data. It is up to the database system to decide how to break the query into different parts, what methods to use to execute each part of the query, and the order in which different parts of the query should be executed.

With certain added features, SQL can be Turing-complete (*https://oreil.ly/npL5B*), which means that, in theory, SQL can be used to solve any computation problem (without making any guarantee about the time or memory required). However, in practice, it's not always easy to write a query to solve a specific task, and it's not always feasible or tractable to execute a query. Anyone working with SQL databases might have nightmarish memories of painfully long SQL queries that are impossible to understand and nobody dares to touch for fear that things might break.[12]

Figuring out how to execute an arbitrary query is the hard part, which is the job of query optimizers. A query optimizer examines all possible ways to execute a query and finds the fastest way to do so.[13] It's possible to use ML to improve query optimizers based on learning from incoming queries.[14] Query optimization is one of the most challenging problems in database systems, and normalization means that data is spread out on multiple relations, which makes joining it together even harder. Even though developing a query optimizer is hard, the good news is that you generally only need one query optimizer and all your applications can leverage it.

12 Greg Kemnitz, a coauthor of the original Postgres paper, shared on Quora (*https://oreil.ly/W0gQa*) that he once wrote a reporting SQL query that was 700 lines long and visited 27 different tables in lookups or joins. The query had about 1,000 lines of comments to help him remember what he was doing. It took him three days to compose, debug, and tune.

13 Yannis E. Ioannidis, "Query Optimization," *ACM Computing Surveys* (CSUR) 28, no. 1 (1996): 121–23, *https://oreil.ly/omXMg*

14 Ryan Marcus et al., "Neo: A Learned Query Optimizer," *arXiv* preprint arXiv:1904.03711 (2019), *https://oreil.ly/wHy6p*.

From Declarative Data Systems to Declarative ML Systems

Possibly inspired by the success of declarative data systems, many people have looked forward to declarative ML.[15] With a declarative ML system, users only need to declare the features' schema and the task, and the system will figure out the best model to perform that task with the given features. Users won't have to write code to construct, train, and tune models. Popular frameworks for declarative ML are Ludwig (*https://oreil.ly/28VWI*), developed at Uber, and H2O AutoML (*https://oreil.ly/sA70M*). In Ludwig, users can specify the model structure—such as the number of fully connected layers and the number of hidden units—on top of the features' schema and output. In H2O AutoML, you don't need to specify the model structure or hyperparameters. It experiments with multiple model architectures and picks out the best model given the features and the task.

Here is an example to show how H2O AutoML works. You give the system your data (inputs and outputs) and specify the number of models you want to experiment. It'll experiment with that number of models and show you the best-performing model:

```
# Identify predictors and response
x = train.columns
y = "response"
x.remove(y)

# For binary classification, response should be a factor
train[y] = train[y].asfactor()
test[y] = test[y].asfactor()

# Run AutoML for 20 base models
aml = H2OAutoML(max_models=20, seed=1)
aml.train(x=x, y=y, training_frame=train)

# Show the best-performing models on the AutoML Leaderboard
lb = aml.leaderboard

# Get the best-performing model
aml.leader
```

While declarative ML can be useful in many cases, it leaves unanswered the biggest challenges with ML in production. Declarative ML systems today abstract away the model development part, and as we'll cover in the next six chapters, with models being increasingly commoditized, model development is often the easier part. The hard part lies in feature engineering, data processing, model evaluation, data shift detection, continual learning, and so on.

15 Matthias Boehm, Alexandre V. Evfimievski, Niketan Pansare, and Berthold Reinwald, "Declarative Machine Learning—A Classification of Basic Properties and Types," *arXiv*, May 19, 2016, *https://oreil.ly/OvW07*.

NoSQL

The relational data model has been able to generalize to a lot of use cases, from ecommerce to finance to social networks. However, for certain use cases, this model can be restrictive. For example, it demands that your data follows a strict schema, and schema management is painful. In a survey by Couchbase in 2014, frustration with schema management was the #1 reason for the adoption of their nonrelational database.[16] It can also be difficult to write and execute SQL queries for specialized applications.

The latest movement against the relational data model is NoSQL. Originally started as a hashtag for a meetup to discuss nonrelational databases, NoSQL has been retroactively reinterpreted as Not Only SQL,[17] as many NoSQL data systems also support relational models. Two major types of nonrelational models are the document model and the graph model. The document model targets use cases where data comes in self-contained documents and relationships between one document and another are rare. The graph model goes in the opposite direction, targeting use cases where relationships between data items are common and important. We'll examine each of these two models, starting with the document model.

Document model

The document model is built around the concept of "document." A document is often a single continuous string, encoded as JSON, XML, or a binary format like BSON (Binary JSON). All documents in a document database are assumed to be encoded in the same format. Each document has a unique key that represents that document, which can be used to retrieve it.

A collection of documents could be considered analogous to a table in a relational database, and a document analogous to a row. In fact, you can convert a relation into a collection of documents that way. For example, you can convert the book data in Tables 3-3 and 3-4 into three JSON documents as shown in Examples 3-1, 3-2, and 3-3. However, a collection of documents is much more flexible than a table. All rows in a table must follow the same schema (e.g., have the same sequence of columns), while documents in the same collection can have completely different schemas.

16 James Phillips, "Surprises in Our NoSQL Adoption Survey," *Couchbase*, December 16, 2014, *https://oreil.ly/ueyEX*.

17 Martin Kleppmann, *Designing Data-Intensive Applications* (Sebastopol, CA: O'Reilly, 2017).

Example 3-1. Document 1: harry_potter.json

```
{
  "Title": "Harry Potter",
  "Author": "J .K. Rowling",
  "Publisher": "Banana Press",
  "Country": "UK",
  "Sold as": [
    {"Format": "Paperback", "Price": "$20"},
    {"Format": "E-book", "Price": "$10"}
  ]
}
```

Example 3-2. Document 2: sherlock_holmes.json

```
{
  "Title": "Sherlock Holmes",
  "Author": "Conan Doyle",
  "Publisher": "Guava Press",
  "Country": "US",
  "Sold as": [
    {"Format": "Paperback", "Price": "$30"},
    {"Format": "E-book", "Price": "$15"}
  ]
}
```

Example 3-3. Document 3: the_hobbit.json

```
{
  "Title": "The Hobbit",
  "Author": "J.R.R. Tolkien",
  "Publisher": "Banana Press",
  "Country": "UK",
  "Sold as": [
    {"Format": "Paperback", "Price": "$30"},
  ]
}
```

Because the document model doesn't enforce a schema, it's often referred to as schemaless. This is misleading because, as discussed previously, data stored in documents will be read later. The application that reads the documents usually assumes some kind of structure of the documents. Document databases just shift the responsibility of assuming structures from the application that writes the data to the application that reads the data.

The document model has better locality than the relational model. Consider the book data example in Tables 3-3 and 3-4 where the information about a book is spread across both the Book table and the Publisher table (and potentially also the Format table). To retrieve information about a book, you'll have to query multiple tables.

In the document model, all information about a book can be stored in a document, making it much easier to retrieve.

However, compared to the relational model, it's harder and less efficient to execute joins across documents compared to across tables. For example, if you want to find all books whose prices are below $25, you'll have to read all documents, extract the prices, compare them to $25, and return all the documents containing the books with prices below $25.

Because of the different strengths of the document and relational data models, it's common to use both models for different tasks in the same database systems. More and more database systems, such as PostgreSQL and MySQL, support them both.

Graph model

The graph model is built around the concept of a "graph." A graph consists of nodes and edges, where the edges represent the relationships between the nodes. A database that uses graph structures to store its data is called a graph database. If in document databases, the content of each document is the priority, then in graph databases, the relationships between data items are the priority.

Because the relationships are modeled explicitly in graph models, it's faster to retrieve data based on relationships. Consider an example of a graph database in Figure 3-5. The data from this example could potentially come from a simple social network. In this graph, nodes can be of different data types: person, city, country, company, etc.

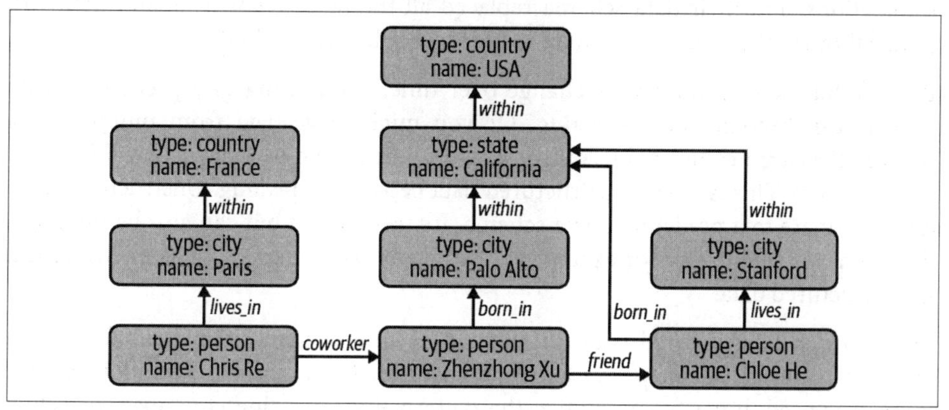

Figure 3-5. An example of a simple graph database

Imagine you want to find everyone who was born in the USA. Given this graph, you can start from the node USA and traverse the graph following the edges "within" and "born_in" to find all the nodes of the type "person." Now, imagine that instead of using the graph model to represent this data, we use the relational model. There'd be no easy way to write an SQL query to find everyone who was born in the USA, especially given that there are an unknown number of hops between *country* and *person*—there are three hops between Zhenzhong Xu and USA while there are only two hops between Chloe He and USA. Similarly, there'd be no easy way for this type of query with a document database.

Many queries that are easy to do in one data model are harder to do in another data model. Picking the right data model for your application can make your life so much easier.

Structured Versus Unstructured Data

Structured data follows a predefined data model, also known as a data schema. For example, the data model might specify that each data item consists of two values: the first value, "name," is a string of at most 50 characters, and the second value, "age," is an 8-bit integer in the range between 0 and 200. The predefined structure makes your data easier to analyze. If you want to know the average age of people in the database, all you have to do is to extract all the age values and average them out.

The disadvantage of structured data is that you have to commit your data to a predefined schema. If your schema changes, you'll have to retrospectively update all your data, often causing mysterious bugs in the process. For example, you've never kept your users' email addresses before but now you do, so you have to retrospectively update email information to all previous users. One of the strangest bugs one of my colleagues encountered was when they could no longer use users' ages with their transactions, and their data schema replaced all the null ages with 0, and their ML model thought the transactions were made by people 0 years old.[18]

Because business requirements change over time, committing to a predefined data schema can become too restricting. Or you might have data from multiple data sources that are beyond your control, and it's impossible to make them follow the same schema. This is where unstructured data becomes appealing. Unstructured data doesn't adhere to a predefined data schema. It's usually text but can also be numbers, dates, images, audio, etc. For example, a text file of logs generated by your ML model is unstructured data.

Even though unstructured data doesn't adhere to a schema, it might still contain intrinsic patterns that help you extract structures. For example, the following text is unstructured, but you can notice the pattern that each line contains two values separated by a comma, the first value is textual, and the second value is numerical. However, there is no guarantee that all lines must follow this format. You can add a new line to that text even if that line doesn't follow this format.

---

18 In this specific example, replacing the null age values with −1 solved the problem.

```
Lisa, 43
Jack, 23
Huyen, 59
```

Unstructured data also allows for more flexible storage options. For example, if your storage follows a schema, you can only store data following that schema. But if your storage doesn't follow a schema, you can store any type of data. You can convert all your data, regardless of types and formats, into bytestrings and store them together.

A repository for storing structured data is called a data warehouse. A repository for storing unstructured data is called a data lake. Data lakes are usually used to store raw data before processing. Data warehouses are used to store data that has been processed into formats ready to be used. Table 3-5 shows a summary of the key differences between structured and unstructured data.

Table 3-5. The key differences between structured and unstructured data

Structured data	Unstructured data
Schema clearly defined	Data doesn't have to follow a schema
Easy to search and analyze	Fast arrival
Can only handle data with a specific schema	Can handle data from any source
Schema changes will cause a lot of troubles	No need to worry about schema changes (yet), as the worry is shifted to the downstream applications that use this data
Stored in data warehouses	Stored in data lakes

Data Storage Engines and Processing

Data formats and data models specify the interface for how users can store and retrieve data. Storage engines, also known as databases, are the implementation of how data is stored and retrieved on machines. It's useful to understand different types of databases as your team or your adjacent team might need to select a database appropriate for your application.

Typically, there are two types of workloads that databases are optimized for, transactional processing and analytical processing, and there's a big difference between them, which we'll cover in this section. We will then cover the basics of the ETL (extract, transform, load) process that you will inevitably encounter when building an ML system in production.

Transactional and Analytical Processing

Traditionally, a transaction refers to the action of buying or selling something. In the digital world, a transaction refers to any kind of action: tweeting, ordering a ride through a ride-sharing service, uploading a new model, watching a YouTube video, and so on. Even though these different transactions involve different types of data, the way they're processed is similar across applications. The transactions are inserted as they are generated, and occasionally updated when something changes, or deleted when they are no longer needed.[19] This type of processing is known as *online transaction processing* (OLTP).

Because these transactions often involve users, they need to be processed fast (low latency) so that they don't keep users waiting. The processing method needs to have high availability—that is, the processing system needs to be available any time a user wants to make a transaction. If your system can't process a transaction, that transaction won't go through.

Transactional databases are designed to process online transactions and satisfy the low latency, high availability requirements. When people hear transactional databases, they usually think of ACID (atomicity, consistency, isolation, durability). Here are their definitions for those needing a quick reminder:

Atomicity
: To guarantee that all the steps in a transaction are completed successfully as a group. If any step in the transaction fails, all other steps must fail also. For example, if a user's payment fails, you don't want to still assign a driver to that user.

Consistency
: To guarantee that all the transactions coming through must follow predefined rules. For example, a transaction must be made by a valid user.

Isolation
: To guarantee that two transactions happen at the same time as if they were isolated. Two users accessing the same data won't change it at the same time. For example, you don't want two users to book the same driver at the same time.

Durability
: To guarantee that once a transaction has been committed, it will remain committed even in the case of a system failure. For example, after you've ordered a ride and your phone dies, you still want your ride to come.

However, transactional databases don't necessarily need to be ACID, and some developers find ACID to be too restrictive. According to Martin Kleppmann, "systems that do not meet the ACID criteria are sometimes called BASE, which stands for *B*asically *A*vailable, *S*oft state, and *E*ventual consistency. This is even more vague than the definition of ACID."[20]

---

19 This paragraph, as well as many parts of this chapter, is inspired by Martin Kleppmann's *Designing Data-Intensive Applications*.

20 Kleppmann, *Designing Data-Intensive Applications*.

Because each transaction is often processed as a unit separately from other transactions, transactional databases are often row-major. This also means that transactional databases might not be efficient for questions such as "What's the average price for all the rides in September in San Francisco?" This kind of analytical question requires aggregating data in columns across multiple rows of data. Analytical databases are designed for this purpose. They are efficient with queries that allow you to look at data from different viewpoints. We call this type of processing *online analytical processing* (OLAP).

However, both the terms OLTP and OLAP have become outdated, as shown in Figure 3-6, for three reasons. First, the separation of transactional and analytical databases was due to limitations of technology—it was hard to have databases that could handle both transactional and analytical queries efficiently. However, this separation is being closed. Today, we have transactional databases that can handle analytical queries, such as CockroachDB (*https://oreil.ly/UsPCr*). We also have analytical databases that can handle transactional queries, such as Apache Iceberg (*https://oreil.ly/pgAfK*) and DuckDB (*https://oreil.ly/jVTHZ*).

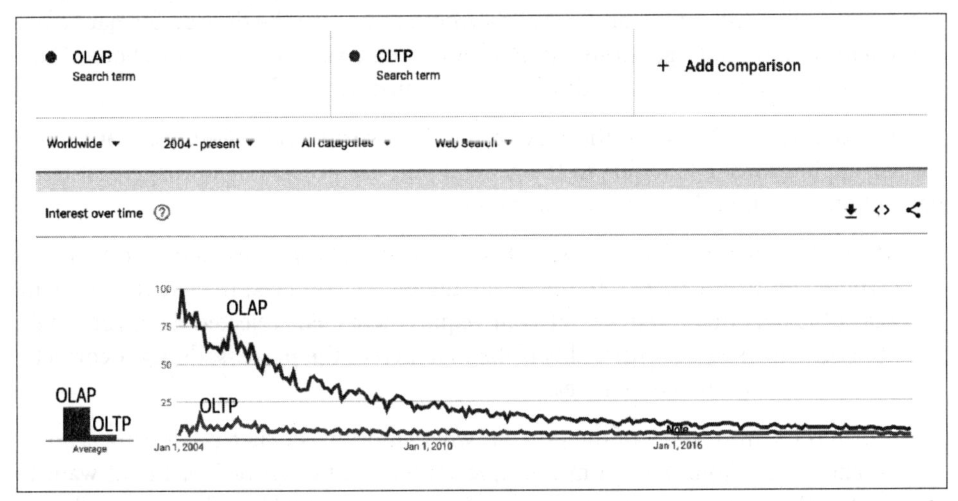

Figure 3-6. OLAP and OLTP are outdated terms, as of 2021, according to Google Trends (https://oreil.ly/O8gAH)

Second, in the traditional OLTP or OLAP paradigms, storage and processing are tightly coupled—how data is stored is also how data is processed. This may result in the same data being stored in multiple databases and using different processing engines to solve different types of queries. An interesting paradigm in the last decade has been to decouple storage from processing (also known as compute), as adopted by many data vendors including Google's BigQuery, Snowflake, IBM, and Teradata.[21] In this paradigm, the data can be stored in the same place, with a processing layer on top that can be optimized for different types of queries.

Third, "online" has become an overloaded term that can mean many different things. Online used to just mean "connected to the internet." Then, it grew to also mean "in production"—we say a feature is online after that feature has been deployed in production.

In the data world today, *online* might refer to the speed at which your data is processed and made available: online, nearline, or offline. According to Wikipedia, online processing means data is immediately available for input/output. Nearline, which is short for near-online, means data is not immediately available but can be made online quickly without human intervention. *Offline* means data is not immediately available and requires some human intervention to become online.[22]

ETL: Extract, Transform, and Load

In the early days of the relational data model, data was mostly structured. When data is *extracted* from different sources, it's first *transformed* into the desired format before being *loaded* into the target destination such as a database or a data warehouse. This process is called *ETL*, which stands for extract, transform, and load.

Even before ML, ETL was all the rage in the data world, and it's still relevant today for ML applications. ETL refers to the general purpose processing and aggregating of data into the shape and the format that you want.

Extract is extracting the data you want from all your data sources. Some of them will be corrupted or malformatted. In the extracting phase, you need to validate your data and reject the data that doesn't meet your requirements. For rejected data, you might have to notify the sources. Since this is the first step of the process, doing it correctly can save you a lot of time downstream.

Transform is the meaty part of the process, where most of the data processing is done. You might want to join data from multiple sources and clean it. You might want to standardize the value ranges (e.g., one data source might use "Male" and "Female" for genders, but another uses "M" and "F" or "1" and "2"). You can apply operations such as transposing, deduplicating, sorting, aggregating, deriving new features, more data validating, etc.

21 Tino Tereshko, "Separation of Storage and Compute in BigQuery," Google Cloud blog, November 29, 2017, *https://oreil.ly/utf7z*; Suresh H., "Snowflake Architecture and Key Concepts: A Comprehensive Guide," Hevo blog, January 18, 2019, *https://oreil.ly/GyvKl*; Preetam Kumar, "Cutting the Cord: Separating Data from Compute in Your Data Lake with Object Storage," IBM blog, September 21, 2017, *https://oreil.ly/Nd3xD*; "The Power of Separating Cloud Compute and Cloud Storage," Teradata, last accessed April 2022, *https://oreil.ly/f82gP*.

22 Wikipedia, s.v. "Nearline storage," last accessed April 2022, *https://oreil.ly/OCmiB*.

Load is deciding how and how often to load your transformed data into the target destination, which can be a file, a database, or a data warehouse.

The idea of ETL sounds simple but powerful, and it's the underlying structure of the data layer at many organizations. An overview of the ETL process is shown in Figure 3-7.

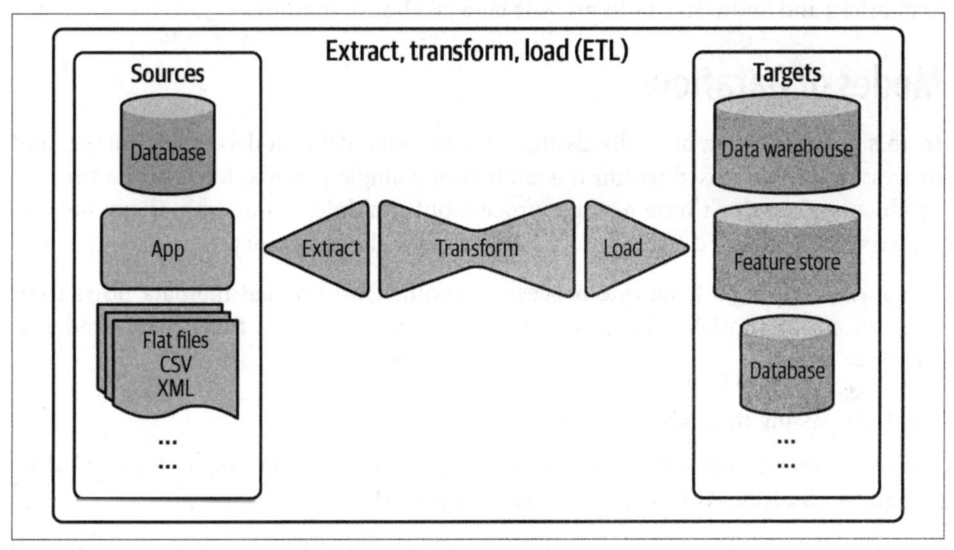

Figure 3-7. An overview of the ETL process

When the internet first became ubiquitous and hardware had just become so much more powerful, collecting data suddenly became so much easier. The amount of data grew rapidly. Not only that, but the nature of data also changed. The number of data sources expanded, and data schemas evolved.

Finding it difficult to keep data structured, some companies had this idea: "Why not just store all data in a data lake so we don't have to deal with schema changes? Whichever application needs data can just pull out raw data from there and process it." This process of loading data into storage first then processing it later is sometimes called *ELT* (extract, load, transform). This paradigm allows for the fast arrival of data since there's little processing needed before data is stored.

However, as data keeps on growing, this idea becomes less attractive. It's inefficient to search through a massive amount of raw data for the data that you want.[23] At the same time, as companies switch to running applications on the cloud

23 In the first draft of this book, I had cost as a reason why you shouldn't store everything. However, as of today, storage has become so cheap that the storage cost is rarely a problem.

and infrastructures become standardized, data structures also become standardized. Committing data to a predefined schema becomes more feasible.

As companies weigh the pros and cons of storing structured data versus storing unstructured data, vendors evolve to offer hybrid solutions that combine the flexibility of data lakes and the data management aspect of data warehouses. For example, Databricks and Snowflake both provide data lakehouse solutions.

Modes of Dataflow

In this chapter, we've been discussing data formats, data models, data storage, and processing for data used within the context of a single process. Most of the time, in production, you don't have a single process but multiple. A question arises: how do we pass data between different processes that don't share memory?

When data is passed from one process to another, we say that the data flows from one process to another, which gives us a dataflow. There are three main modes of dataflow:

- Data passing through databases
- Data passing through services using requests such as the requests provided by REST and RPC APIs (e.g., POST/GET requests)
- Data passing through a real-time transport like Apache Kafka and Amazon Kinesis

We'll go over each of them in this section.

Data Passing Through Databases

The easiest way to pass data between two processes is through databases, which we've discussed in the section "Data Storage Engines and Processing" on page 67. For example, to pass data from process A to process B, process A can write that data into a database, and process B simply reads from that database.

This mode, however, doesn't always work because of two reasons. First, it requires that both processes must be able to access the same database. This might be infeasible, especially if the two processes are run by two different companies.

Second, it requires both processes to access data from databases, and read/write from databases can be slow, making it unsuitable for applications with strict latency requirements—e.g., almost all consumer-facing applications.

Data Passing Through Services

One way to pass data between two processes is to send data directly through a network that connects these two processes. To pass data from process B to process A, process A first sends a request to process B that specifies the data A needs, and B returns the requested data through the same network. Because processes communicate through requests, we say that this is *request-driven*.

This mode of data passing is tightly coupled with the service-oriented architecture. A service is a process that can be accessed remotely, e.g., through a network. In this example, B is exposed to A as a service that A can send requests to. For B to be able to request data from A, A will also need to be exposed to B as a service.

Two services in communication with each other can be run by different companies in different applications. For example, a service might be run by a stock exchange that keeps track of the current stock prices. Another service might be run by an investment firm that requests the current stock prices and uses them to predict future stock prices.

Two services in communication with each other can also be parts of the same application. Structuring different components of your application as separate services allows each component to be developed, tested, and maintained independently of one another. Structuring an application as separate services gives you a microservice architecture.

To put the microservice architecture in the context of ML systems, imagine you're an ML engineer working on the price optimization problem for a company that owns a ride-sharing application like Lyft. In reality, Lyft has hundreds of services (*https://oreil.ly/6fl8f*) in its microservice architecture, but for the sake of simplicity, let's consider only three services:

Driver management service
: Predicts how many drivers will be available in the next minute in a given area.

Ride management service
: Predicts how many rides will be requested in the next minute in a given area.

Price optimization service
: Predicts the optimal price for each ride. The price for a ride should be low enough for riders to be willing to pay, yet high enough for drivers to be willing to drive and for the company to make a profit.

Because the price depends on supply (the available drivers) and demand (the requested rides), the price optimization service needs data from both the driver management and ride management services. Each time a user requests a ride, the price optimization service requests the predicted number of rides and predicted number of drivers to predict the optimal price for this ride.[24]

The most popular styles of requests used for passing data through networks are REST (representational state transfer) and RPC (remote procedure call). Their detailed analysis is beyond the scope of this book, but one major difference is that REST was designed for requests over networks, whereas RPC "tries to make a request to a remote network service look the same as calling a function or method in your programming language." Because of this, "REST seems to be the predominant style for public APIs. The main focus of RPC frameworks is on requests between services owned by the same organization, typically within the same data center."[25]

Implementations of a REST architecture are said to be RESTful. Even though many people think of REST as HTTP, REST doesn't exactly mean HTTP because HTTP is just an implementation of REST.[26]

Data Passing Through Real-Time Transport

To understand the motivation for real-time transports, let's go back to the preceding example of the ride-sharing app with three simple services: driver management, ride management, and price optimization. In the last section, we discussed how the price optimization service needs data from the ride and driver management services to predict the optimal price for each ride.

Now, imagine that the driver management service also needs to know the number of rides from the ride management service to know how many drivers to mobilize. It also wants to know the predicted prices from the price optimization service to use them as incentives for potential drivers (e.g., if you get on the road now you can get a 2x surge charge). Similarly, the ride management service might also want data from the driver management and price optimization services. If we pass data through services as discussed in the previous section, each of these services needs to send requests to the other two services, as shown in Figure 3-8.

---

24 In practice, the price optimization might not have to request the predicted number of rides/drivers every time it has to make a price prediction. It's a common practice to use the cached predicted number of rides/drivers and request new predictions every minute or so.

25 Kleppmann, *Designing Data-Intensive Applications*.

26 Tyson Trautmann, "Debunking the Myths of RPC and REST," *Ethereal Bits*, December 4, 2012 (accessed via the Internet Archive), *https://oreil.ly/4sUrL*.

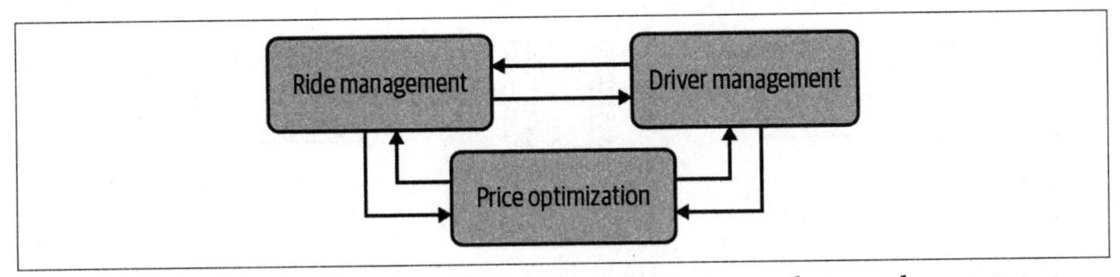

Figure 3-8. In the request-driven architecture, each service needs to send requests to two other services

With only three services, data passing is already getting complicated. Imagine having hundreds, if not thousands of services like what major internet companies have. Interservice data passing can blow up and become a bottleneck, slowing down the entire system.

Request-driven data passing is synchronous: the target service has to listen to the request for the request to go through. If the price optimization service requests data from the driver management service and the driver management service is down, the price optimization service will keep resending the request until it times out. And if the price optimization service is down before it receives a response, the response will be lost. A service that is down can cause all services that require data from it to be down.

What if there's a broker that coordinates data passing among services? Instead of having services request data directly from each other and creating a web of complex interservice data passing, each service only has to communicate with the broker, as shown in Figure 3-9. For example, instead of having other services request the driver management services for the predicted number of drivers for the next minute, what if whenever the driver management service makes a prediction, this prediction is broadcast to a broker? Whichever service wants data from the driver management service can check that broker for the most recent predicted number of drivers. Similarly, whenever the price optimization service makes a prediction about the surge charge for the next minute, this prediction is broadcast to the broker.

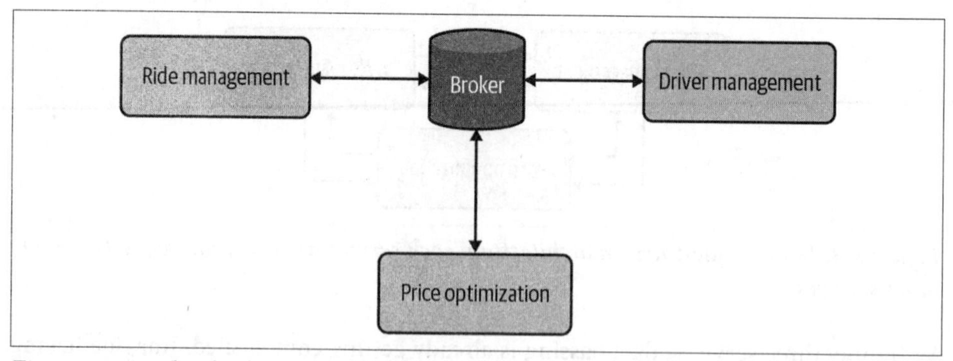

Figure 3-9. With a broker, a service only has to communicate with the broker instead of with other services

Technically, a database can be a broker—each service can write data to a database and other services that need the data can read from that database. However, as mentioned in the section "Data Passing Through Databases" on page 72, reading and writing from databases are too slow for applications with strict latency requirements. Instead of using databases to broker data, we use in-memory storage to broker data. Real-time transports can be thought of as in-memory storage for data passing among services.

A piece of data broadcast to a real-time transport is called an event. This architecture is, therefore, also called *event-driven*. A real-time transport is sometimes called an event bus.

Request-driven architecture works well for systems that rely more on logic than on data. Event-driven architecture works better for systems that are data-heavy.

The two most common types of real-time transports are pubsub, which is short for publish-subscribe, and message queue. In the pubsub model, any service can publish to different topics in a real-time transport, and any service that subscribes to a topic can read all the events in that topic. The services that produce data don't care about what services consume their data. Pubsub solutions often have a retention policy—data will be retained in the real-time transport for a certain period of time (e.g., seven days) before being deleted or moved to a permanent storage (like Amazon S3). See Figure 3-10.

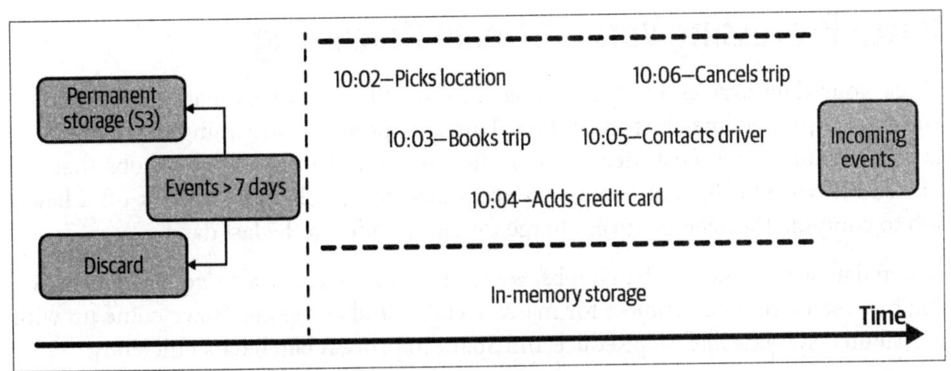

Figure 3-10. Incoming events are stored in in-memory storage before being discarded or moved to more permanent storage

In a message queue model, an event often has intended consumers (an event with intended consumers is called a message), and the message queue is responsible for getting the message to the right consumers.

Examples of pubsub solutions are Apache Kafka and Amazon Kinesis.[27] Examples of message queues are Apache RocketMQ and RabbitMQ. Both paradigms have gained a lot of traction in the last few years. Figure 3-11 shows some of the companies that use Apache Kafka and RabbitMQ.

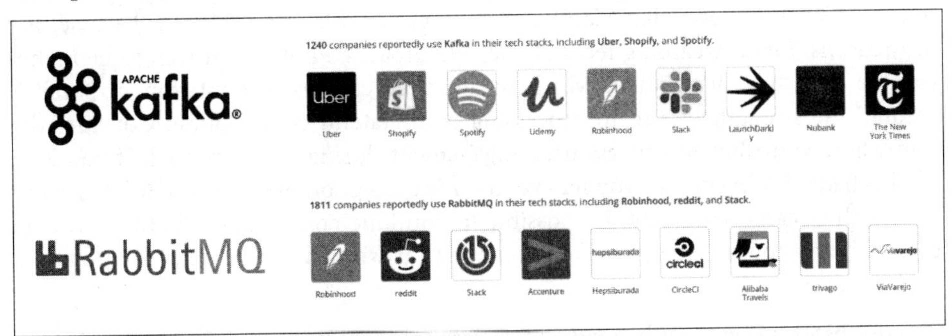

Figure 3-11. Companies that use Apache Kafka and RabbitMQ. Source: Screenshot from Stackshare (https://oreil.ly/OqAgL)

27 If you want to learn more about how Apache Kafka works, Mitch Seymour has a great animation (*https://oreil.ly/kBZzU*) to explain it using otters!

Batch Processing Versus Stream Processing

Once your data arrives in data storage engines like databases, data lakes, or data warehouses, it becomes historical data. This is opposed to streaming data (data that is still streaming in). Historical data is often processed in batch jobs—jobs that are kicked off periodically. For example, once a day, you might want to kick off a batch job to compute the average surge charge for all the rides in the last day.

When data is processed in batch jobs, we refer to it as *batch processing*. Batch processing has been a research subject for many decades, and companies have come up with distributed systems like MapReduce and Spark to process batch data efficiently.

When you have data in real-time transports like Apache Kafka and Amazon Kinesis, we say that you have streaming data. *Stream processing* refers to doing computation on streaming data. Computation on streaming data can also be kicked off periodically, but the periods are usually much shorter than the periods for batch jobs (e.g., every five minutes instead of every day). Computation on streaming data can also be kicked off whenever the need arises. For example, whenever a user requests a ride, you process your data stream to see what drivers are currently available.

Stream processing, when done right, can give low latency because you can process data as soon as data is generated, without having to first write it into databases. Many people believe that stream processing is less efficient than batch processing because you can't leverage tools like MapReduce or Spark. This is not always the case, for two reasons. First, streaming technologies like Apache Flink are proven to be highly scalable and fully distributed, which means they can do computation in parallel. Second, the strength of stream processing is in stateful computation. Consider the case where you want to process user engagement during a 30-day trial. If you kick off this batch job every day, you'll have to do computation over the last 30 days every day. With stream processing, it's possible to continue computing only the new data each day and joining the new data computation with the older data computation, preventing redundancy.

Because batch processing happens much less frequently than stream processing, in ML, batch processing is usually used to compute features that change less often, such as drivers' ratings (if a driver has had hundreds of rides, their rating is less likely to change significantly from one day to the next). *Batch features*—features extracted through batch processing—are also known as *static features*.

Stream processing is used to compute features that change quickly, such as how many drivers are available right now, how many rides have been requested in the last minute, how many rides will be finished in the next two minutes, the median price of the last 10 rides in this area, etc. Features about the current state of the system like these are important to make the optimal price predictions. *Streaming features*—features extracted through stream processing—are also known as *dynamic features*.

For many problems, you need not only batch features or streaming features, but both. You need infrastructure that allows you to process streaming data as well as batch data and join them together to feed into your ML models. We'll discuss more on how batch features and streaming features can be used together to generate predictions in Chapter 7.

To do computation on data streams, you need a stream computation engine (the way Spark and MapReduce are batch computation engines). For simple streaming computation, you might be able to get away with the built-in stream computation capacity of real-time transports like Apache Kafka, but Kafka stream processing is limited in its ability to deal with various data sources.

For ML systems that leverage streaming features, the streaming computation is rarely simple. The number of stream features used in an application such as fraud detection and credit scoring can be in the hundreds, if not thousands. The stream feature extraction logic can require complex queries with join and aggregation along different dimensions. To extract these features requires efficient stream processing engines. For this purpose, you might want to look into tools like Apache Flink, KSQL, and Spark Streaming. Of these three engines, Apache Flink and KSQL are more recognized in the industry and provide a nice SQL abstraction for data scientists.

Stream processing is more difficult because the data amount is unbounded and the data comes in at variable rates and speeds. It's easier to make a stream processor do batch processing than to make a batch processor do stream processing. Apache Flink's core maintainers have been arguing for years that batch processing is a special case of stream processing.[28]

Summary

This chapter is built on the foundations established in Chapter 2 around the importance of data in developing ML systems. In this chapter, we learned it's important to choose the right format to store our data to make it easier to use the data in the future. We discussed different data formats and the pros and cons of row-major versus column-major formats as well as text versus binary formats.

We continued to cover three major data models: relational, document, and graph. Even though the relational model is the most well known given the popularity of SQL, all three models are widely used today, and each is good for a certain set of tasks.

When talking about the relational model compared to the document model, many people think of the former as structured and the latter as unstructured. The division

28 Kostas Tzoumas, "Batch Is a Special Case of Streaming," *Ververica*, September 15, 2015, *https://oreil.ly/IcIl2*.

between structured and unstructured data is quite fluid—the main question is who has to shoulder the responsibility of assuming the structure of data. Structured data means that the code that writes the data has to assume the structure. Unstructured data means that the code that reads the data has to assume the structure.

We continued the chapter with data storage engines and processing. We studied databases optimized for two distinct types of data processing: transactional processing and analytical processing. We studied data storage engines and processing together because traditionally storage is coupled with processing: transactional databases for transactional processing and analytical databases for analytical processing. However, in recent years, many vendors have worked on decoupling storage and processing. Today, we have transactional databases that can handle analytical queries and analytical databases that can handle transactional queries.

When discussing data formats, data models, data storage engines, and processing, data is assumed to be within a process. However, while working in production, you'll likely work with multiple processes, and you'll likely need to transfer data between them. We discussed three modes of data passing. The simplest mode is passing through databases. The most popular mode of data passing for processes is data passing through services. In this mode, a process is exposed as a service that another process can send requests for data. This mode of data passing is tightly coupled with microservice architectures, where each component of an application is set up as a service.

A mode of data passing that has become increasingly popular over the last decade is data passing through a real-time transport like Apache Kafka and RabbitMQ. This mode of data passing is somewhere between passing through databases and passing through services: it allows for asynchronous data passing with reasonably low latency.

As data in real-time transports have different properties from data in databases, they require different processing techniques, as discussed in the section "Batch Processing Versus Stream Processing" on page 78. Data in databases is often processed in batch jobs and produces static features, whereas data in real-time transports is often processed using stream computation engines and produces dynamic features. Some people argue that batch processing is a special case of stream processing, and stream computation engines can be used to unify both processing pipelines.

Once we have our data systems figured out, we can collect data and create training data, which will be the focus of the next chapter.

CHAPTER 4

Training Data

In Chapter 3, we covered how to handle data from the systems perspective. In this chapter, we'll go over how to handle data from the data science perspective. Despite the importance of training data in developing and improving ML models, ML curricula are heavily skewed toward modeling, which is considered by many practitioners the "fun" part of the process. Building a state-of-the-art model is interesting. Spending days wrangling with a massive amount of malformatted data that doesn't even fit into your machine's memory is frustrating.

Data is messy, complex, unpredictable, and potentially treacherous. If not handled properly, it can easily sink your entire ML operation. But this is precisely the reason why data scientists and ML engineers should learn how to handle data well, saving us time and headache down the road.

In this chapter, we will go over techniques to obtain or create good training data. Training data, in this chapter, encompasses all the data used in the developing phase of ML models, including the different splits used for training, validation, and testing (the train, validation, test splits). This chapter starts with different sampling techniques to select data for training. We'll then address common challenges in creating training data, including the label multiplicity problem, the lack of labels problem, the class imbalance problem, and techniques in data augmentation to address the lack of data problem.

We use the term "training data" instead of "training dataset" because "dataset" denotes a set that is finite and stationary. Data in production is neither finite nor stationary, a phenomenon that we will cover in the section "Data Distribution Shifts" on page 237. Like other steps in building ML systems, creating training data is an iterative process. As your model evolves through a project lifecycle, your training data will likely also evolve.

Before we move forward, I just want to echo a word of caution that has been said many times yet is still not enough. Data is full of potential biases. These biases have many possible causes. There are biases caused during collecting, sampling, or labeling. Historical data might be embedded with human biases, and ML models, trained on this data, can perpetuate them. Use data but don't trust it too much!

Sampling

Sampling is an integral part of the ML workflow that is, unfortunately, often overlooked in typical ML coursework. Sampling happens in many steps of an ML project lifecycle, such as sampling from all possible real-world data to create training data; sampling from a given dataset to create splits for training, validation, and testing; or sampling from all possible events that happen within your ML system for monitoring purposes. In this section, we'll focus on sampling methods for creating training data, but these sampling methods can also be used for other steps in an ML project lifecycle.

In many cases, sampling is necessary. One case is when you don't have access to all possible data in the real world, the data that you use to train your model is a subset of real-world data, created by one sampling method or another. Another case is when it's infeasible to process all the data that you have access to—because it requires too much time or resources—so you have to sample that data to create a subset that is feasible to process. In many other cases, sampling is helpful as it allows you to accomplish a task faster and cheaper. For example, when considering a new model, you might want to do a quick experiment with a small subset of your data to see if the new model is promising first before training this new model on all your data.[1]

Understanding different sampling methods and how they are being used in our workflow can, first, help us avoid potential sampling biases, and second, help us choose the methods that improve the efficiency of the data we sample.

There are two families of sampling: nonprobability sampling and random sampling. We'll start with nonprobability sampling methods, followed by several common random sampling methods.

1 Some readers might argue that this approach might not work with large models, as certain large models don't work for small datasets but work well with a lot more data. In this case, it's still important to experiment with datasets of different sizes to figure out the effect of the dataset size on your model.

Nonprobability Sampling

Nonprobability sampling is when the selection of data isn't based on any probability criteria. Here are some of the criteria for nonprobability sampling:

Convenience sampling
: Samples of data are selected based on their availability. This sampling method is popular because, well, it's convenient.

Snowball sampling
: Future samples are selected based on existing samples. For example, to scrape legitimate Twitter accounts without having access to Twitter databases, you start with a small number of accounts, then you scrape all the accounts they follow, and so on.

Judgment sampling
: Experts decide what samples to include.

Quota sampling
: You select samples based on quotas for certain slices of data without any randomization. For example, when doing a survey, you might want 100 responses from each of the age groups: under 30 years old, between 30 and 60 years old, and above 60 years old, regardless of the actual age distribution.

The samples selected by nonprobability criteria are not representative of the real-world data and therefore are riddled with selection biases.[2] Because of these biases, you might think that it's a bad idea to select data to train ML models using this family of sampling methods. You're right. Unfortunately, in many cases, the selection of data for ML models is still driven by convenience.

One example of these cases is language modeling. Language models are often trained not with data that is representative of all possible texts but with data that can be easily collected—Wikipedia, Common Crawl, Reddit.

Another example is data for sentiment analysis of general text. Much of this data is collected from sources with natural labels (ratings) such as IMDB reviews and Amazon reviews. These datasets are then used for other sentiment analysis tasks. IMDB reviews and Amazon reviews are biased toward users who are willing to leave reviews online, and not necessarily representative of people who don't have access to the internet or people who aren't willing to put reviews online.

2 James J. Heckman, "Sample Selection Bias as a Specification Error," *Econometrica* 47, no. 1 (January 1979): 153–61, *https://oreil.ly/I5AhM*.

A third example is data for training self-driving cars. Initially, data collected for self-driving cars came largely from two areas: Phoenix, Arizona (because of its lax regulations), and the Bay Area in California (because many companies that build self-driving cars are located here). Both areas have generally sunny weather. In 2016, Waymo expanded its operations to Kirkland, Washington, specially for Kirkland's rainy weather,[3] but there's still a lot more self-driving car data for sunny weather than for rainy or snowy weather.

Nonprobability sampling can be a quick and easy way to gather your initial data to get your project off the ground. However, for reliable models, you might want to use probability-based sampling, which we will cover next.

Simple Random Sampling

In the simplest form of random sampling, you give all samples in the population equal probabilities of being selected.[4] For example, you randomly select 10% of the population, giving all members of this population an equal 10% chance of being selected.

The advantage of this method is that it's easy to implement. The drawback is that rare categories of data might not appear in your selection. Consider the case where a class appears only in 0.01% of your data population. If you randomly select 1% of your data, samples of this rare class will unlikely be selected. Models trained on this selection might think that this rare class doesn't exist.

Stratified Sampling

To avoid the drawback of simple random sampling, you can first divide your population into the groups that you care about and sample from each group separately. For example, to sample 1% of data that has two classes, A and B, you can sample 1% of class A and 1% of class B. This way, no matter how rare class A or B is, you'll ensure that samples from it will be included in the selection. Each group is called a stratum, and this method is called stratified sampling.

3 Rachel Lerman, "Google Is Testing Its Self-Driving Car in Kirkland," *Seattle Times*, February 3, 2016, *https://oreil.ly/3IA1V*.

4 Population here refers to a "statistical population" (*https://oreil.ly/w7GDX*), a (potentially infinite) set of all possible samples that can be sampled.

One drawback of this sampling method is that it isn't always possible, such as when it's impossible to divide all samples into groups. This is especially challenging when one sample might belong to multiple groups, as in the case of multilabel tasks.[5] For instance, a sample can be both class A and class B.

Weighted Sampling

In weighted sampling, each sample is given a weight, which determines the probability of it being selected. For example, if you have three samples, A, B, and C, and want them to be selected with the probabilities of 50%, 30%, and 20% respectively, you can give them the weights 0.5, 0.3, and 0.2.

This method allows you to leverage domain expertise. For example, if you know that a certain subpopulation of data, such as more recent data, is more valuable to your model and want it to have a higher chance of being selected, you can give it a higher weight.

This also helps with the case when the data you have comes from a different distribution compared to the true data. For example, if in your data, red samples account for 25% and blue samples account for 75%, but you know that in the real world, red and blue have equal probability to happen, you can give red samples weights three times higher than blue samples.

In Python, you can do weighted sampling with `random.choices` as follows:

```
# Choose two items from the list such that 1, 2, 3, 4 each has
# 20% chance of being selected, while 100 and 1000 each have only 10% chance.
import random
random.choices(population=[1, 2, 3, 4, 100, 1000],
               weights=[0.2, 0.2, 0.2, 0.2, 0.1, 0.1],
               k=2)
# This is equivalent to the following
random.choices(population=[1, 1, 2, 2, 3, 3, 4, 4, 100, 1000],
               k=2)
```

A common concept in ML that is closely related to weighted sampling is sample weights. Weighted sampling is used to select samples to train your model with, whereas sample weights are used to assign "weights" or "importance" to training samples. Samples with higher weights affect the loss function more. Changing sample weights can change your model's decision boundaries significantly, as shown in Figure 4-1.

5 Multilabel tasks are tasks where one example can have multiple labels.

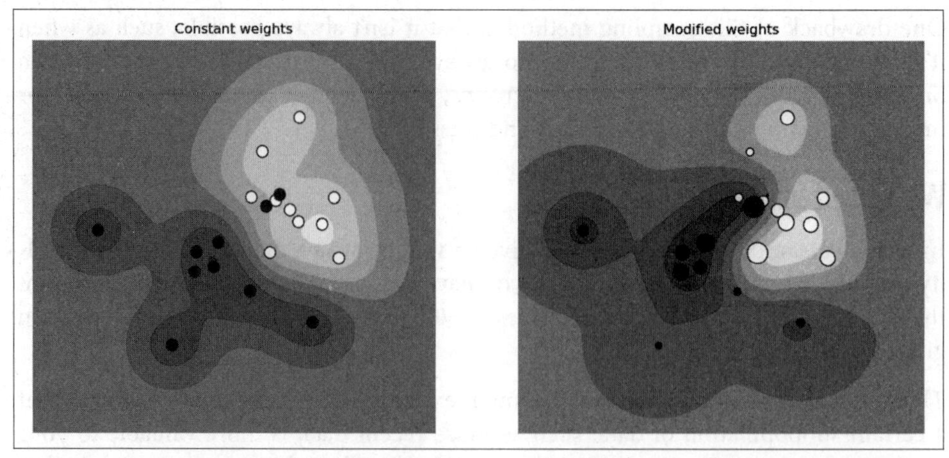

Figure 4-1. Sample weights can affect the decision boundary. On the left is when all samples are given equal weights. On the right is when samples are given different weights. Source: scikit-learn[6]

Reservoir Sampling

Reservoir sampling is a fascinating algorithm that is especially useful when you have to deal with streaming data, which is usually what you have in production.

Imagine you have an incoming stream of tweets and you want to sample a certain number, *k*, of tweets to do analysis or train a model on. You don't know how many tweets there are, but you know you can't fit them all in memory, which means you don't know in advance the probability at which a tweet should be selected. You want to ensure that:

- Every tweet has an equal probability of being selected.
- You can stop the algorithm at any time and the tweets are sampled with the correct probability.

One solution for this problem is reservoir sampling. The algorithm involves a reservoir, which can be an array, and consists of three steps:

1. Put the first k elements into the reservoir.
2. For each incoming n^{th} element, generate a random number i such that $1 \leq i \leq n$.
3. If $1 \leq i \leq k$: replace the ith element in the reservoir with the nth element. Else, do nothing.

6 "SVM: Weighted Samples," scikit-learn, *https://oreil.ly/BDqbk*.

This means that each incoming n^{th} element has $\frac{k}{n}$ probability of being in the reservoir. You can also prove that each element in the reservoir has $\frac{k}{n}$ probability of being there. This means that all samples have an equal chance of being selected. If we stop the algorithm at any time, all samples in the reservoir have been sampled with the correct probability. Figure 4-2 shows an illustrative example of how reservoir sampling works.

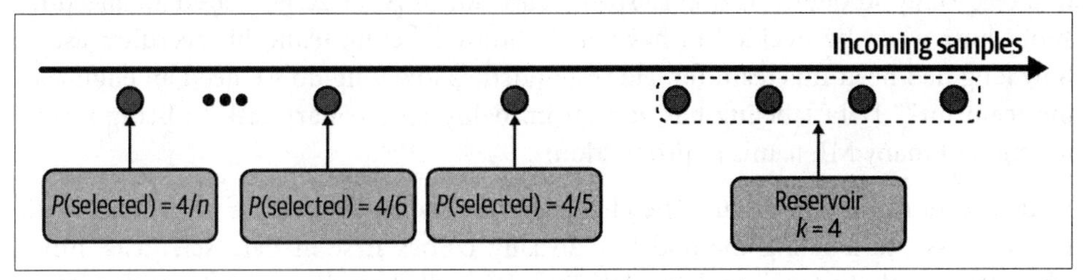

Figure 4-2. A visualization of how reservoir sampling works

Importance Sampling

Importance sampling is one of the most important sampling methods, not just in ML. It allows us to sample from a distribution when we only have access to another distribution.

Imagine you have to sample x from a distribution $P(x)$, but $P(x)$ is really expensive, slow, or infeasible to sample from. However, you have a distribution $Q(x)$ that is a lot easier to sample from. So you sample x from $Q(x)$ instead and weigh this sample by $\frac{P(x)}{Q(x)}$. $Q(x)$ is called the *proposal distribution* or the *importance distribution*. $Q(x)$ can be any distribution as long as $Q(x) > 0$ whenever $P(x) \neq 0$. The following equation shows that in expectation, x sampled from $P(x)$ is equal to x sampled from $Q(x)$ weighted by $\frac{P(x)}{Q(x)}$:

$$E_{P(x)}[x] = \sum_x P(x)x = \sum_x Q(x)x\frac{P(x)}{Q(x)} = E_{Q(x)}\left[x\frac{P(x)}{Q(x)}\right]$$

One example where importance sampling is used in ML is policy-based reinforcement learning. Consider the case when you want to update your policy. You want to estimate the value functions of the new policy, but calculating the total rewards of taking an action can be costly because it requires considering all possible outcomes until the end of the time horizon after that action. However, if the new policy is relatively close to the old policy, you can calculate the total rewards based on the old policy instead and reweight them according to the new policy. The rewards from the old policy make up the proposal distribution.

Labeling

Despite the promise of unsupervised ML, most ML models in production today are supervised, which means that they need labeled data to learn from. The performance of an ML model still depends heavily on the quality and quantity of the labeled data it's trained on.

In a talk to my students, Andrej Karpathy, director of AI at Tesla, shared an anecdote about how when he decided to have an in-house labeling team, his recruiter asked how long he'd need this team for. He responded: "How long do we need an engineering team for?" Data labeling has gone from being an auxiliary task to being a core function of many ML teams in production.

In this section, we will discuss the challenge of obtaining labels for your data. We'll first discuss the labeling method that usually comes first in data scientists' mind when talking about labeling: hand-labeling. We will then discuss tasks with natural labels, which are tasks where labels can be inferred from the system without requiring human annotations, followed by what to do when natural and hand labels are lacking.

Hand Labels

Anyone who has ever had to work with data in production has probably felt this at a visceral level: acquiring hand labels for your data is difficult for many, many reasons. First, hand-labeling data can be expensive, especially if subject matter expertise is required. To classify whether a comment is spam, you might be able to find 20 annotators on a crowdsourcing platform and train them in 15 minutes to label your data. However, if you want to label chest X-rays, you'd need to find board-certified radiologists, whose time is limited and expensive.

Second, hand labeling poses a threat to data privacy. Hand labeling means that someone has to look at your data, which isn't always possible if your data has strict privacy requirements. For example, you can't just ship your patients' medical records or your company's confidential financial information to a third-party service for labeling. In many cases, your data might not even be allowed to leave your organization, and you might have to hire or contract annotators to label your data on premises.

Third, hand labeling is slow. For example, accurate transcription of speech utterance at the phonetic level can take 400 times longer than the utterance duration.[7] So if you want to annotate 1 hour of speech, it'll take 400 hours or almost 3 months for a person to do so. In a study to use ML to help classify lung cancers from X-rays, my colleagues had to wait almost a year to obtain sufficient labels.

---

7 Xiaojin Zhu, "Semi-Supervised Learning with Graphs" (doctoral diss., Carnegie Mellon University, 2005), *https://oreil.ly/VYy4C*.

Slow labeling leads to slow iteration speed and makes your model less adaptive to changing environments and requirements. If the task changes or data changes, you'll have to wait for your data to be relabeled before updating your model. Imagine the scenario when you have a sentiment analysis model to analyze the sentiment of every tweet that mentions your brand. It has only two classes: NEGATIVE and POSITIVE. However, after deployment, your PR team realizes that the most damage comes from angry tweets and they want to attend to angry messages faster. So you have to update your sentiment analysis model to have three classes: NEGATIVE, POSITIVE, and ANGRY. To do so, you will need to look at your data again to see which existing training examples should be relabeled ANGRY. If you don't have enough ANGRY examples, you will have to collect more data. The longer the process takes, the more your existing model performance will degrade.

Label multiplicity

Often, to obtain enough labeled data, companies have to use data from multiple sources and rely on multiple annotators who have different levels of expertise. These different data sources and annotators also have different levels of accuracy. This leads to the problem of label ambiguity or label multiplicity: what to do when there are multiple conflicting labels for a data instance.

Consider this simple task of entity recognition. You give three annotators the following sample and ask them to annotate all entities they can find:

> Darth Sidious, known simply as the Emperor, was a Dark Lord of the Sith who reigned over the galaxy as Galactic Emperor of the First Galactic Empire.

You receive back three different solutions, as shown in Table 4-1. Three annotators have identified different entities. Which one should your model train on? A model trained on data labeled by annotator 1 will perform very differently from a model trained on data labeled by annotator 2.

Table 4-1. Entities identified by different annotators might be very different

Annotator	# entities	Annotation
1	3	[*Darth Sidious*], known simply as the Emperor, was a [*Dark Lord of the Sith*] who reigned over the galaxy as [*Galactic Emperor of the First Galactic Empire*].
2	6	[*Darth Sidious*], known simply as the [*Emperor*], was a [*Dark Lord*] of the [*Sith*] who reigned over the galaxy as [*Galactic Emperor*] of the [*First Galactic Empire*].
3	4	[*Darth Sidious*], known simply as the [*Emperor*], was a [*Dark Lord of the Sith*] who reigned over the galaxy as [*Galactic Emperor of the First Galactic Empire*].

Disagreements among annotators are extremely common. The higher the level of domain expertise required, the higher the potential for annotating disagreement.[8] If one human expert thinks the label should be A while another believes it should be B, how do we resolve this conflict to obtain one single ground truth? If human experts can't agree on a label, what does human-level performance even mean?

To minimize the disagreement among annotators, it's important to first have a clear problem definition. For example, in the preceding entity recognition task, some disagreements could have been eliminated if we clarify that in case of multiple possible entities, pick the entity that comprises the longest substring. This means *Galactic Emperor of the First Galactic Empire* instead of *Galactic Emperor* and *First Galactic Empire*. Second, you need to incorporate that definition into the annotators' training to make sure that all annotators understand the rules.

Data lineage

Indiscriminately using data from multiple sources, generated with different annotators, without examining their quality can cause your model to fail mysteriously. Consider a case when you've trained a moderately good model with 100K data samples. Your ML engineers are confident that more data will improve the model performance, so you spend a lot of money to hire annotators to label another million data samples.

However, the model performance actually decreases after being trained on the new data. The reason is that the new million samples were crowdsourced to annotators who labeled data with much less accuracy than the original data. It can be especially difficult to remedy this if you've already mixed your data and can't differentiate new data from old data.

It's good practice to keep track of the origin of each of your data samples as well as its labels, a technique known as *data lineage*. Data lineage helps you both flag potential biases in your data and debug your models. For example, if your model fails mostly on the recently acquired data samples, you might want to look into how the new data was acquired. On more than one occasion, we've discovered that the problem wasn't with our model, but because of the unusually high number of wrong labels in the data that we'd acquired recently.

8 If something is so obvious to label, you wouldn't need domain expertise.

Natural Labels

Hand-labeling isn't the only source for labels. You might be lucky enough to work on tasks with natural ground truth labels. Tasks with natural labels are tasks where the model's predictions can be automatically evaluated or partially evaluated by the system. An example is the model that estimates time of arrival for a certain route on Google Maps. If you take that route, by the end of your trip, Google Maps knows how long the trip actually took, and thus can evaluate the accuracy of the predicted time of arrival. Another example is stock price prediction. If your model predicts a stock's price in the next two minutes, then after two minutes, you can compare the predicted price with the actual price.

The canonical example of tasks with natural labels is recommender systems. The goal of a recommender system is to recommend to users items relevant to them. Whether a user clicks on the recommended item or not can be seen as the feedback for that recommendation. A recommendation that gets clicked on can be presumed to be good (i.e., the label is POSITIVE) and a recommendation that doesn't get clicked on after a period of time, say 10 minutes, can be presumed to be bad (i.e., the label is NEGATIVE).

Many tasks can be framed as recommendation tasks. For example, you can frame the task of predicting ads' click-through rates as recommending the most relevant ads to users based on their activity histories and profiles. Natural labels that are inferred from user behaviors like clicks and ratings are also known as behavioral labels.

Even if your task doesn't inherently have natural labels, it might be possible to set up your system in a way that allows you to collect some feedback on your model. For example, if you're building a machine translation system like Google Translate, you can have the option for the community to submit alternative translations for bad translations—these alternative translations can be used to train the next iteration of your models (though you might want to review these suggested translations first). Newsfeed ranking is not a task with inherent labels, but by adding the Like button and other reactions to each newsfeed item, Facebook is able to collect feedback on their ranking algorithm.

Tasks with natural labels are fairly common in the industry. In a survey of 86 companies in my network, I found that 63% of them work with tasks with natural labels, as shown in Figure 4-3. This doesn't mean that 63% of tasks that can benefit from ML solutions have natural labels. What is more likely is that companies find it easier and cheaper to first start on tasks that have natural labels.

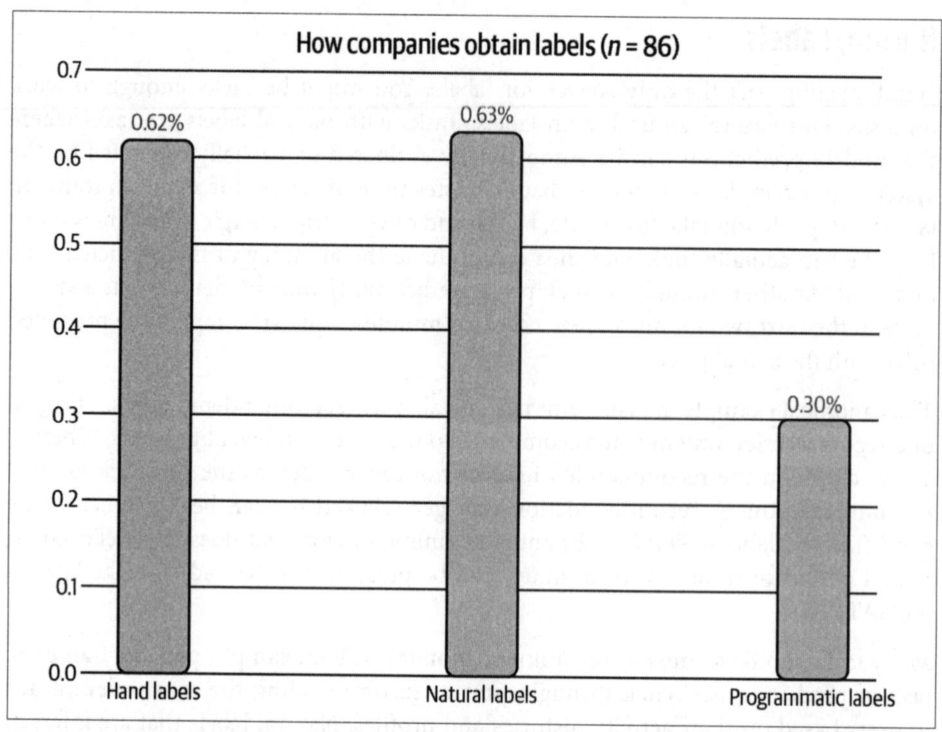

Figure 4-3. Sixty-three percent of companies in my network work on tasks with natural labels. The percentages don't sum to 1 because a company can work with tasks with different label sources.[9]

In the previous example, a recommendation that doesn't get clicked on after a period of time can be presumed to be bad. This is called an *implicit label*, as this negative label is presumed from the lack of a positive label. It's different from *explicit labels* where users explicitly demonstrate their feedback on a recommendation by giving it a low rating or downvoting it.

Feedback loop length

For tasks with natural ground truth labels, the time it takes from when a prediction is served until when the feedback on it is provided is the feedback loop length. Tasks with short feedback loops are tasks where labels are generally available within minutes. Many recommender systems have short feedback loops. If the recommended items are related products on Amazon or people to follow on Twitter, the time between when the item is recommended until it's clicked on, if it's clicked on at all, is short.

9 We'll cover programmatic labels in the section "Weak supervision" on page 95.

However, not all recommender systems have minute-long feedback loops. If you work with longer content types like blog posts or articles or YouTube videos, the feedback loop can be hours. If you build a system to recommend clothes for users like the one Stitch Fix has, you wouldn't get feedback until users have received the items and tried them on, which could be weeks later.

Different Types of User Feedback

If you want to extract labels from user feedback, it's important to note that there are different types of user feedback. They can occur at different stages during a user journey on your app and differ by volume, strength of signal, and feedback loop length.

For example, consider an ecommerce application similar to what Amazon has. Types of feedback a user on this application can provide might include clicking on a product recommendation, adding a product to cart, buying a product, rating, leaving a review, and returning a previously bought product.

Clicking on a product happens much faster and more frequently (and therefore incurs a higher volume) than purchasing a product. However, buying a product is a much stronger signal on whether a user likes that product compared to just clicking on it.

When building a product recommender system, many companies focus on optimizing for clicks, which give them a higher volume of feedback to evaluate their models. However, some companies focus on purchases, which gives them a stronger signal that is also more correlated to their business metrics (e.g., revenue from product sales). Both approaches are valid. There's no definite answer to what type of feedback you should optimize for your use case, and it merits serious discussions between all stakeholders involved.

Choosing the right window length requires thorough consideration, as it involves the speed and accuracy trade-off. A short window length means that you can capture labels faster, which allows you to use these labels to detect issues with your model and address those issues as soon as possible. However, a short window length also means that you might prematurely label a recommendation as bad before it's clicked on.

No matter how long you set your window length to be, there might still be premature negative labels. In early 2021, a study by the Ads team at Twitter found that even though the majority of clicks on ads happen within the first five minutes, some clicks happen hours after when the ad is shown.[10] This means that this type of label tends to give an underestimate of the actual click-through rate. If you only record 1,000 POSITIVE labels, the actual number of clicks might be a bit over 1,000.

For tasks with long feedback loops, natural labels might not arrive for weeks or even months. Fraud detection is an example of a task with long feedback loops. For a certain period of time after a transaction, users can dispute whether that transaction is fraudulent or not. For example, when a customer read their credit card statement and saw a transaction they didn't recognize, they might dispute it with their bank, giving the bank the feedback to label that transaction as fraudulent. A typical dispute window is one to three months. After the dispute window has passed, if there's no dispute from the user, you might presume the transaction to be legitimate.

Labels with long feedback loops are helpful for reporting a model's performance on quarterly or yearly business reports. However, they are not very helpful if you want to detect issues with your models as soon as possible. If there's a problem with your fraud detection model and it takes you months to catch, by the time the problem is fixed, all the fraudulent transactions your faulty model let through might have caused a small business to go bankrupt.

Handling the Lack of Labels

Because of the challenges in acquiring sufficient high-quality labels, many techniques have been developed to address the problems that result. In this section, we will cover four of them: weak supervision, semi-supervision, transfer learning, and active learning. A summary of these methods is shown in Table 4-2.

Table 4-2. Summaries of four techniques for handling the lack of hand-labeled data

Method	How	Ground truths required?
Weak supervision	Leverages (often noisy) heuristics to generate labels	No, but a small number of labels are recommended to guide the development of heuristics
Semi-supervision	Leverages structural assumptions to generate labels	Yes, a small number of initial labels as seeds to generate more labels
Transfer learning	Leverages models pretrained on another task for your new task	No for zero-shot learning Yes for fine-tuning, though the number of ground truths required is often much smaller than what would be needed if you train the model from scratch
Active learning	Labels data samples that are most useful to your model	Yes

10 Sofia Ira Ktena, Alykhan Tejani, Lucas Theis, Pranay Kumar Myana, Deepak Dilipkumar, Ferenc Huszar, Steven Yoo, and Wenzhe Shi, "Addressing Delayed Feedback for Continuous Training with Neural Networks in CTR Prediction," *arXiv*, July 15, 2019, *https://oreil.ly/5y2WA*.

Weak supervision

If hand labeling is so problematic, what if we don't use hand labels altogether? One approach that has gained popularity is weak supervision. One of the most popular open source tools for weak supervision is Snorkel, developed at the Stanford AI Lab.[11] The insight behind weak supervision is that people rely on heuristics, which can be developed with subject matter expertise, to label data. For example, a doctor might use the following heuristics to decide whether a patient's case should be prioritized as emergent:

> If the nurse's note mentions a serious condition like pneumonia, the patient's case should be given priority consideration.

Libraries like Snorkel are built around the concept of a *labeling function* (LF): a function that encodes heuristics. The preceding heuristics can be expressed by the following function:

```
def labeling_function(note):
   if "pneumonia" in note:
     return "EMERGENT"
```

LFs can encode many different types of heuristics. Here are some of them:

Keyword heuristic
: Such as the preceding example

Regular expressions
: Such as if the note matches or fails to match a certain regular expression

Database lookup
: Such as if the note contains the disease listed in the dangerous disease list

The outputs of other models
: Such as if an existing system classifies this as `EMERGENT`

After you've written LFs, you can apply them to the samples you want to label.

Because LFs encode heuristics, and heuristics are noisy, labels produced by LFs are noisy. Multiple LFs might apply to the same data examples, and they might give conflicting labels. One function might think a nurse's note is `EMERGENT` but another function might think it's not. One heuristic might be much more accurate than another heuristic, which you might not know because you don't have ground truth labels to compare them to. It's important to combine, denoise, and reweight all LFs to

11 Alexander Ratner, Stephen H. Bach, Henry Ehrenberg, Jason Fries, Sen Wu, and Christopher Ré, "Snorkel: Rapid Training Data Creation with Weak Supervision," *Proceedings of the VLDB Endowment* 11, no. 3 (2017): 269–82, *https://oreil.ly/vFPjk*.

get a set of most likely to be correct labels. Figure 4-4 shows at a high level how LFs work.

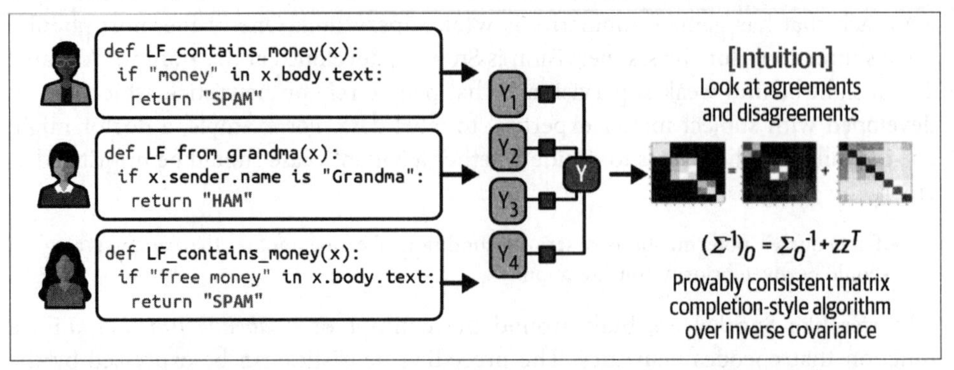

Figure 4-4. A high-level overview of how labeling functions are combined. Source: Adapted from an image by Ratner et al.[12]

In theory, you don't need any hand labels for weak supervision. However, to get a sense of how accurate your LFs are, a small number of hand labels is recommended. These hand labels can help you discover patterns in your data to write better LFs.

Weak supervision can be especially useful when your data has strict privacy requirements. You only need to see a small, cleared subset of data to write LFs, which can be applied to the rest of your data without anyone looking at it.

With LFs, subject matter expertise can be versioned, reused, and shared. Expertise owned by one team can be encoded and used by another team. If your data changes or your requirements change, you can just reapply LFs to your data samples. The approach of using LFs to generate labels for your data is also known as programmatic labeling. Table 4-3 shows some of the advantages of programmatic labeling over hand labeling.

Table 4-3. The advantages of programmatic labeling over hand labeling

Hand labeling	Programmatic labeling
Expensive: Especially when subject matter expertise required	**Cost saving**: Expertise can be versioned, shared, and reused across an organization
Lack of privacy: Need to ship data to human annotators	**Privacy**: Create LFs using a cleared data subsample and then apply LFs to other data without looking at individual samples
Slow: Time required scales linearly with number of labels needed	**Fast**: Easily scale from 1K to 1M samples
Nonadaptive: Every change requires relabeling the data	**Adaptive**: When changes happen, just reapply LFs!

12 Ratner et al., "Snorkel: Rapid Training Data Creation with Weak Supervision."

Here is a case study to show how well weak supervision works in practice. In a study with Stanford Medicine,[13] models trained with weakly supervised labels obtained by a single radiologist after eight hours of writing LFs had comparable performance with models trained on data obtained through almost a year of hand labeling, as shown in Figure 4-5. There are two interesting facts about the results of the experiment. First, the models continued improving with more unlabeled data even without more LFs. Second, LFs were being reused across tasks. The researchers were able to reuse six LFs between the CXR (chest X-rays) task and EXR (extremity X-rays) task.[14]

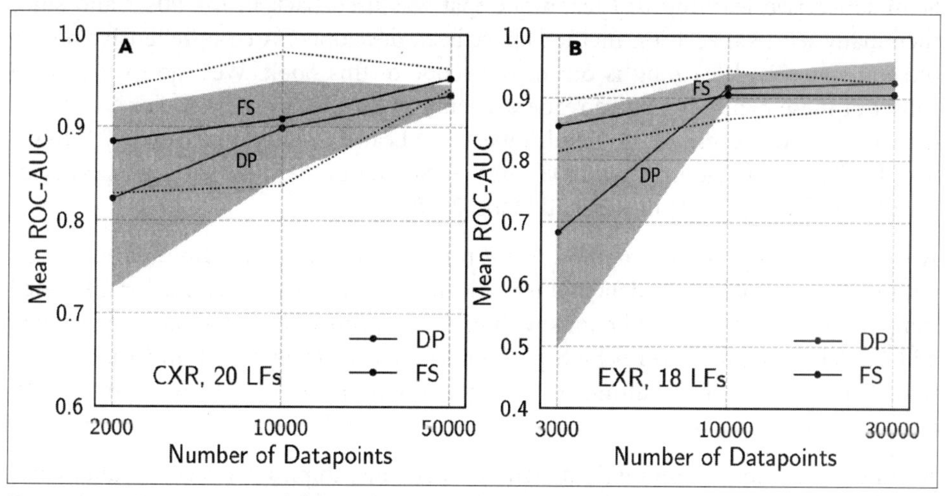

Figure 4-5. Comparison of the performance of a model trained on fully supervised labels (FS) and a model trained with programmatic labels (DP) on CXR and EXR tasks. Source: Dunnmon et al.[15]

My students often ask that if heuristics work so well to label data, why do we need ML models? One reason is that LFs might not cover all data samples, so we can train ML models on data programmatically labeled with LFs and use this trained model to generate predictions for samples that aren't covered by any LF.

Weak supervision is a simple but powerful paradigm. However, it's not perfect. In some cases, the labels obtained by weak supervision might be too noisy to be useful. But even in these cases, weak supervision can be a good way to get you started when

13 Jared A. Dunnmon, Alexander J. Ratner, Khaled Saab, Matthew P. Lungren, Daniel L. Rubin, and Christopher Ré, "Cross-Modal Data Programming Enables Rapid Medical Machine Learning," *Patterns* 1, no. 2 (2020): 100019, *https://oreil.ly/nKt8E*.

14 The two tasks in this study use only 18 and 20 LFs respectively. In practice, I've seen teams using hundreds of LFs for each task.

15 Dummon et al., "Cross-Modal Data Programming."

you want to explore the effectiveness of ML without wanting to invest too much in hand labeling up front.

Semi-supervision

If weak supervision leverages heuristics to obtain noisy labels, semi-supervision leverages structural assumptions to generate new labels based on a small set of initial labels. Unlike weak supervision, semi-supervision requires an initial set of labels.

Semi-supervised learning is a technique that was used back in the 90s,[16] and since then many semi-supervision methods have been developed. A comprehensive review of semi-supervised learning is out of the scope of this book. We'll go over a small subset of these methods to give readers a sense of how they are used. For a comprehensive review, I recommend "Semi-Supervised Learning Literature Survey" (*https://oreil.ly/ULeWD*) (Xiaojin Zhu, 2008) and "A Survey on Semi-Supervised Learning" (*https://oreil.ly/JYgCH*) (Engelen and Hoos, 2018).

A classic semi-supervision method is *self-training*. You start by training a model on your existing set of labeled data and use this model to make predictions for unlabeled samples. Assuming that predictions with high raw probability scores are correct, you add the labels predicted with high probability to your training set and train a new model on this expanded training set. This goes on until you're happy with your model performance.

Another semi-supervision method assumes that data samples that share similar characteristics share the same labels. The similarity might be obvious, such as in the task of classifying the topic of Twitter hashtags. You can start by labeling the hashtag "#AI" as Computer Science. Assuming that hashtags that appear in the same tweet or profile are likely about the same topic, given the profile of MIT CSAIL in Figure 4-6, you can also label the hashtags "#ML" and "#BigData" as Computer Science.

Figure 4-6. Because #ML and #BigData appear in the same Twitter profile as #AI, we can assume that they belong to the same topic

16 Avrim Blum and Tom Mitchell, "Combining Labeled and Unlabeled Data with Co-Training," in *Proceedings of the Eleventh Annual Conference on Computational Learning Theory* (July 1998): 92–100, *https://oreil.ly/T79AE*.

In most cases, the similarity can only be discovered by more complex methods. For example, you might need to use a clustering method or a *k*-nearest neighbors algorithm to discover samples that belong to the same cluster.

A semi-supervision method that has gained popularity in recent years is the perturbation-based method. It's based on the assumption that small perturbations to a sample shouldn't change its label. So you apply small perturbations to your training instances to obtain new training instances. The perturbations might be applied directly to the samples (e.g., adding white noise to images) or to their representations (e.g., adding small random values to embeddings of words). The perturbed samples have the same labels as the unperturbed samples. We'll discuss more about this in the section "Perturbation" on page 114.

In some cases, semi-supervision approaches have reached the performance of purely supervised learning, even when a substantial portion of the labels in a given dataset has been discarded.[17]

Semi-supervision is the most useful when the number of training labels is limited. One thing to consider when doing semi-supervision with limited data is how much of this limited data should be used to evaluate multiple candidate models and select the best one. If you use a small amount, the best performing model on this small evaluation set might be the one that overfits the most to this set. On the other hand, if you use a large amount of data for evaluation, the performance boost gained by selecting the best model based on this evaluation set might be less than the boost gained by adding the evaluation set to the limited training set. Many companies overcome this trade-off by using a reasonably large evaluation set to select the best model, then continuing training the champion model on the evaluation set.

Transfer learning

Transfer learning refers to the family of methods where a model developed for a task is reused as the starting point for a model on a second task. First, the base model is trained for a base task. The base task is usually a task that has cheap and abundant training data. Language modeling is a great candidate because it doesn't require labeled data. Language models can be trained on any body of text—books, Wikipedia articles, chat histories—and the task is: given a sequence of tokens,[18] predict the next token. When given the sequence "I bought NVIDIA shares because I believe in the importance of," a language model might output "hardware" or "GPU" as the next token.

---

17 Avital Oliver, Augustus Odena, Colin Raffel, Ekin D. Cubuk, and Ian J. Goodfellow, "Realistic Evaluation of Deep Semi-Supervised Learning Algorithms," *NeurIPS 2018 Proceedings*, *https://oreil.ly/dRmPV*.

18 A token can be a word, a character, or part of a word.

The trained model can then be used for the task that you're interested in—a downstream task—such as sentiment analysis, intent detection, or question answering. In some cases, such as in zero-shot learning scenarios, you might be able to use the base model on a downstream task directly. In many cases, you might need to *fine-tune* the base model. Fine-tuning means making small changes to the base model, such as continuing to train the base model or a part of the base model on data from a given downstream task.[19]

Sometimes, you might need to modify the inputs using a template to prompt the base model to generate the outputs you want.[20] For example, to use a language model as the base model for a question answering task, you might want to use this prompt:

> *Q: When was the United States founded?*
>
> *A: July 4, 1776.*
>
> *Q: Who wrote the Declaration of Independence?*
>
> *A: Thomas Jefferson.*
>
> *Q: What year was Alexander Hamilton born?*
>
> *A:*

When you input this prompt into a language model such as GPT-3 (*https://oreil.ly/qT0r3*), it might output the year Alexander Hamilton was born.

Transfer learning is especially appealing for tasks that don't have a lot of labeled data. Even for tasks that have a lot of labeled data, using a pretrained model as the starting point can often boost the performance significantly compared to training from scratch.

Transfer learning has gained a lot of interest in recent years for the right reasons. It has enabled many applications that were previously impossible due to the lack of training samples. A nontrivial portion of ML models in production today are the results of transfer learning, including object detection models that leverage models pretrained on ImageNet and text classification models that leverage pretrained language models such as BERT or GPT-3.[21] Transfer learning also lowers the entry

---

19 Jeremy Howard and Sebastian Ruder, "Universal Language Model Fine-tuning for Text Classification," *arXiv*, January 18, 2018, *https://oreil.ly/DBEbw*.

20 Pengfei Liu, Weizhe Yuan, Jinlan Fu, Zhengbao Jiang, Hiroaki Hayashi, and Graham Neubig, "Pre-train, Prompt, and Predict: A Systematic Survey of Prompting Methods in Natural Language Processing," *arXiv*, July 28, 2021, *https://oreil.ly/0lBgn*.

21 Jacob Devlin, Ming-Wei Chang, Kenton Lee, and Kristina Toutanova, "BERT: Pre-training of Deep Bidirectional Transformers for Language Understanding," *arXiv*, October 11, 2018, *https://oreil.ly/RdIGU*; Tom B. Brown, Benjamin Mann, Nick Ryder, Melanie Subbiah, Jared Kaplan, Prafulla Dhariwal, Arvind Neelakantan, et al., "Language Models Are Few-Shot Learners," OpenAI, 2020, *https://oreil.ly/YVmrr*.

barriers into ML, as it helps reduce the up-front cost needed for labeling data to build ML applications.

A trend that has emerged in the last five years is that (usually) the larger the pretrained base model, the better its performance on downstream tasks. Large models are expensive to train. Based on the configuration of GPT-3, it's estimated that the cost of training this model is in the tens of millions USD. Many have hypothesized that in the future only a handful of companies will be able to afford to train large pretrained models. The rest of the industry will use these pretrained models directly or fine-tune them for their specific needs.

Active learning

Active learning is a method for improving the efficiency of data labels. The hope here is that ML models can achieve greater accuracy with fewer training labels if they can choose which data samples to learn from. Active learning is sometimes called query learning—though this term is getting increasingly unpopular—because a model (active learner) sends back queries in the form of unlabeled samples to be labeled by annotators (usually humans).

Instead of randomly labeling data samples, you label the samples that are most helpful to your models according to some metrics or heuristics. The most straightforward metric is uncertainty measurement—label the examples that your model is the least certain about, hoping that they will help your model learn the decision boundary better. For example, in the case of classification problems where your model outputs raw probabilities for different classes, it might choose the data samples with the lowest probabilities for the predicted class. Figure 4-7 illustrates how well this method works on a toy example.

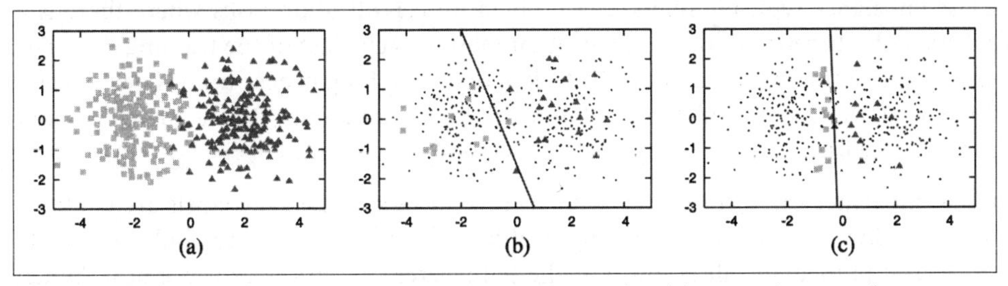

Figure 4-7. How uncertainty-based active learning works. (a) A toy dataset of 400 instances, evenly sampled from two class Gaussians. (b) A model trained on 30 samples randomly labeled gives an accuracy of 70%. (c) A model trained on 30 samples chosen by active learning gives an accuracy of 90%. Source: Burr Settles[22]

22 Burr Settles, *Active Learning* (Williston, VT: Morgan & Claypool, 2012).

Another common heuristic is based on disagreement among multiple candidate models. This method is called query-by-committee, an example of an ensemble method.[23] You need a committee of several candidate models, which are usually the same model trained with different sets of hyperparameters or the same model trained on different slices of data. Each model can make one vote for which samples to label next, and it might vote based on how uncertain it is about the prediction. You then label the samples that the committee disagrees on the most.

There are other heuristics such as choosing samples that, if trained on them, will give the highest gradient updates or will reduce the loss the most. For a comprehensive review of active learning methods, check out "Active Learning Literature Survey" (*https://oreil.ly/4RuBo*) (Settles 2010).

The samples to be labeled can come from different data regimes. They can be synthesized where your model generates samples in the region of the input space that it's most uncertain about.[24] They can come from a stationary distribution where you've already collected a lot of unlabeled data and your model chooses samples from this pool to label. They can come from the real-world distribution where you have a stream of data coming in, as in production, and your model chooses samples from this stream of data to label.

I'm most excited about active learning when a system works with real-time data. Data changes all the time, a phenomenon we briefly touched on in Chapter 1 and will further detail in Chapter 8. Active learning in this data regime will allow your model to learn more effectively in real time and adapt faster to changing environments.

Class Imbalance

Class imbalance typically refers to a problem in classification tasks where there is a substantial difference in the number of samples in each class of the training data. For example, in a training dataset for the task of detecting lung cancer from X-ray images, 99.99% of the X-rays might be of normal lungs, and only 0.01% might contain cancerous cells.

Class imbalance can also happen with regression tasks where the labels are continuous. Consider the task of estimating health-care bills.[25] Health-care bills are highly skewed—the median bill is low, but the 95th percentile bill is astronomical. When predicting hospital bills, it might be more important to predict accurately the bills at the 95th percentile than the median bills. A 100% difference in a $250 bill is acceptable (actual $500, predicted $250), but a 100% difference on a $10k bill is not

23 We'll cover ensembles in Chapter 6.

24 Dana Angluin, "Queries and Concept Learning," *Machine Learning* 2 (1988): 319–42, *https://oreil.ly/0uKs4*.

25 Thanks to Eugene Yan for this wonderful example!

(actual $20k, predicted $10k). Therefore, we might have to train the model to be better at predicting 95th percentile bills, even if it reduces the overall metrics.

Challenges of Class Imbalance

ML, especially deep learning, works well in situations when the data distribution is more balanced, and usually not so well when the classes are heavily imbalanced, as illustrated in Figure 4-8. Class imbalance can make learning difficult for the following three reasons.

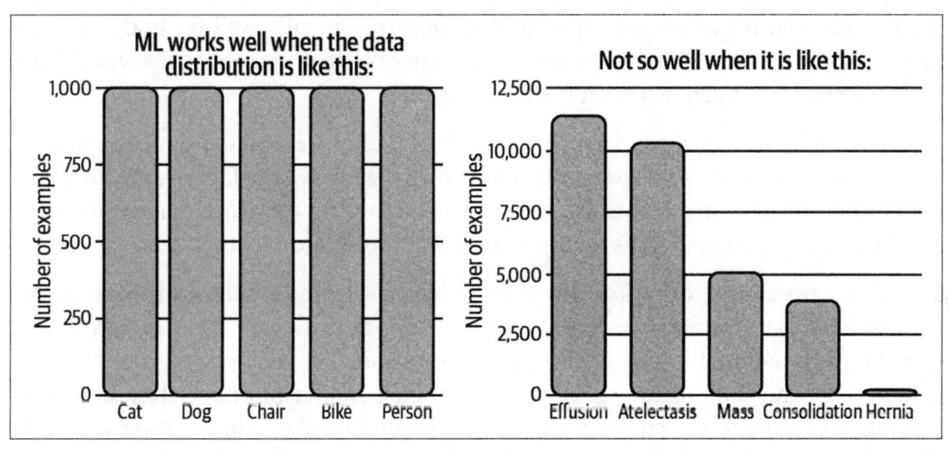

Figure 4-8. ML works well in situations where the classes are balanced. Source: Adapted from an image by Andrew Ng[26]

The first reason is that class imbalance often means there's insufficient signal for your model to learn to detect the minority classes. In the case where there is a small number of instances in the minority class, the problem becomes a few-shot learning problem where your model only gets to see the minority class a few times before having to make a decision on it. In the case where there is no instance of the rare classes in your training set, your model might assume these rare classes don't exist.

The second reason is that class imbalance makes it easier for your model to get stuck in a nonoptimal solution by exploiting a simple heuristic instead of learning anything useful about the underlying pattern of the data. Consider the preceding lung cancer detection example. If your model learns to always output the majority class, its accuracy is already 99.99%.[27] This heuristic can be very hard for gradient descent

26 Andrew Ng, "Bridging AI's Proof-of-Concept to Production Gap" (HAI Seminar, September 22, 2020), video, 1:02:07, *https://oreil.ly/FSFWS*.

27 And this is why accuracy is a bad metric for tasks with class imbalance, as we'll explore more in the section "Handling Class Imbalance" on page 105.

algorithms to beat because a small amount of randomness added to this heuristic might lead to worse accuracy.

The third reason is that class imbalance leads to asymmetric costs of error—the cost of a wrong prediction on a sample of the rare class might be much higher than a wrong prediction on a sample of the majority class.

For example, misclassification on an X-ray with cancerous cells is much more dangerous than misclassification on an X-ray of a normal lung. If your loss function isn't configured to address this asymmetry, your model will treat all samples the same way. As a result, you might obtain a model that performs equally well on both majority and minority classes, while you much prefer a model that performs less well on the majority class but much better on the minority one.

When I was in school, most datasets I was given had more or less balanced classes.[28] It was a shock for me to start working and realize that class imbalance is the norm. In real-world settings, rare events are often more interesting (or more dangerous) than regular events, and many tasks focus on detecting those rare events.

The classical example of tasks with class imbalance is fraud detection. Most credit card transactions are not fraudulent. As of 2018, 6.8¢ for every $100 in cardholder spending is fraudulent.[29] Another is churn prediction. The majority of your customers are probably not planning on canceling their subscription. If they are, your business has more to worry about than churn prediction algorithms. Other examples include disease screening (most people, fortunately, don't have terminal illness) and resume screening (98% of job seekers are eliminated at the initial resume screening[30]).

A less obvious example of a task with class imbalance is object detection (*https://oreil.ly/CGEf5*). Object detection algorithms currently work by generating a large number of bounding boxes over an image then predicting which boxes are most likely to have objects in them. Most bounding boxes do not contain a relevant object.

Outside the cases where class imbalance is inherent in the problem, class imbalance can also be caused by biases during the sampling process. Consider the case when you want to create training data to detect whether an email is spam or not. You decide to use all the anonymized emails from your company's email database. According to Talos Intelligence, as of May 2021, nearly 85% of all emails are spam.[31] But most spam

28 I imagined that it'd be easier to learn ML theory if I didn't have to figure out how to deal with class imbalance.

29 The Nilson Report, "Payment Card Fraud Losses Reach $27.85 Billion," PR Newswire, November 21, 2019, *https://oreil.ly/NM5zo*.

30 "Job Market Expert Explains Why Only 2% of Job Seekers Get Interviewed," WebWire, January 7, 2014, *https://oreil.ly/UpL8S*.

31 "Email and Spam Data," Talos Intelligence, last accessed May 2021, *https://oreil.ly/lI5Jr*.

emails were filtered out before they reached your company's database, so in your dataset, only a small percentage is spam.

Another cause for class imbalance, though less common, is due to labeling errors. Annotators might have read the instructions wrong or followed the wrong instructions (thinking there are only two classes, POSITIVE and NEGATIVE, while there are actually three), or simply made errors. Whenever faced with the problem of class imbalance, it's important to examine your data to understand the causes of it.

Handling Class Imbalance

Because of its prevalence in real-world applications, class imbalance has been thoroughly studied over the last two decades.[32] Class imbalance affects tasks differently based on the level of imbalance. Some tasks are more sensitive to class imbalance than others. Japkowicz showed that sensitivity to imbalance increases with the complexity of the problem, and that noncomplex, linearly separable problems are unaffected by all levels of class imbalance.[33] Class imbalance in binary classification problems is a much easier problem than class imbalance in multiclass classification problems. Ding et al. showed that very deep neural networks—with "very deep" meaning over 10 layers back in 2017—performed much better on imbalanced data than shallower neural networks.[34]

There have been many techniques suggested to mitigate the effect of class imbalance. However, as neural networks have grown to be much larger and much deeper, with more learning capacity, some might argue that you shouldn't try to "fix" class imbalance if that's how the data looks in the real world. A good model should learn to model that imbalance. However, developing a model good enough for that can be challenging, so we still have to rely on special training techniques.

In this section, we will cover three approaches to handling class imbalance: choosing the right metrics for your problem; data-level methods, which means changing the data distribution to make it less imbalanced; and algorithm-level methods, which means changing your learning method to make it more robust to class imbalance.

These techniques might be necessary but not sufficient. For a comprehensive survey, I recommend "Survey on Deep Learning with Class Imbalance" (*https://oreil.ly/9QvBr*) (Johnson and Khoshgoftaar 2019).

---

32 Nathalie Japkowciz and Shaju Stephen, "The Class Imbalance Problem: A Systematic Study," 2002, *https://oreil.ly/d7lVu*.

33 Nathalie Japkowicz, "The Class Imbalance Problem: Significance and Strategies," 2000, *https://oreil.ly/Ma50Z*.

34 Wan Ding, Dong-Yan Huang, Zhuo Chen, Xinguo Yu, and Weisi Lin, "Facial Action Recognition Using Very Deep Networks for Highly Imbalanced Class Distribution," *2017 Asia-Pacific Signal and Information Processing Association Annual Summit and Conference (APSIPA ASC)*, 2017, *https://oreil.ly/WeW6J*.

Using the right evaluation metrics

The most important thing to do when facing a task with class imbalance is to choose the appropriate evaluation metrics. Wrong metrics will give you the wrong ideas of how your models are doing and, subsequently, won't be able to help you develop or choose models good enough for your task.

The overall accuracy and error rate are the most frequently used metrics to report the performance of ML models. However, these are insufficient metrics for tasks with class imbalance because they treat all classes equally, which means the performance of your model on the majority class will dominate these metrics. This is especially bad when the majority class isn't what you care about.

Consider a task with two labels: CANCER (the positive class) and NORMAL (the negative class), where 90% of the labeled data is NORMAL. Consider two models, A and B, with the confusion matrices shown in Tables 4-4 and 4-5.

Table 4-4. Model A's confusion matrix; model A can detect 10 out of 100 CANCER cases

Model A	Actual CANCER	Actual NORMAL
Predicted CANCER	10	10
Predicted NORMAL	90	890

Table 4-5. Model B's confusion matrix; model B can detect 90 out of 100 CANCER cases

Model B	Actual CANCER	Actual NORMAL
Predicted CANCER	90	90
Predicted NORMAL	10	810

If you're like most people, you'd probably prefer model B to make predictions for you since it has a better chance of telling you if you actually have cancer. However, they both have the same accuracy of 0.9.

Metrics that help you understand your model's performance with respect to specific classes would be better choices. Accuracy can still be a good metric if you use it for each class individually. The accuracy of model A on the CANCER class is 10% and the accuracy of model B on the CANCER class is 90%.

F1, precision, and recall are metrics that measure your model's performance with respect to the positive class in binary classification problems, as they rely on true positive—an outcome where the model correctly predicts the positive class.[35]

Precision, Recall, and F1

For readers needing a refresh, precision, recall, and F1 scores, for binary tasks, are calculated using the count of true positives, true negatives, false positives, and false negatives. Definitions for these terms are shown in Table 4-6.

Table 4-6. Definitions of True Positive, False Positive, False Negative, and True Negative in a binary classification task

	Predicted Positive	Predicted Negative
Positive label	True Positive (hit)	False Negative (type II error, miss)
Negative label	False Positive (type I error, false alarm)	True Negative (correct rejection)

Precision = True Positive / (True Positive + False Positive)

Recall = True Positive / (True Positive + False Negative)

F1 = 2 × Precision × Recall / (Precision + Recall)

F1, precision, and recall are asymmetric metrics, which means that their values change depending on which class is considered the positive class. In our case, if we consider CANCER the positive class, model A's F1 is 0.17. However, if we consider NORMAL the positive class, model A's F1 is 0.95. Accuracy, precision, recall, and F1 scores of model A and model B when CANCER is the positive class are shown in Table 4-7.

Table 4-7. Both models have the same accuracy even though one model is clearly superior

	CANCER (1)	NORMAL (0)	Accuracy	Precision	Recall	F1
Model A	10/100	890/900	0.9	0.5	0.1	0.17
Model B	90/100	810/900	0.9	0.5	0.9	0.64

Many classification problems can be modeled as regression problems. Your model can output a probability, and based on that probability, you classify the sample. For example, if the value is greater than 0.5, it's a positive label, and if it's less than or

35 As of July 2021, when you use `scikit-learn.metrics.f1_score`, `pos_label` is set to 1 by default, but you can change it to 0 if you want 0 to be your positive label.

equal to 0.5, it's a negative label. This means that you can tune the threshold to increase the *true positive rate* (also known as *recall*) while decreasing the *false positive rate* (also known as the *probability of false alarm*), and vice versa. We can plot the true positive rate against the false positive rate for different thresholds. This plot is known as the *ROC curve* (receiver operating characteristics). When your model is perfect, the recall is 1.0, and the curve is just a line at the top. This curve shows you how your model's performance changes depending on the threshold, and helps you choose the threshold that works best for you. The closer to the perfect line, the better your model's performance.

The area under the curve (AUC) measures the area under the ROC curve. Since the closer to the perfect line the better, the larger this area the better, as shown in Figure 4-9.

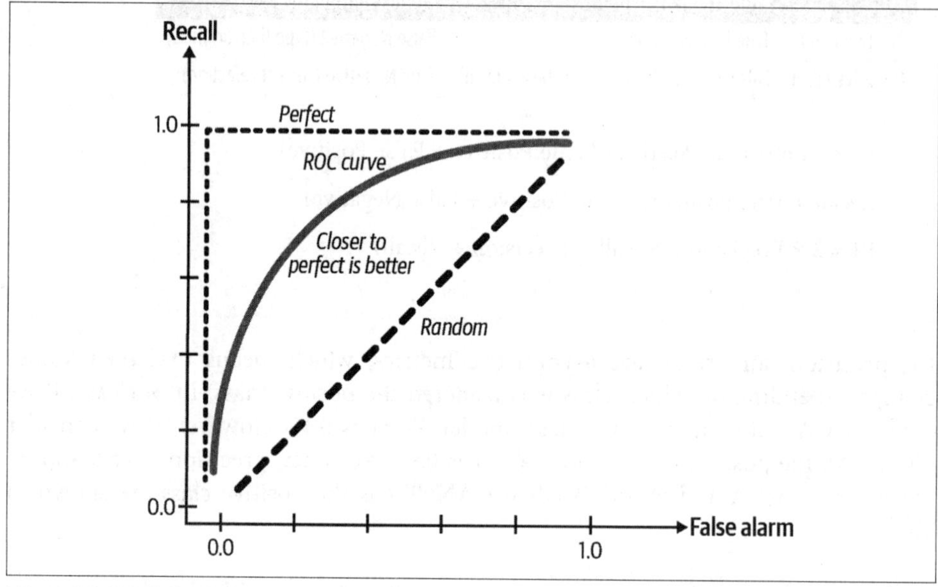

Figure 4-9. ROC curve

Like F1 and recall, the ROC curve focuses only on the positive class and doesn't show how well your model does on the negative class. Davis and Goadrich suggested that we should plot precision against recall instead, in what they termed the Precision-Recall Curve. They argued that this curve gives a more informative picture of an algorithm's performance on tasks with heavy class imbalance.[36]

36 Jesse Davis and Mark Goadrich, "The Relationship Between Precision-Recall and ROC Curves," *Proceedings of the 23rd International Conference on Machine Learning*, 2006, *https://oreil.ly/s40F3*.

Data-level methods: Resampling

Data-level methods modify the distribution of the training data to reduce the level of imbalance to make it easier for the model to learn. A common family of techniques is resampling. Resampling includes oversampling, adding more instances from the minority classes, and undersampling, removing instances of the majority classes. The simplest way to undersample is to randomly remove instances from the majority class, whereas the simplest way to oversample is to randomly make copies of the minority class until you have a ratio that you're happy with. Figure 4-10 shows a visualization of oversampling and undersampling.

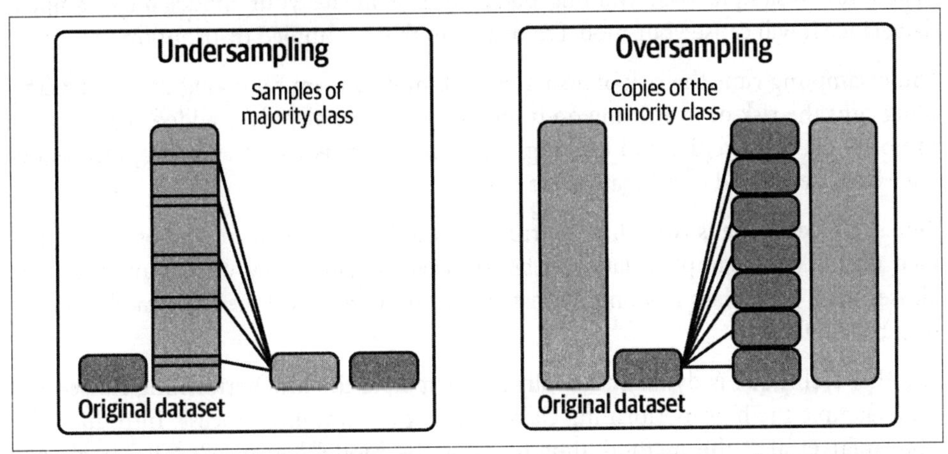

Figure 4-10. Illustrations of how undersampling and oversampling work. Source: Adapted from an image by Rafael Alencar[37]

A popular method of undersampling low-dimensional data that was developed back in 1976 is Tomek links.[38] With this technique, you find pairs of samples from opposite classes that are close in proximity and remove the sample of the majority class in each pair.

While this makes the decision boundary more clear and arguably helps models learn the boundary better, it may make the model less robust because the model doesn't get to learn from the subtleties of the true decision boundary.

A popular method of oversampling low-dimensional data is SMOTE (synthetic minority oversampling technique).[39] It synthesizes novel samples of the minority class

37 Rafael Alencar, "Resampling Strategies for Imbalanced Datasets," Kaggle, *https://oreil.ly/p8Whs*.

38 Ivan Tomek, "An Experiment with the Edited Nearest-Neighbor Rule," *IEEE Transactions on Systems, Man, and Cybernetics* (June 1976): 448–52, *https://oreil.ly/JCxHZ*.

39 N.V. Chawla, K.W. Bowyer, L.O. Hall, and W.P. Kegelmeyer, "SMOTE: Synthetic Minority Over-sampling Technique, *Journal of Artificial Intelligence Research* 16 (2002): 341–78, *https://oreil.ly/f6y46*.

through sampling convex combinations of existing data points within the minority class.[40]

Both SMOTE and Tomek links have only been proven effective in low-dimensional data. Many of the sophisticated resampling techniques, such as Near-Miss and one-sided selection,[41] require calculating the distance between instances or between instances and the decision boundaries, which can be expensive or infeasible for high-dimensional data or in high-dimensional feature space, such as the case with large neural networks.

When you resample your training data, never evaluate your model on resampled data, since it will cause your model to overfit to that resampled distribution.

Undersampling runs the risk of losing important data from removing data. Oversampling runs the risk of overfitting on training data, especially if the added copies of the minority class are replicas of existing data. Many sophisticated sampling techniques have been developed to mitigate these risks.

One such technique is two-phase learning.[42] You first train your model on the resampled data. This resampled data can be achieved by randomly undersampling large classes until each class has only *N* instances. You then fine-tune your model on the original data.

Another technique is dynamic sampling: oversample the low-performing classes and undersample the high-performing classes during the training process. Introduced by Pouyanfar et al.,[43] the method aims to show the model less of what it has already learned and more of what it has not.

Algorithm-level methods

If data-level methods mitigate the challenge of class imbalance by altering the distribution of your training data, algorithm-level methods keep the training data distribution intact but alter the algorithm to make it more robust to class imbalance.

---

40 "Convex" here approximately means "linear."

41 Jianping Zhang and Inderjeet Mani, "kNN Approach to Unbalanced Data Distributions: A Case Study involving Information Extraction" (Workshop on Learning from Imbalanced Datasets II, ICML, Washington, DC, 2003), *https://oreil.ly/qnpra*; Miroslav Kubat and Stan Matwin, "Addressing the Curse of Imbalanced Training Sets: One-Sided Selection," 2000, *https://oreil.ly/8pheJ*.

42 Hansang Lee, Minseok Park, and Junmo Kim, "Plankton Classification on Imbalanced Large Scale Database via Convolutional Neural Networks with Transfer Learning," *2016 IEEE International Conference on Image Processing (ICIP)*, 2016, *https://oreil.ly/YiA8p*.

43 Samira Pouyanfar, Yudong Tao, Anup Mohan, Haiman Tian, Ahmed S. Kaseb, Kent Gauen, Ryan Dailey, et al., "Dynamic Sampling in Convolutional Neural Networks for Imbalanced Data Classification," *2018 IEEE Conference on Multimedia Information Processing and Retrieval (MIPR)*, 2018, *https://oreil.ly/D3Ak5*.

Because the loss function (or the cost function) guides the learning process, many algorithm-level methods involve adjustment to the loss function. The key idea is that if there are two instances, x_1 and x_2, and the loss resulting from making the wrong prediction on x_1 is higher than x_2, the model will prioritize making the correct prediction on x_1 over making the correct prediction on x_2. By giving the training instances we care about higher weight, we can make the model focus more on learning these instances.

Let $L(x; \theta)$ be the loss caused by the instance x for the model with the parameter set θ. The model's loss is often defined as the average loss caused by all instances. N denotes the total number of training samples.

$$L(X; \theta) = \Sigma_x \frac{1}{N} L(x; \theta)$$

This loss function values the loss caused by all instances equally, even though wrong predictions on some instances might be much costlier than wrong predictions on other instances. There are many ways to modify this cost function. In this section, we will focus on three of them, starting with cost-sensitive learning.

Cost-sensitive learning. Back in 2001, based on the insight that misclassification of different classes incurs different costs, Elkan proposed cost-sensitive learning in which the individual loss function is modified to take into account this varying cost.[44] The method started by using a cost matrix to specify C_{ij}: the cost if class i is classified as class j. If $i = j$, it's a correct classification, and the cost is usually 0. If not, it's a misclassification. If classifying POSITIVE examples as NEGATIVE is twice as costly as the other way around, you can make C_{10} twice as high as C_{01}.

For example, if you have two classes, POSITIVE and NEGATIVE, the cost matrix can look like that in Table 4-8.

Table 4-8. Example of a cost matrix

	Actual NEGATIVE	Actual POSITIVE
Predicted NEGATIVE	$C(0, 0) = C_{00}$	$C(1, 0) = C_{10}$
Predicted POSITIVE	$C(0, 1) = C_{01}$	$C(1, 1) = C_{11}$

44 Charles Elkan, "The Foundations of Cost-Sensitive Learning," *Proceedings of the Seventeenth International Joint Conference on Artificial Intelligence* (IJCAI'01), 2001, *https://oreil.ly/WGq5M*.

The loss caused by instance x of class i will become the weighted average of all possible classifications of instance x.

$$L(x; \theta) = \Sigma_j C_{ij} P(j \mid x; \theta)$$

The problem with this loss function is that you have to manually define the cost matrix, which is different for different tasks at different scales.

Class-balanced loss. What might happen with a model trained on an imbalanced dataset is that it'll bias toward majority classes and make wrong predictions on minority classes. What if we punish the model for making wrong predictions on minority classes to correct this bias?

In its vanilla form, we can make the weight of each class inversely proportional to the number of samples in that class, so that the rarer classes have higher weights. In the following equation, N denotes the total number of training samples:

$$W_i = \frac{N}{\text{number of samples of class i}}$$

The loss caused by instance x of class i will become as follows, with $\text{Loss}(x, j)$ being the loss when x is classified as class j. It can be cross entropy or any other loss function.

$$L(x; \theta) = W_i \Sigma_j P(j \mid x; \theta) \text{Loss}(x, j)$$

A more sophisticated version of this loss can take into account the overlap among existing samples, such as class-balanced loss based on effective number of samples.[45]

Focal loss. In our data, some examples are easier to classify than others, and our model might learn to classify them quickly. We want to incentivize our model to focus on learning the samples it still has difficulty classifying. What if we adjust the loss so that if a sample has a lower probability of being right, it'll have a higher weight? This is exactly what focal loss does.[46] The equation for focal loss and its performance compared to cross entropy loss is shown in Figure 4-11.

45 Yin Cui, Menglin Jia, Tsung-Yi Lin, Yang Song, and Serge Belongie, "Class-Balanced Loss Based on Effective Number of Samples," *Proceedings of the Conference on Computer Vision and Pattern*, 2019, *https://oreil.ly/jCzGH*.

46 Tsung-Yi Lin, Priya Goyal, Ross Girshick, Kaiming He, and Piotr Dollár, "Focal Loss for Dense Object Detection," *arXiv*, August 7, 2017, *https://oreil.ly/Km2dF*.

In practice, ensembles have shown to help with the class imbalance problem.[47] However, we don't include ensembling in this section because class imbalance isn't usually why ensembles are used. Ensemble techniques will be covered in Chapter 6.

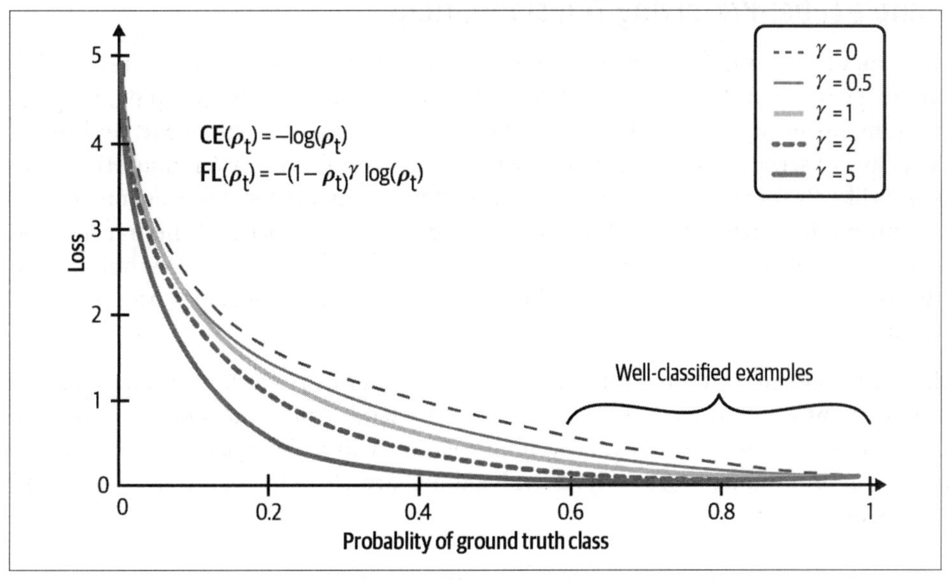

Figure 4-11. The model trained with focal loss (FL) shows reduced loss values compared to the model trained with cross entropy loss (CE). Source: Adapted from an image by Lin et al.

Data Augmentation

Data augmentation is a family of techniques that are used to increase the amount of training data. Traditionally, these techniques are used for tasks that have limited training data, such as in medical imaging. However, in the last few years, they have shown to be useful even when we have a lot of data—augmented data can make our models more robust to noise and even adversarial attacks.

Data augmentation has become a standard step in many computer vision tasks and is finding its way into natural language processing (NLP) tasks. The techniques depend heavily on the data format, as image manipulation is different from text manipulation. In this section, we will cover three main types of data augmentation: simple label-preserving transformations; perturbation, which is a term for "adding

47 Mikel Galar, Alberto Fernandez, Edurne Barrenechea, Humberto Bustince, and Francisco Herrera, "A Review on Ensembles for the Class Imbalance Problem: Bagging-, Boosting-, and Hybrid-Based Approaches," *IEEE Transactions on Systems, Man, and Cybernetics, Part C (Applications and Reviews)* 42, no. 4 (July 2012): 463–84, *https://oreil.ly/1ND4g*.

noises"; and data synthesis. In each type, we'll go over examples for both computer vision and NLP.

Simple Label-Preserving Transformations

In computer vision, the simplest data augmentation technique is to randomly modify an image while preserving its label. You can modify the image by cropping, flipping, rotating, inverting (horizontally or vertically), erasing part of the image, and more. This makes sense because a rotated image of a dog is still a dog. Common ML frameworks like PyTorch, TensorFlow, and Keras all have support for image augmentation. According to Krizhevsky et al. in their legendary AlexNet paper, "The transformed images are generated in Python code on the CPU while the GPU is training on the previous batch of images. So these data augmentation schemes are, in effect, computationally free."[48]

In NLP, you can randomly replace a word with a similar word, assuming that this replacement wouldn't change the meaning or the sentiment of the sentence, as shown in Table 4-9. Similar words can be found either with a dictionary of synonymous words or by finding words whose embeddings are close to each other in a word embedding space.

Table 4-9. Three sentences generated from an original sentence

Original sentence	I'm so happy to see you.
Generated sentences	I'm so *glad* to see you. I'm so happy to see *y'all*. I'm *very* happy to see you.

This type of data augmentation is a quick way to double or triple your training data.

Perturbation

Perturbation is also a label-preserving operation, but because sometimes it's used to trick models into making wrong predictions, I thought it deserves its own section.

Neural networks, in general, are sensitive to noise. In the case of computer vision, this means that adding a small amount of noise to an image can cause a neural network to misclassify it. Su et al. showed that 67.97% of the natural images in the Kaggle CIFAR-10 test dataset and 16.04% of the ImageNet test images can be misclassified by changing just one pixel (see Figure 4-12).[49]

48 Alex Krizhevsky, Ilya Sutskever, and Geoffrey E. Hinton, "ImageNet Classification with Deep Convolutional Neural Networks, 2012, *https://oreil.ly/aphzA*.

49 Jiawei Su, Danilo Vasconcellos Vargas, and Sakurai Kouichi, "One Pixel Attack for Fooling Deep Neural Networks," *IEEE Transactions on Evolutionary Computation* 23, no. 5 (2019): 828–41, *https://oreil.ly/LzN9D*.

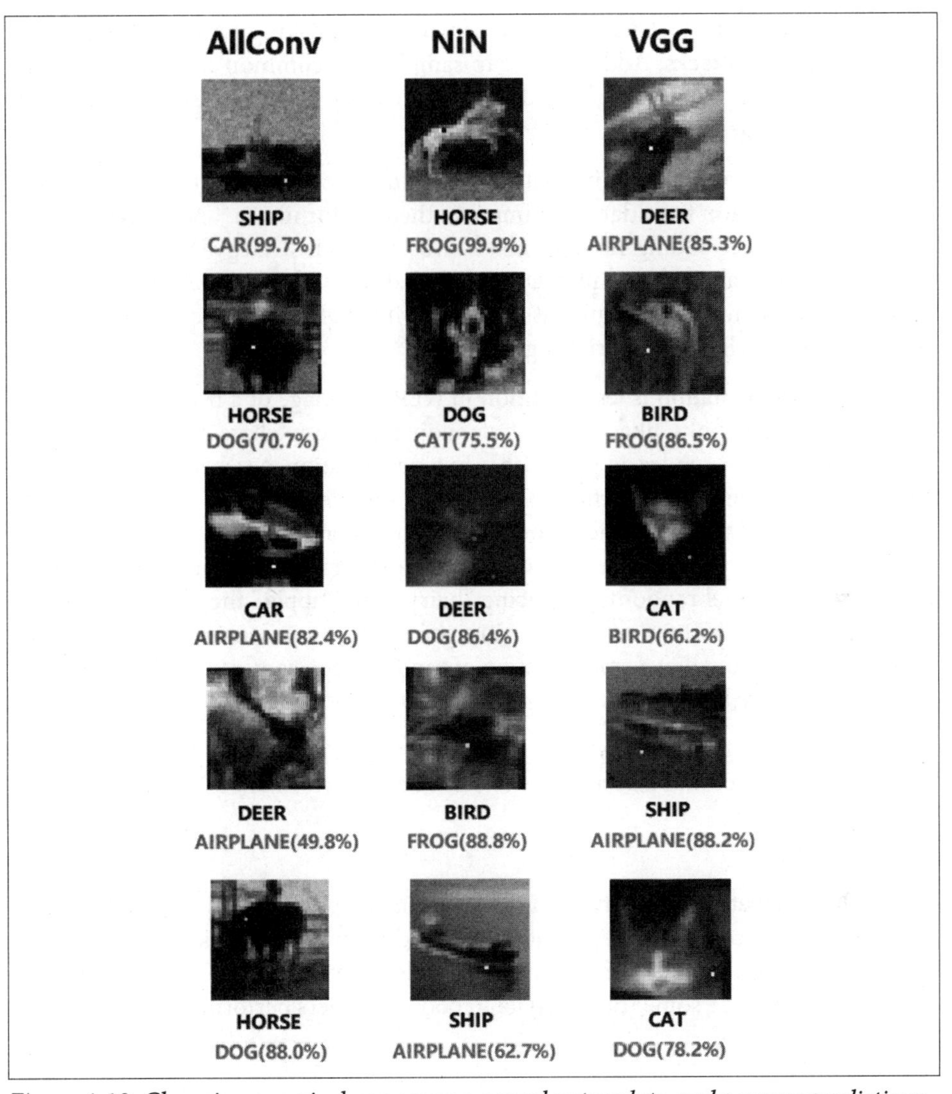

Figure 4-12. Changing one pixel can cause a neural network to make wrong predictions. The three models used are AllConv, NiN, and VGG. The original labels made by those models are above the labels made after one pixel was changed. Source: Su et al.[50]

50 Su et al., "One Pixel Attack."

Using deceptive data to trick a neural network into making wrong predictions is called adversarial attacks. Adding noise to samples is a common technique to create adversarial samples. The success of adversarial attacks is especially exaggerated as the resolution of images increases.

Adding noisy samples to training data can help models recognize the weak spots in their learned decision boundary and improve their performance.[51] Noisy samples can be created by either adding random noise or by a search strategy. Moosavi-Dezfooli et al. proposed an algorithm, called DeepFool, that finds the minimum possible noise injection needed to cause a misclassification with high confidence.[52] This type of augmentation is called adversarial augmentation.[53]

Adversarial augmentation is less common in NLP (an image of a bear with randomly added pixels still looks like a bear, but adding random characters to a random sentence will likely render it gibberish), but perturbation has been used to make models more robust. One of the most notable examples is BERT, where the model chooses 15% of all tokens in each sequence at random and chooses to replace 10% of the chosen tokens with random words. For example, given the sentence "My dog is hairy," and the model randomly replacing "hairy" with "apple," the sentence becomes "My dog is apple." So 1.5% of all tokens might result in nonsensical meaning. Their ablation studies show that a small fraction of random replacement gives their model a small performance boost.[54]

In Chapter 6, we'll go over how to use perturbation not just as a way to improve your model's performance, but also as a way to evaluate its performance.

Data Synthesis

Since collecting data is expensive and slow, with many potential privacy concerns, it'd be a dream if we could sidestep it altogether and train our models with synthesized data. Even though we're still far from being able to synthesize all training data, it's possible to synthesize some training data to boost a model's performance.

---

51 Ian J. Goodfellow, Jonathon Shlens, and Christian Szegedy, "Explaining and Harnessing Adversarial Examples," *arXiv*, March 20, 2015, *https://oreil.ly/9v2No*; Ian J. Goodfellow, David Warde-Farley, Mehdi Mirza, Aaron Courville, and Yoshua Bengio, "Maxout Networks, *arXiv*, February 18, 2013, *https://oreil.ly/L8mch*.

52 Seyed-Mohsen Moosavi-Dezfooli, Alhussein Fawzi, and Pascal Frossard, "DeepFool: A Simple and Accurate Method to Fool Deep Neural Networks," in *Proceedings of IEEE Conference on Computer Vision and Pattern Recognition (CVPR)*, 2016, *https://oreil.ly/dYVL8*.

53 Takeru Miyato, Shin-ichi Maeda, Masanori Koyama, and Shin Ishii, "Virtual Adversarial Training: A Regularization Method for Supervised and Semi-Supervised Learning," *IEEE Transactions on Pattern Analysis and Machine Intelligence*, 2017, *https://oreil.ly/MBQeu*.

54 Devlin et al., "BERT: Pre-training of Deep Bidirectional Transformers for Language Understanding."

In NLP, templates can be a cheap way to bootstrap your model. One team I worked with used templates to bootstrap training data for their conversational AI (chatbot). A template might look like: "Find me a [CUISINE] restaurant within [NUMBER] miles of [LOCATION]" (see Table 4-10). With lists of all possible cuisines, reasonable numbers (you would probably never want to search for restaurants beyond 1,000 miles), and locations (home, office, landmarks, exact addresses) for each city, you can generate thousands of training queries from a template.

Table 4-10. Three sentences generated from a template

Template	Find me a [CUISINE] restaurant within [NUMBER] miles of [LOCATION].
Generated queries	Find me a *Vietnamese* restaurant within *2* miles of *my office*.
	Find me a *Thai* restaurant within *5* miles of *my home*.
	Find me a *Mexican* restaurant within *3* miles of *Google headquarters*.

In computer vision, a straightforward way to synthesize new data is to combine existing examples with discrete labels to generate continuous labels. Consider a task of classifying images with two possible labels: DOG (encoded as 0) and CAT (encoded as 1). From example x_1 of label DOG and example x_2 of label CAT, you can generate x' such as:

$$x' = \gamma x_1 + (1 - \gamma) x_2$$

The label of x' is a combination of the labels of x_1 and x_2: $\gamma \times 0 + (1 - \gamma) \times 1$. This method is called mixup. The authors showed that mixup improves models' generalization, reduces their memorization of corrupt labels, increases their robustness to adversarial examples, and stabilizes the training of generative adversarial networks.[55]

Using neural networks to synthesize training data is an exciting approach that is actively being researched but not yet popular in production. Sandfort et al. showed that by adding images generated using CycleGAN to their original training data, they were able to improve their model's performance significantly on computed tomography (CT) segmentation tasks.[56]

If you're interested in learning more about data augmentation for computer vision, "A Survey on Image Data Augmentation for Deep Learning" (*https://oreil.ly/3TUpK*) (Shorten and Khoshgoftaar 2019) is a comprehensive review.

---

55 Hongyi Zhang, Moustapha Cisse, Yann N. Dauphin, and David Lopez-Paz, "*mixup*: Beyond Empirical Risk Minimization," *ICLR 2018*, *https://oreil.ly/lIM5E*.

56 Veit Sandfort, Ke Yan, Perry J. Pickhardt, and Ronald M. Summers, "Data Augmentation Using Generative Adversarial Networks (CycleGAN) to Improve Generalizability in CT Segmentation Tasks," *Scientific Reports* 9, no. 1 (2019): 16884, *https://oreil.ly/TDUwm*.

Summary

Training data still forms the foundation of modern ML algorithms. No matter how clever your algorithms might be, if your training data is bad, your algorithms won't be able to perform well. It's worth it to invest time and effort to curate and create training data that will enable your algorithms to learn something meaningful.

In this chapter, we've discussed the multiple steps to create training data. We first covered different sampling methods, both nonprobability sampling and random sampling, that can help us sample the right data for our problem.

Most ML algorithms in use today are supervised ML algorithms, so obtaining labels is an integral part of creating training data. Many tasks, such as delivery time estimation or recommender systems, have natural labels. Natural labels are usually delayed, and the time it takes from when a prediction is served until when the feedback on it is provided is the feedback loop length. Tasks with natural labels are fairly common in the industry, which might mean that companies prefer to start on tasks that have natural labels over tasks without natural labels.

For tasks that don't have natural labels, companies tend to rely on human annotators to annotate their data. However, hand labeling comes with many drawbacks. For example, hand labels can be expensive and slow. To combat the lack of hand labels, we discussed alternatives including weak supervision, semi-supervision, transfer learning, and active learning.

ML algorithms work well in situations when the data distribution is more balanced, and not so well when the classes are heavily imbalanced. Unfortunately, problems with class imbalance are the norm in the real world. In the following section, we discussed why class imbalance made it hard for ML algorithms to learn. We also discussed different techniques to handle class imbalance, from choosing the right metrics to resampling data to modifying the loss function to encourage the model to pay attention to certain samples.

We ended the chapter with a discussion on data augmentation techniques that can be used to improve a model's performance and generalization for both computer vision and NLP tasks.

Once you have your training data, you will want to extract features from it to train your ML models, which we will cover in the next chapter.

CHAPTER 5

Feature Engineering

In 2014, the paper "Practical Lessons from Predicting Clicks on Ads at Facebook" (*https://oreil.ly/oS16J*) claimed that having the right features is the most important thing in developing their ML models. Since then, many of the companies that I've worked with have discovered time and time again that once they have a workable model, having the right features tends to give them the biggest performance boost compared to clever algorithmic techniques such as hyperparameter tuning. State-of-the-art model architectures can still perform poorly if they don't use a good set of features.

Due to its importance, a large part of many ML engineering and data science jobs is to come up with new useful features. In this chapter, we will go over common techniques and important considerations with respect to feature engineering. We will dedicate a section to go into detail about a subtle yet disastrous problem that has derailed many ML systems in production: data leakage and how to detect and avoid it.

We will end the chapter discussing how to engineer good features, taking into account both the feature importance and feature generalization. Talking about feature engineering, some people might think of feature stores. Since feature stores are closer to infrastructure to support multiple ML applications, we'll cover feature stores in Chapter 10.

Learned Features Versus Engineered Features

When I cover this topic in class, my students frequently ask: "Why do we have to worry about feature engineering? Doesn't deep learning promise us that we no longer have to engineer features?"

They are right. The promise of deep learning is that we won't have to handcraft features. For this reason, deep learning is sometimes called feature learning.[1] Many features can be automatically learned and extracted by algorithms. However, we're still far from the point where all features can be automated. This is not to mention that, as of this writing, the majority of ML applications in production aren't deep learning. Let's go over an example to understand what features can be automatically extracted and what features still need to be handcrafted.

Imagine that you want to build a sentiment analysis classifier to classify whether a comment is spam or not. Before deep learning, when given a piece of text, you would have to manually apply classical text processing techniques such as lemmatization, expanding contractions, removing punctuation, and lowercasing everything. After that, you might want to split your text into n-grams with *n* values of your choice.

For those unfamiliar, an n-gram is a contiguous sequence of *n* items from a given sample of text. The items can be phonemes, syllables, letters, or words. For example, given the post "I like food," its word-level 1-grams are ["I", "like", "food"] and its word-level 2-grams are ["I like", "like food"]. This sentence's set of n-gram features, if we want *n* to be 1 and 2, is: ["I", "like", "food", "I like", "like food"].

Figure 5-1 shows an example of classical text processing techniques you can use to handcraft n-gram features for your text.

1 Loris Nanni, Stefano Ghidoni, and Sheryl Brahnam, "Handcrafted vs. Non-handcrafted Features for Computer Vision Classification," *Pattern Recognition* 71 (November 2017): 158–72, *https://oreil.ly/CGfYQ*; Wikipedia, s.v. "Feature learning," *https://oreil.ly/fJmwN*.

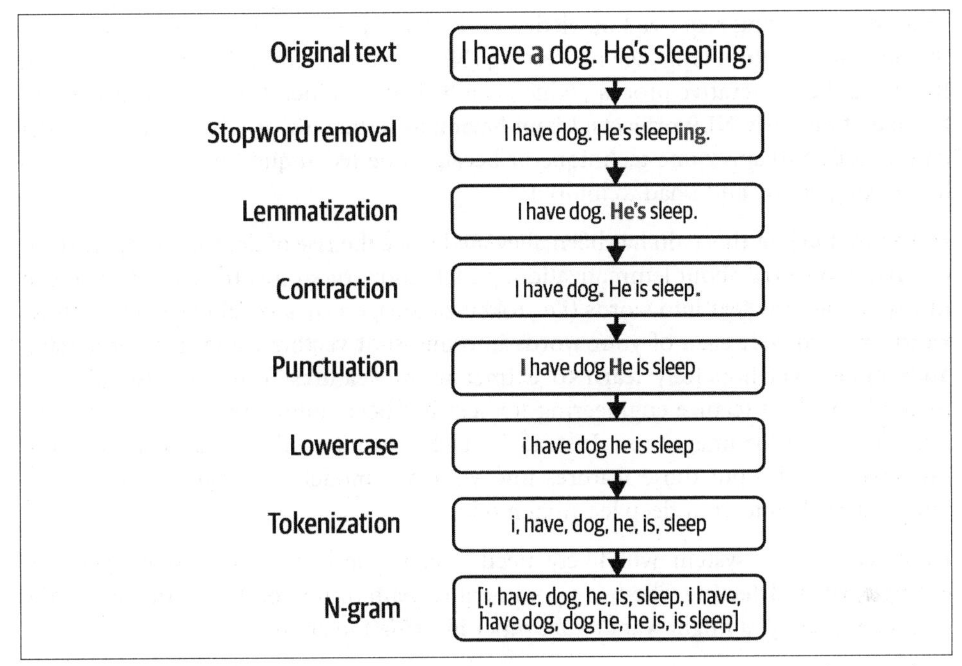
Original text
I have a dog. He's sleeping.
Stopword removal
I have dog. He's sleeping.
Lemmatization
I have dog. He's sleep.
Contraction
I have dog. He is sleep.
Punctuation
I have dog He is sleep
Lowercase
i have dog he is sleep
Tokenization
i, have, dog, he, is, sleep
N-gram
[i, have, dog, he, is, sleep, i have, have dog, dog he, he is, is sleep]

Figure 5-1. An example of techniques that you can use to handcraft n-gram features for your text

Once you've generated n-grams for your training data, you can create a vocabulary that maps each n-gram to an index. Then you can convert each post into a vector based on its n-grams' indices. For example, if we have a vocabulary of seven n-grams as shown in Table 5-1, each post can be a vector of seven elements. Each element corresponds to the number of times the n-gram at that index appears in the post. "I like food" will be encoded as the vector [1, 1, 0, 1, 1, 0, 1]. This vector can then be used as an input into an ML model.

Table 5-1. Example of a 1-gram and 2-gram vocabulary

I	like	good	food	I like	good food	like food
0	1	2	3	4	5	6

Feature engineering requires knowledge of domain-specific techniques—in this case, the domain is natural language processing (NLP) and the native language of the text. It tends to be an iterative process, which can be brittle. When I followed this method for one of my early NLP projects, I kept having to restart my process either because I had forgotten to apply one technique or because one technique I used turned out to be working poorly and I had to undo it.

However, much of this pain has been alleviated since the rise of deep learning. Instead of having to worry about lemmatization, punctuation, or stopword removal, you can just split your raw text into words (i.e., tokenization), create a vocabulary out of those words, and convert each of your words into one-shot vectors using this vocabulary. Your model will hopefully learn to extract useful features from this. In this new method, much of feature engineering for text has been automated. Similar progress has been made for images too. Instead of having to manually extract features from raw images and input those features into your ML models, you can just input raw images directly into your deep learning models.

However, an ML system will likely need data beyond just text and images. For example, when detecting whether a comment is spam or not, on top of the text in the comment itself, you might want to use other information about:

The comment
: How many upvotes/downvotes does it have?

The user who posted this comment
: When was this account created, how often do they post, and how many upvotes/downvotes do they have?

The thread in which the comment was posted
: How many views does it have? Popular threads tend to attract more spam.

There are many possible features to use in your model. Some of them are shown in Figure 5-2. The process of choosing what information to use and how to extract this information into a format usable by your ML models is feature engineering. For complex tasks such as recommending videos for users to watch next on TikTok, the number of features used can go up to millions. For domain-specific tasks such as predicting whether a transaction is fraudulent, you might need subject matter expertise with banking and frauds to be able to come up with useful features.

Comment ID	Time	User	Text	# ▲	# ▼	Link	# img	Thread ID	Reply to	# replies	...
93880839	2020-10-30 T 10:45 UTC	gitrekt	Your mom is a nice lady.	1	0	0	0	2332332	n0tab0t	1	...

User ID	Created	User	Subs	# ▲	# ▼	# replies	Karma	# threads	Verified email	Awards	...
4402903	2015-01-57 T 3:09 PST	gitrekt	[r/ml, r/memes, r/socialist]	15	90	28	304	776	No		...

Thread ID	Time	User	Text	# ▲	# ▼	Link	# img	# replies	# views	Awards	...
93883208	2020-10-30 T 2:45 PST	doge	Human is temporary, AGI is forever	120	50	1	0	32	2405	1	...

Figure 5-2. Some of the possible features about a comment, a thread, or a user to be included in your model

Common Feature Engineering Operations

Because of the importance and the ubiquity of feature engineering in ML projects, there have been many techniques developed to streamline the process. In this section, we will discuss several of the most important operations that you might want to consider while engineering features from your data. They include handling missing values, scaling, discretization, encoding categorical features, and generating the old-school but still very effective cross features as well as the newer and exciting positional features. This list is nowhere near being comprehensive, but it does comprise some of the most common and useful operations to give you a good starting point. Let's dive in!

Handling Missing Values

One of the first things you might notice when dealing with data in production is that some values are missing. However, one thing that many ML engineers I've interviewed don't know is that not all types of missing values are equal.[2] To illustrate this point, consider the task of predicting whether someone is going to buy a house in the next 12 months. A portion of the data we have is in Table 5-2.

2 In my experience, how well a person handles missing values for a given dataset during interviews strongly correlates with how well they will do in their day-to-day jobs.

Table 5-2. Example data for predicting house buying in the next 12 months

ID	Age	Gender	Annual income	Marital status	Number of children	Job	Buy?
1		A	150,000		1	Engineer	No
2	27	B	50,000			Teacher	No
3		A	100,000	Married	2		Yes
4	40	B			2	Engineer	Yes
5	35	B		Single	0	Doctor	Yes
6		A	50,000		0	Teacher	No
7	33	B	60,000	Single		Teacher	No
8	20	B	10,000			Student	No

There are three types of missing values. The official names for these types are a little bit confusing, so we'll go into detailed examples to mitigate the confusion.

Missing not at random (MNAR)
: This is when the reason a value is missing is because of the true value itself. In this example, we might notice that some respondents didn't disclose their income. Upon investigation it may turn out that the income of respondents who failed to report tends to be higher than that of those who did disclose. *The income values are missing for reasons related to the values themselves.*

Missing at random (MAR)
: This is when the reason a *value is missing is not due to the value itself, but due to another observed variable.* In this example, we might notice that age values are often missing for respondents of the gender "A," which might be because the people of gender A in this survey don't like disclosing their age.

Missing completely at random (MCAR)
: This is when *there's no pattern in when the value is missing.* In this example, we might think that the missing values for the column "Job" might be completely random, not because of the job itself and not because of any other variable. People just forget to fill in that value sometimes for no particular reason. However, this type of missing is very rare. There are usually reasons why certain values are missing, and you should investigate.

When encountering missing values, you can either fill in the missing values with certain values (imputation) or remove the missing values (deletion). We'll go over both.

Deletion

When I ask candidates about how to handle missing values during interviews, many tend to prefer deletion, not because it's a better method, but because it's easier to do.

One way to delete is *column deletion*: if a variable has too many missing values, just remove that variable. For example, in the example above, over 50% of the values for the variable "Marital status" are missing, so you might be tempted to remove this variable from your model. The drawback of this approach is that you might remove important information and reduce the accuracy of your model. Marital status might be highly correlated to buying houses, as married couples are much more likely to be homeowners than single people.[3]

Another way to delete is *row deletion*: if a sample has missing value(s), just remove that sample. This method can work when the missing values are completely at random (MCAR) and the number of examples with missing values is small, such as less than 0.1%. You don't want to do row deletion if that means 10% of your data samples are removed.

However, removing rows of data can also remove important information that your model needs to make predictions, especially if the missing values are not at random (MNAR). For example, you don't want to remove samples of gender B respondents with missing income because the fact that income is missing is information itself (missing income might mean higher income, and thus, more correlated to buying a house) and can be used to make predictions.

On top of that, removing rows of data can create biases in your model, especially if the missing values are at random (MAR). For example, if you remove all examples missing age values in the data in Table 5-2, you will remove all respondents with gender A from your data, and your model won't be able to make good predictions for respondents with gender A.

Imputation

Even though deletion is tempting because it's easy to do, deleting data can lead to losing important information and introduce biases into your model. If you don't want to delete missing values, you will have to impute them, which means "fill them with certain values." Deciding which "certain values" to use is the hard part.

3 Rachel Bogardus Drew, "3 Facts About Marriage and Homeownership," Joint Center for Housing Studies of Harvard University, December 17, 2014, *https://oreil.ly/MWxFp*.

One common practice is to fill in missing values with their defaults. For example, if the job is missing, you might fill it with an empty string "". Another common practice is to fill in missing values with the mean, median, or mode (the most common value). For example, if the temperature value is missing for a data sample whose month value is July, it's not a bad idea to fill it with the median temperature of July.

Both practices work well in many cases, but sometimes they can cause hair-pulling bugs. One time, in one of the projects I was helping with, we discovered that the model was spitting out garbage because the app's frontend no longer asked users to enter their age, so age values were missing, and the model filled them with 0. But the model never saw the age value of 0 during training, so it couldn't make reasonable predictions.

In general, you want to avoid filling missing values with possible values, such as filling the missing number of children with 0—0 is a possible value for the number of children. It makes it hard to distinguish between people whose information is missing and people who don't have children.

Multiple techniques might be used at the same time or in sequence to handle missing values for a particular set of data. Regardless of what techniques you use, one thing is certain: there is no perfect way to handle missing values. With deletion, you risk losing important information or accentuating biases. With imputation, you risk injecting your own bias into and adding noise to your data, or worse, data leakage. If you don't know what data leakage is, don't panic, we'll cover it in the section "Data Leakage" on page 135.

Scaling

Consider the task of predicting whether someone will buy a house in the next 12 months, and the data shown in Table 5-2. The values of the variable Age in our data range from 20 to 40, whereas the values of the variable Annual Income range from 10,000 to 150,000. When we input these two variables into an ML model, it won't understand that 150,000 and 40 represent different things. It will just see them both as numbers, and because the number 150,000 is much bigger than 40, it might give it more importance, regardless of which variable is actually more useful for generating predictions.

Before inputting features into models, it's important to scale them to be similar ranges. This process is called *feature scaling*. This is one of the simplest things you can do that often results in a performance boost for your model. Neglecting to do so can cause your model to make gibberish predictions, especially with classical algorithms like gradient-boosted trees and logistic regression.[4]

4 Feature scaling once boosted my model's performance by almost 10%.

An intuitive way to scale your features is to get them to be in the range [0, 1]. Given a variable x, its values can be rescaled to be in this range using the following formula:

$$x' = \frac{x - \min(x)}{\max(x) - \min(x)}$$

You can validate that if x is the maximum value, the scaled value x' will be 1. If x is the minimum value, the scaled value x' will be 0.

If you want your feature to be in an arbitrary range [a, b]—empirically, I find the range [–1, 1] to work better than the range [0, 1]—you can use the following formula:

$$x' = a + \frac{(x - \min(x))(b - a)}{\max(x) - \min(x)}$$

Scaling to an arbitrary range works well when you don't want to make any assumptions about your variables. If you think that your variables might follow a normal distribution, it might be helpful to normalize them so that they have zero mean and unit variance. This process is called *standardization*:

$$x' = \frac{x - \bar{x}}{\sigma},$$

with $\bar{x}$ being the mean of variable x, and σ being its standard deviation.

In practice, ML models tend to struggle with features that follow a skewed distribution. To help mitigate the skewness, a technique commonly used is log transformation (*https://oreil.ly/RMwEy*): apply the log function to your feature. An example of how the log transformation can make your data less skewed is shown in Figure 5-3. While this technique can yield performance gain in many cases, it doesn't work for all cases, and you should be wary of the analysis performed on log-transformed data instead of the original data.[5]

5 Changyong Feng, Hongyue Wang, Naiji Lu, Tian Chen, Hua He, Ying Lu, and Xin M. Tu, "Log-Transformation and Its Implications for Data Analysis," *Shanghai Archives of Psychiatry* 26, no. 2 (April 2014): 105–9, *https://oreil.ly/hHJjt*.

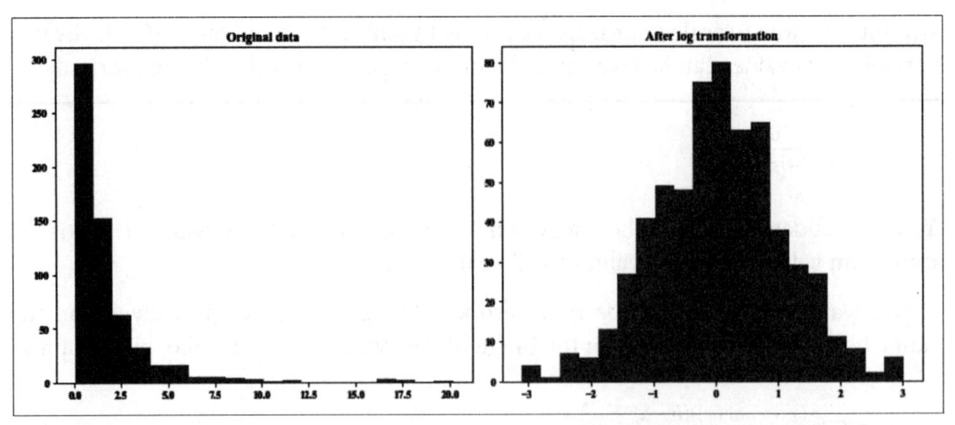

Figure 5-3. In many cases, the log transformation can help reduce the skewness of your data

There are two important things to note about scaling. One is that it's a common source of data leakage (this will be covered in greater detail in the section "Data Leakage" on page 135). Another is that it often requires global statistics—you have to look at the entire or a subset of training data to calculate its min, max, or mean. During inference, you reuse the statistics you had obtained during training to scale new data. If the new data has changed significantly compared to the training, these statistics won't be very useful. Therefore, it's important to retrain your model often to account for these changes.

Discretization

This technique is included in this book for completeness, though in practice, I've rarely found discretization to help. Imagine that we've built a model with the data in Table 5-2. During training, our model has seen the annual income values of "150,000," "50,000," "100,000," and so on. During inference, our model encounters an example with an annual income of "9,000.50."

Intuitively, we know that $9,000.50 a year isn't much different from $10,000/year, and we want our model to treat both of them the same way. But the model doesn't know that. Our model only knows that 9,000.50 is different from 10,000, and it will treat them differently.

Discretization is the process of turning a continuous feature into a discrete feature. This process is also known as quantization or binning. This is done by creating buckets for the given values. For annual income, you might want to group them into three buckets as follows:

- Lower income: less than $35,000/year
- Middle income: between $35,000 and $100,000/year
- Upper income: more than $100,000/year

Instead of having to learn an infinite number of possible incomes, our model can focus on learning only three categories, which is a much easier task to learn. This technique is supposed to be more helpful with limited training data.

Even though, by definition, discretization is meant for continuous features, it can be used for discrete features too. The age variable is discrete, but it might still be useful to group the values into buckets such as follows:

- Less than 18
- Between 18 and 22
- Between 22 and 30
- Between 30 and 40
- Between 40 and 65
- Over 65

The downside is that this categorization introduces discontinuities at the category boundaries—$34,999 is now treated as completely different from $35,000, which is treated the same as $100,000. Choosing the boundaries of categories might not be all that easy. You can try to plot the histograms of the values and choose the boundaries that make sense. In general, common sense, basic quantiles, and sometimes subject matter expertise can help.

Encoding Categorical Features

We've talked about how to turn continuous features into categorical features. In this section, we'll discuss how to best handle categorical features.

People who haven't worked with data in production tend to assume that categories are *static*, which means the categories don't change over time. This is true for many categories. For example, age brackets and income brackets are unlikely to change, and you know exactly how many categories there are in advance. Handling these categories is straightforward. You can just give each category a number and you're done.

However, in production, categories change. Imagine you're building a recommender system to predict what products users might want to buy from Amazon. One of the features you want to use is the product brand. When looking at Amazon's historical data, you realize that there are a lot of brands. Even back in 2019, there were already over two million brands on Amazon![6]

The number of brands is overwhelming, but you think: "I can still handle this." You encode each brand as a number, so now you have two million numbers, from 0 to 1,999,999, corresponding to two million brands. Your model does spectacularly on the historical test set, and you get approval to test it on 1% of today's traffic.

In production, your model crashes because it encounters a brand it hasn't seen before and therefore can't encode. New brands join Amazon all the time. To address this, you create a category UNKNOWN with the value of 2,000,000 to catch all the brands your model hasn't seen during training.

Your model doesn't crash anymore, but your sellers complain that their new brands are not getting any traffic. It's because your model didn't see the category UNKNOWN in the train set, so it just doesn't recommend any product of the UNKNOWN brand. You fix this by encoding only the top 99% most popular brands and encode the bottom 1% brand as UNKNOWN. This way, at least your model knows how to deal with UNKNOWN brands.

Your model seems to work fine for about one hour, then the click-through rate on product recommendations plummets. Over the last hour, 20 new brands joined your site; some of them are new luxury brands, some of them are sketchy knockoff brands, some of them are established brands. However, your model treats them all the same way it treats unpopular brands in the training data.

This isn't an extreme example that only happens if you work at Amazon. This problem happens quite a lot. For example, if you want to predict whether a comment is spam, you might want to use the account that posted this comment as a feature, and new accounts are being created all the time. The same goes for new product types, new website domains, new restaurants, new companies, new IP addresses, and so on. If you work with any of them, you'll have to deal with this problem.

Finding a way to solve this problem turns out to be surprisingly difficult. You don't want to put them into a set of buckets because it can be really hard—how would you even go about putting new user accounts into different groups?

One solution to this problem is the *hashing trick*, popularized by the package Vowpal Wabbit developed at Microsoft.[7] The gist of this trick is that you use a hash function to generate a hashed value of each category. The hashed value will become the index of that category. Because you can specify the hash space, you can fix the number of encoded values for a feature in advance, without having to know how many

6 "Two Million Brands on Amazon," *Marketplace Pulse*, June 11, 2019, *https://oreil.ly/zrqtd*.

7 Wikipedia, s.v. "Feature hashing," *https://oreil.ly/tINTc*.

categories there will be. For example, if you choose a hash space of 18 bits, which corresponds to 2^{18} = 262,144 possible hashed values, all the categories, even the ones that your model has never seen before, will be encoded by an index between 0 and 262,143.

One problem with hashed functions is collision: two categories being assigned the same index. However, with many hash functions, the collisions are random; new brands can share an index with any of the existing brands instead of always sharing an index with unpopular brands, which is what happens when we use the preceding UNKNOWN category. The impact of colliding hashed features is, fortunately, not that bad. In research done by Booking.com, even for 50% colliding features, the performance loss is less than 0.5%, as shown in Figure 5-4.[8]

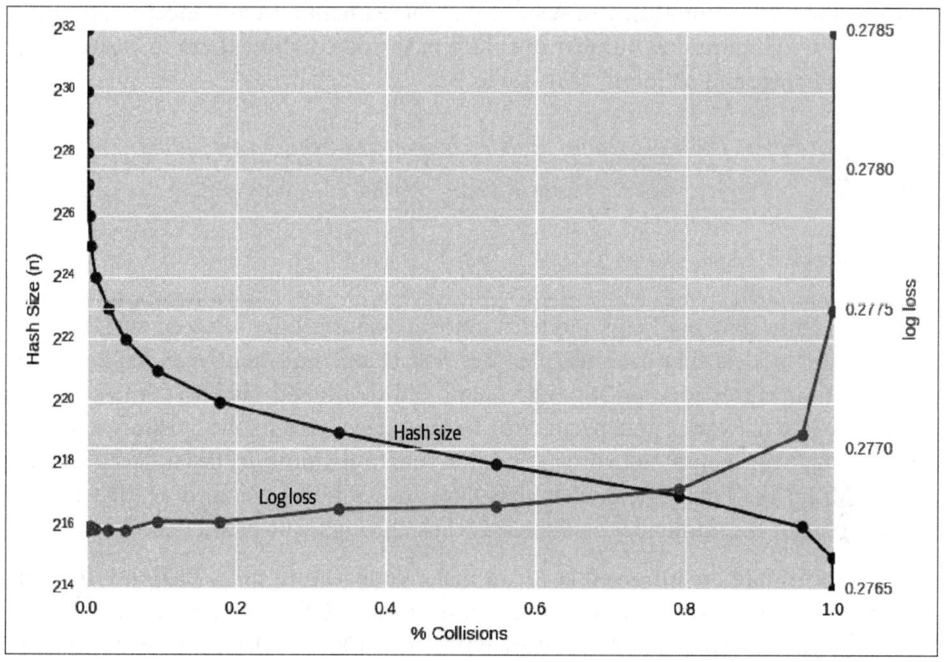

Figure 5-4. A 50% collision rate only causes the log loss to increase less than 0.5%. Source: Lucas Bernardi

You can choose a hash space large enough to reduce the collision. You can also choose a hash function with properties that you want, such as a locality-sensitive hashing function where similar categories (such as websites with similar names) are hashed into values close to each other.

8 Lucas Bernardi, "Don't Be Tricked by the Hashing Trick," Booking.com, January 10, 2018, *https://oreil.ly/VZmaY*.

Because it's a trick, it's often considered hacky by academics and excluded from ML curricula. But its wide adoption in the industry is a testimonial to how effective the trick is. It's essential to Vowpal Wabbit and it's part of the frameworks of scikit-learn, TensorFlow, and gensim. It can be especially useful in continual learning settings where your model learns from incoming examples in production. We'll cover continual learning in Chapter 9.

Feature Crossing

Feature crossing is the technique to combine two or more features to generate new features. This technique is useful to model the nonlinear relationships between features. For example, for the task of predicting whether someone will want to buy a house in the next 12 months, you suspect that there might be a nonlinear relationship between marital status and number of children, so you combine them to create a new feature "marriage and children" as in Table 5-3.

Table 5-3. Example of how two features can be combined to create a new feature

Marriage	Single	Married	Single	Single	Married
Children	0	2	1	0	1
Marriage and children	Single, 0	Married, 2	Single, 1	Single, 0	Married, 1

Because feature crossing helps model nonlinear relationships between variables, it's essential for models that can't learn or are bad at learning nonlinear relationships, such as linear regression, logistic regression, and tree-based models. It's less important in neural networks, but it can still be useful because explicit feature crossing occasionally helps neural networks learn nonlinear relationships faster. DeepFM and xDeepFM are the family of models that have successfully leveraged explicit feature interactions for recommender systems and click-through-rate prediction.[9]

A caveat of feature crossing is that it can make your feature space blow up. Imagine feature A has 100 possible values and feature B has 100 possible features; crossing these two features will result in a feature with 100 × 100 = 10,000 possible values. You will need a lot more data for models to learn all these possible values. Another caveat is that because feature crossing increases the number of features models use, it can make models overfit to the training data.

9 Huifeng Guo, Ruiming Tang, Yunming Ye, Zhenguo Li, and Xiuqiang He, "DeepFM: A Factorization-Machine Based Neural Network for CTR Prediction," *Proceedings of the Twenty-Sixth International Joint Conference on Artificial Intelligence* (IJCAI, 2017), *https://oreil.ly/1Vs3v*; Jianxun Lian, Xiaohuan Zhou, Fuzheng Zhang, Zhongxia Chen, Xing Xie, and Guangzhong Sun, "xDeepFM: Combining Explicit and Implicit Feature Interactions for Recommender Systems," *arXiv*, 2018, *https://oreil.ly/WFmFt*.

Discrete and Continuous Positional Embeddings

First introduced to the deep learning community in the paper "Attention Is All You Need" (*https://oreil.ly/eXk16*) (Vaswani et al. 2017), positional embedding has become a standard data engineering technique for many applications in both computer vision and NLP. We'll walk through an example to show why positional embedding is necessary and how to do it.

Consider the task of language modeling where you want to predict the next token (e.g., a word, character, or subword) based on the previous sequence of tokens. In practice, a sequence length can be up to 512, if not larger. However, for simplicity, let's use words as our tokens and use the sequence length of 8. Given an arbitrary sequence of 8 words, such as "Sometimes all I really want to do is," we want to predict the next word.

> ### Embeddings
>
> An embedding is a vector that represents a piece of data. We call the set of all possible embeddings generated by the same algorithm for a type of data "an embedding space." All embedding vectors in the same space are of the same size.
>
> One of the most common uses of embeddings is word embeddings, where you can represent each word with a vector. However, embeddings for other types of data are increasingly popular. For example, ecommerce solutions like Criteo and Coveo have embeddings for products.[10] Pinterest has embeddings for images, graphs, queries, and even users.[11] Given that there are so many types of data with embeddings, there has been a lot of interest in creating universal embeddings for multimodal data.

If we use a recurrent neural network, it will process words in sequential order, which means the order of words is implicitly inputted. However, if we use a model like a transformer, words are processed in parallel, so words' positions need to be explicitly inputted so that our model knows the order of these words ("a dog bites a child" is very different from "a child bites a dog"). We don't want to input the absolute positions, 0, 1, 2, ..., 7, into our model because empirically, neural networks don't work well with inputs that aren't unit-variance (that's why we scale our features, as discussed previously in the section "Scaling" on page 126).

10 Flavian Vasile, Elena Smirnova, and Alexis Conneau, "Meta-Prod2Vec—Product Embeddings Using Side-Information for Recommendation," *arXiv*, July 25, 2016, *https://oreil.ly/KDaEd*; "Product Embeddings and Vectors," Coveo, *https://oreil.ly/ShaSY*.

11 Andrew Zhai, "Representation Learning for Recommender Systems," August 15, 2021, *https://oreil.ly/OchiL*.

If we rescale the positions to between 0 and 1, so 0, 1, 2, …, 7 become 0, 0.143, 0.286, …, 1, the differences between the two positions will be too small for neural networks to learn to differentiate.

A way to handle position embeddings is to treat it the way we'd treat word embedding. With word embedding, we use an embedding matrix with the vocabulary size as its number of columns, and each column is the embedding for the word at the index of that column. With position embedding, the number of columns is the number of positions. In our case, since we only work with the previous sequence size of 8, the positions go from 0 to 7 (see Figure 5-5).

The embedding size for positions is usually the same as the embedding size for words so that they can be summed. For example, the embedding for the word "food" at position 0 is the sum of the embedding vector for the word "food" and the embedding vector for position 0. This is the way position embeddings are implemented in Hugging Face's BERT as of August 2021. Because the embeddings change as the model weights get updated, we say that the position embeddings are learned.

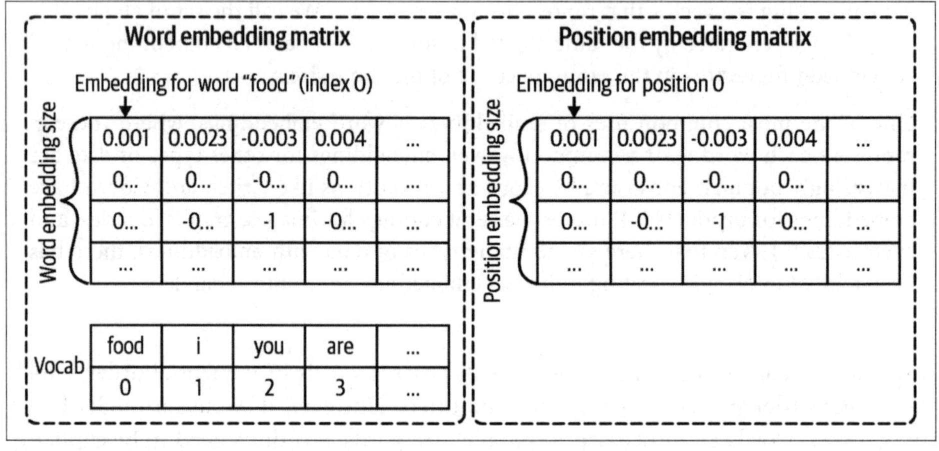

Figure 5-5. One way to embed positions is to treat them the way you'd treat word embeddings

Position embeddings can also be fixed. The embedding for each position is still a vector with *S* elements (*S* is the position embedding size), but each element is predefined using a function, usually sine and cosine. In the original Transformer paper (*https://oreil.ly/hifg6*), if the element is at an even index, use sine. Else, use cosine. See Figure 5-6.

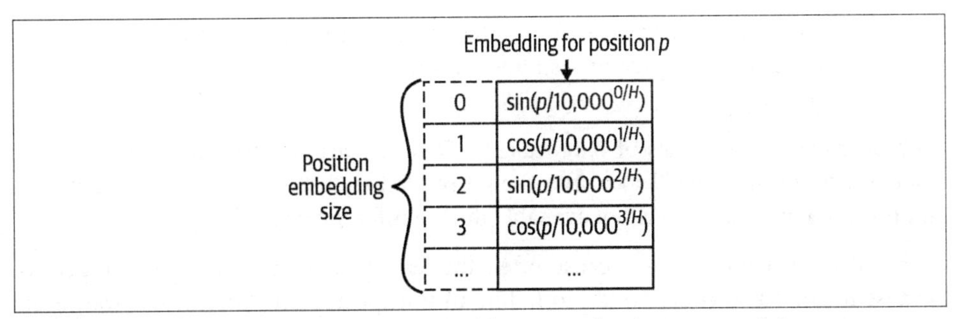

Figure 5-6. Example of fixed position embedding. H is the dimension of the outputs produced by the model.

Fixed positional embedding is a special case of what is known as Fourier features. If positions in positional embeddings are discrete, Fourier features can also be continuous. Consider the task involving representations of 3D objects, such as a teapot. Each position on the surface of the teapot is represented by a three-dimensional coordinate, which is continuous. When positions are continuous, it'd be very hard to build an embedding matrix with continuous column indices, but fixed position embeddings using sine and cosine functions still work.

The following is the generalized format for the embedding vector at coordinate v, also called the Fourier features of coordinate v. Fourier features have been shown to improve models' performance for tasks that take in coordinates (or positions) as inputs. If interested, you might want to read more about it in "Fourier Features Let Networks Learn High Frequency Functions in Low Dimensional Domains" (*https://oreil.ly/cbxr1*) (Tancik et al. 2020).

$$\gamma(v) = \left[a_1 \cos\left(2\pi b_1{}^T v\right), a_1 \sin\left(2\pi b_1{}^T v\right), ..., a_m \cos\left(2\pi b_m{}^T v\right), a_m \sin\left(2\pi b_m{}^T v\right)\right]^T$$

Data Leakage

In July 2021, *MIT Technology Review* ran a provocative article titled "Hundreds of AI Tools Have Been Built to Catch Covid. None of Them Helped." These models were trained to predict COVID-19 risks from medical scans. The article listed multiple examples where ML models that performed well during evaluation failed to be usable in actual production settings.

In one example, researchers trained their model on a mix of scans taken when patients were lying down and standing up. "Because patients scanned while lying down were more likely to be seriously ill, the model learned to predict serious covid risk from a person's position."

In some other cases, models were "found to be picking up on the text font that certain hospitals used to label the scans. As a result, fonts from hospitals with more serious caseloads became predictors of covid risk."[12]

Both of these are examples of data leakage. *Data leakage* refers to the phenomenon when a form of the label "leaks" into the set of features used for making predictions, and this same information is not available during inference.

Data leakage is challenging because often the leakage is nonobvious. It's dangerous because it can cause your models to fail in an unexpected and spectacular way, even after extensive evaluation and testing. Let's go over another example to demonstrate what data leakage is.

Suppose you want to build an ML model to predict whether a CT scan of a lung shows signs of cancer. You obtained the data from hospital A, removed the doctors' diagnosis from the data, and trained your model. It did really well on the test data from hospital A, but poorly on the data from hospital B.

After extensive investigation, you learned that at hospital A, when doctors think that a patient has lung cancer, they send that patient to a more advanced scan machine, which outputs slightly different CT scan images. Your model learned to rely on the information on the scan machine used to make predictions on whether a scan image shows signs of lung cancer. Hospital B sends the patients to different CT scan machines at random, so your model has no information to rely on. We say that labels are leaked into the features during training.

Data leakage can happen not only with newcomers to the field, but has also happened to several experienced researchers whose work I admire, and in one of my own projects. Despite its prevalence, data leakage is rarely covered in ML curricula.

Cautionary Tale: Data Leakage with Kaggle Competition

In 2020, the University of Liverpool launched an Ion Switching competition on Kaggle (*https://oreil.ly/TkvpU*). The task was to identify the number of ion channels open at each time point. They synthesized test data from training data, and some people were able to reverse engineer and obtain test labels from the leak.[13] The two winning teams in this competition are the two teams that were able to exploit the leak, though they might have still been able to win without exploiting the leak.[14]

12 Will Douglas Heaven, "Hundreds of AI Tools Have Been Built to Catch Covid. None of Them Helped," *MIT Technology Review*, July 30, 2021, *https://oreil.ly/Ig1b1*.

13 Zidmie, "The leak explained!" Kaggle, *https://oreil.ly/1JgLj*.

14 Addison Howard, "Competition Recap—Congratulations to our Winners!" Kaggle, *https://oreil.ly/wVUU4*.

Common Causes for Data Leakage

In this section, we'll go over some common causes for data leakage and how to avoid them.

Splitting time-correlated data randomly instead of by time

When I learned ML in college, I was taught to randomly split my data into train, validation, and test splits. This is also how data is often reportedly split in ML research papers. However, this is also one common cause for data leakage.

In many cases, data is time-correlated, which means that the time the data is generated affects its label distribution. Sometimes, the correlation is obvious, as in the case of stock prices. To oversimplify it, the prices of similar stocks tend to move together. If 90% of the tech stocks go down today, it's very likely the other 10% of the tech stocks go down too. When building models to predict the future stock prices, you want to split your training data by time, such as training your model on data from the first six days and evaluating it on data from the seventh day. If you randomly split your data, prices from the seventh day will be included in your train split and leak into your model the condition of the market on that day. We say that the information from the future is leaked into the training process.

However, in many cases, the correlation is nonobvious. Consider the task of predicting whether someone will click on a song recommendation. Whether someone will listen to a song depends not only on their music taste but also on the general music trend that day. If an artist passes away one day, people will be much more likely to listen to that artist. By including samples from a certain day in the train split, information about the music trend that day will be passed into your model, making it easier for it to make predictions on other samples on that same day.

To prevent future information from leaking into the training process and allowing models to cheat during evaluation, split your data by time, instead of splitting randomly, whenever possible. For example, if you have data from five weeks, use the first four weeks for the train split, then randomly split week 5 into validation and test splits as shown in Figure 5-7.

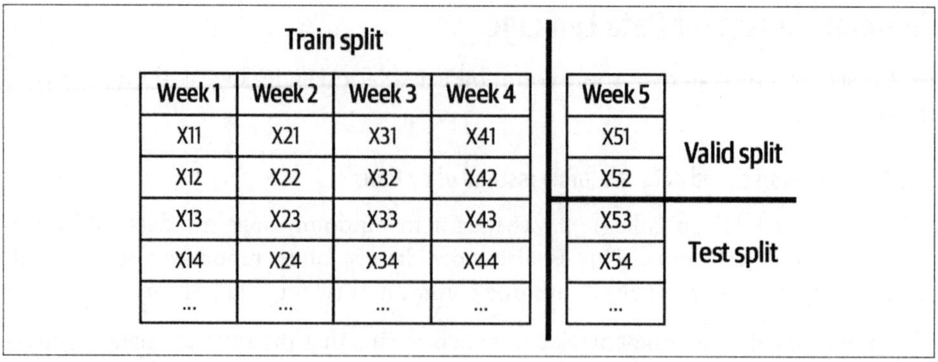

Train split

Week 1	Week 2	Week 3	Week 4	Week 5
X11	X21	X31	X41	X51
X12	X22	X32	X42	X52
X13	X23	X33	X43	X53
X14	X24	X34	X44	X54
...	...	...	...	...

Valid split

Test split

Figure 5-7. Split data by time to prevent future information from leaking into the training process

Scaling before splitting

As discussed in the section "Scaling" on page 126, it's important to scale your features. Scaling requires global statistics—e.g., mean, variance—of your data. One common mistake is to use the entire training data to generate global statistics before splitting it into different splits, leaking the mean and variance of the test samples into the training process, allowing a model to adjust its predictions for the test samples. This information isn't available in production, so the model's performance will likely degrade.

To avoid this type of leakage, always split your data first before scaling, then use the statistics from the train split to scale all the splits. Some even suggest that we split our data before any exploratory data analysis and data processing, so that we don't accidentally gain information about the test split.

Filling in missing data with statistics from the test split

One common way to handle the missing values of a feature is to fill (input) them with the mean or median of all values present. Leakage might occur if the mean or median is calculated using entire data instead of just the train split. This type of leakage is similar to the type of leakage caused by scaling, and it can be prevented by using only statistics from the train split to fill in missing values in all the splits.

Poor handling of data duplication before splitting

If you have duplicates or near-duplicates in your data, failing to remove them before splitting your data might cause the same samples to appear in both train and validation/test splits. Data duplication is quite common in the industry, and has also been found in popular research datasets. For example, CIFAR-10 and CIFAR-100 are two popular datasets used for computer vision research. They were released in 2009, yet it was not until 2019 that Barz and Denzler discovered that 3.3% and 10% of the images from the test sets of the CIFAR-10 and CIFAR-100 datasets have duplicates in the training set.[15]

Data duplication can result from data collection or merging of different data sources. A 2021 *Nature* article listed data duplication as a common pitfall when using ML to detect COVID-19, which happened because "one dataset combined several other datasets without realizing that one of the component datasets already contains another component."[16] Data duplication can also happen because of data processing—for example, oversampling might result in duplicating certain examples.

To avoid this, always check for duplicates before splitting and also after splitting just to make sure. If you oversample your data, do it after splitting.

Group leakage

A group of examples have strongly correlated labels but are divided into different splits. For example, a patient might have two lung CT scans that are a week apart, which likely have the same labels on whether they contain signs of lung cancer, but one of them is in the train split and the second is in the test split. This type of leakage is common for objective detection tasks that contain photos of the same object taken milliseconds apart—some of them landed in the train split while others landed in the test split. It's hard avoiding this type of data leakage without understanding how your data was generated.

15 Björn Barz and Joachim Denzler, "Do We Train on Test Data? Purging CIFAR of Near-Duplicates," *Journal of Imaging* 6, no. 6 (2020): 41.

16 Michael Roberts, Derek Driggs, Matthew Thorpe, Julian Gilbey, Michael Yeung, Stephan Ursprung, Angelica I. Aviles-Rivero, et al. "Common Pitfalls and Recommendations for Using Machine Learning to Detect and Prognosticate for COVID-19 Using Chest Radiographs and CT Scans," *Nature Machine Intelligence* 3 (2021): 199–217, *https://oreil.ly/TzbKJ*.

Leakage from data generation process

The example earlier about how information on whether a CT scan shows signs of lung cancer is leaked via the scan machine is an example of this type of leakage. Detecting this type of data leakage requires a deep understanding of the way data is collected. For example, it would be very hard to figure out that the model's poor performance in hospital B is due to its different scan machine procedure if you don't know about different scan machines or that the procedures at the two hospitals are different.

There's no foolproof way to avoid this type of leakage, but you can mitigate the risk by keeping track of the sources of your data and understanding how it is collected and processed. Normalize your data so that data from different sources can have the same means and variances. If different CT scan machines output images with different resolutions, normalizing all the images to have the same resolution would make it harder for models to know which image is from which scan machine. And don't forget to incorporate subject matter experts, who might have more contexts on how data is collected and used, into the ML design process!

Detecting Data Leakage

Data leakage can happen during many steps, from generating, collecting, sampling, splitting, and processing data to feature engineering. It's important to monitor for data leakage during the entire lifecycle of an ML project.

Measure the predictive power of each feature or a set of features with respect to the target variable (label). If a feature has unusually high correlation, investigate how this feature is generated and whether the correlation makes sense. It's possible that two features independently don't contain leakage, but two features together can contain leakage. For example, when building a model to predict how long an employee will stay at a company, the starting date and the end date separately doesn't tell us much about their tenure, but both together can give us that information.

Do ablation studies to measure how important a feature or a set of features is to your model. If removing a feature causes the model's performance to deteriorate significantly, investigate why that feature is so important. If you have a massive amount of features, say a thousand features, it might be infeasible to do ablation studies on every possible combination of them, but it can still be useful to occasionally do ablation studies with a subset of features that you suspect the most. This is another example of how subject matter expertise can come in handy in feature engineering. Ablation studies can be run offline at your own schedule, so you can leverage your machines during downtime for this purpose.

Keep an eye out for new features added to your model. If adding a new feature significantly improves your model's performance, either that feature is really good or that feature just contains leaked information about labels.

Be very careful every time you look at the test split. If you use the test split in any way other than to report a model's final performance, whether to come up with ideas for new features or to tune hyperparameters, you risk leaking information from the future into your training process.

Engineering Good Features

Generally, adding more features leads to better model performance. In my experience, the list of features used for a model in production only grows over time. However, more features doesn't always mean better model performance. Having too many features can be bad both during training and serving your model for the following reasons:

- The more features you have, the more opportunities there are for data leakage.
- Too many features can cause overfitting.
- Too many features can increase memory required to serve a model, which, in turn, might require you to use a more expensive machine/instance to serve your model.
- Too many features can increase inference latency when doing online prediction, especially if you need to extract these features from raw data for predictions online. We'll go deeper into online prediction in Chapter 7.
- Useless features become technical debts. Whenever your data pipeline changes, all the affected features need to be adjusted accordingly. For example, if one day your application decides to no longer take in information about users' age, all features that use users' age need to be updated.

In theory, if a feature doesn't help a model make good predictions, regularization techniques like L1 regularization should reduce that feature's weight to 0. However, in practice, it might help models learn faster if the features that are no longer useful (and even possibly harmful) are removed, prioritizing good features.

You can store removed features to add them back later. You can also just store general feature definitions to reuse and share across teams in an organization. When talking about feature definition management, some people might think of feature stores as the solution. However, not all feature stores manage feature definitions. We'll discuss feature stores further in Chapter 10.

There are two factors you might want to consider when evaluating whether a feature is good for a model: importance to the model and generalization to unseen data.

Feature Importance

There are many different methods for measuring a feature's importance. If you use a classical ML algorithm like boosted gradient trees, the easiest way to measure the importance of your features is to use built-in feature importance functions implemented by XGBoost.[17] For more model-agnostic methods, you might want to look into SHAP (SHapley Additive exPlanations).[18] InterpretML (*https://oreil.ly/oPllN*) is a great open source package that leverages feature importance to help you understand how your model makes predictions.

The exact algorithm for feature importance measurement is complex, but intuitively, a feature's importance to a model is measured by how much that model's performance deteriorates if that feature or a set of features containing that feature is removed from the model. SHAP is great because it not only measures a feature's importance to an entire model, it also measures each feature's contribution to a model's specific prediction. Figures 5-8 and 5-9 show how SHAP can help you understand the contribution of each feature to a model's predictions.

17 With XGBoost function `get_score` (*https://oreil.ly/8sCfD*).

18 A great open source Python package for calculating SHAP can be found on GitHub (*https://oreil.ly/hGxcF*).

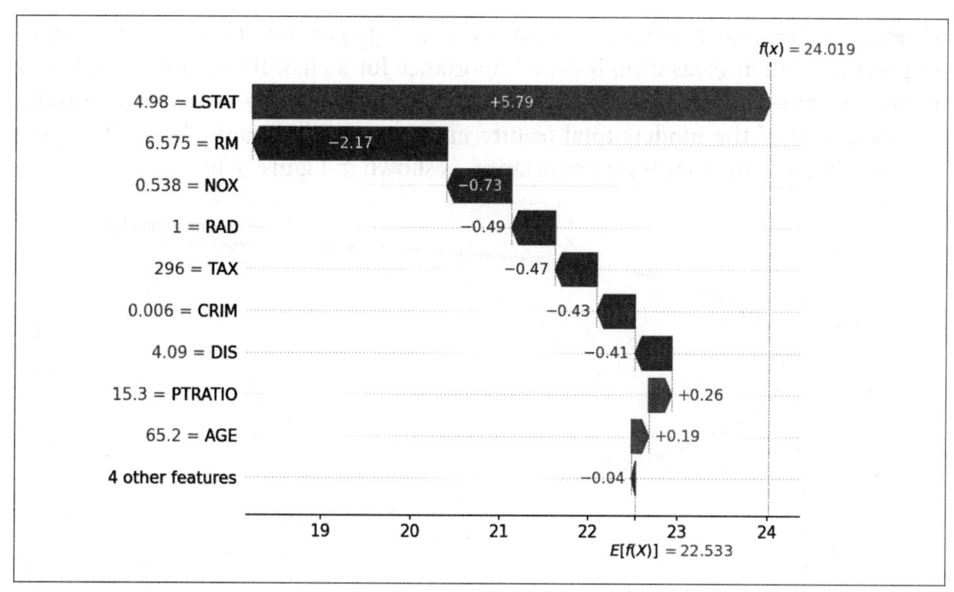

Figure 5-8. How much each feature contributes to a model's single prediction, measured by SHAP. The value LSTAT = 4.98 contributes the most to this specific prediction. Source: Scott Lundberg[19]

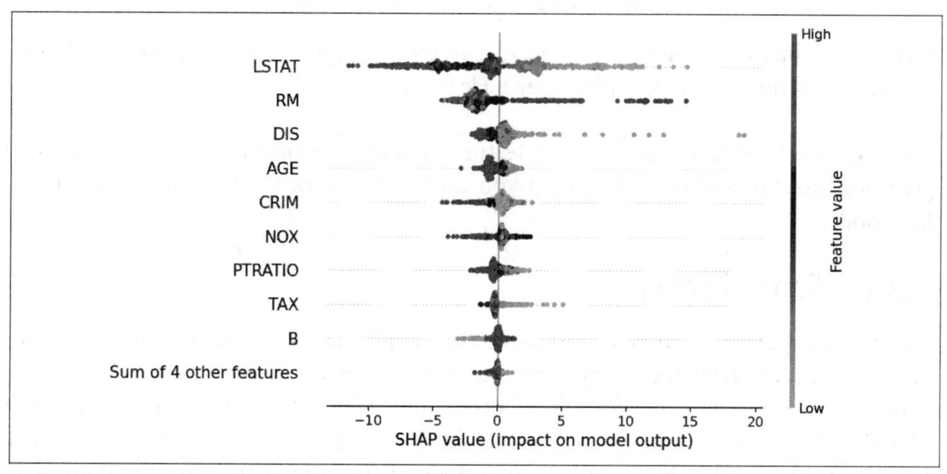

Figure 5-9. How much each feature contributes to a model, measured by SHAP. The feature LSTAT has the highest importance. Source: Scott Lundberg

19 Scott Lundberg, SHAP (SHapley Additive exPlanations), GitHub repository, last accessed 2021, *https://oreil.ly/c8qqE*.

Often, a small number of features accounts for a large portion of your model's feature importance. When measuring feature importance for a click-through rate prediction model, the ads team at Facebook found out that the top 10 features are responsible for about half of the model's total feature importance, whereas the last 300 features contribute less than 1% feature importance, as shown in Figure 5-10.[20]

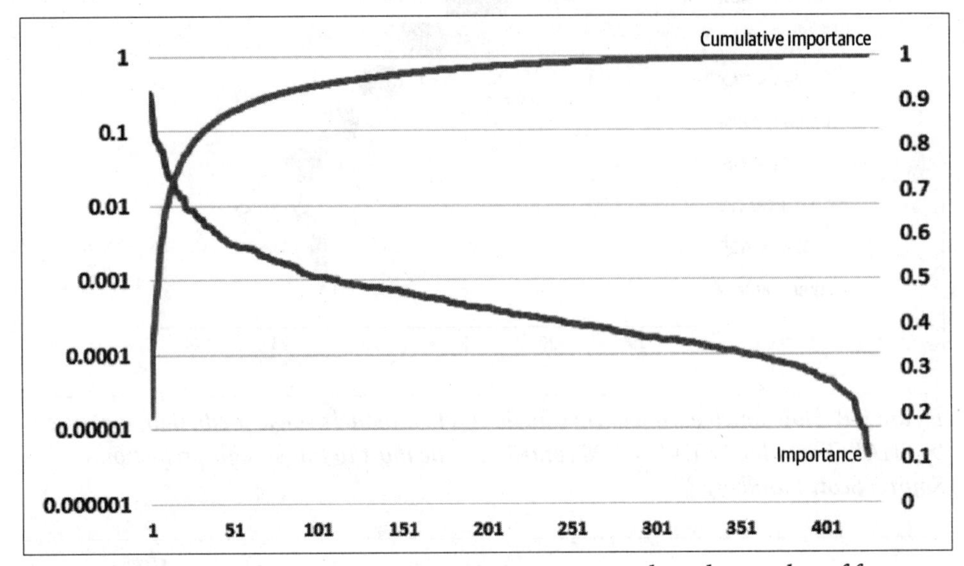

Figure 5-10. Boosting feature importance. X-axis corresponds to the number of features. Feature importance is in log scale. Source: He et al.

Not only good for choosing the right features, feature importance techniques are also great for interpretability as they help you understand how your models work under the hood.

Feature Generalization

Since the goal of an ML model is to make correct predictions on unseen data, features used for the model should generalize to unseen data. Not all features generalize equally. For example, for the task of predicting whether a comment is spam, the identifier of each comment is not generalizable at all and shouldn't be used as a feature for the model. However, the identifier of the user who posts the comment, such as username, might still be useful for a model to make predictions.

20 Xinran He, Junfeng Pan, Ou Jin, Tianbing Xu, Bo Liu, Tao Xu, Yanxin Shi, et al., "Practical Lessons from Predicting Clicks on Ads at Facebook," in *ADKDD '14: Proceedings of the Eighth International Workshop on Data Mining for Online Advertising* (August 2014): 1–9, *https://oreil.ly/dHXeC*.

Measuring feature generalization is a lot less scientific than measuring feature importance, and it requires both intuition and subject matter expertise on top of statistical knowledge. Overall, there are two aspects you might want to consider with regards to generalization: feature coverage and distribution of feature values.

Coverage is the percentage of the samples that has values for this feature in the data—so the fewer values that are missing, the higher the coverage. A rough rule of thumb is that if this feature appears in a very small percentage of your data, it's not going to be very generalizable. For example, if you want to build a model to predict whether someone will buy a house in the next 12 months and you think that the number of children someone has will be a good feature, but you can only get this information for 1% of your data, this feature might not be very useful.

This rule of thumb is rough because some features can still be useful even if they are missing in most of your data. This is especially true when the missing values are not at random, which means having the feature or not might be a strong indication of its value. For example, if a feature appears only in 1% of your data, but 99% of the examples with this feature have POSITIVE labels, this feature is useful and you should use it.

Coverage of a feature can differ wildly between different slices of data and even in the same slice of data over time. If the coverage of a feature differs a lot between the train and test split (such as it appears in 90% of the examples in the train split but only in 20% of the examples in the test split), this is an indication that your train and test splits don't come from the same distribution. You might want to investigate whether the way you split your data makes sense and whether this feature is a cause for data leakage.

For the feature values that are present, you might want to look into their distribution. If the set of values that appears in the seen data (such as the train split) has no overlap with the set of values that appears in the unseen data (such as the test split), this feature might even hurt your model's performance.

As a concrete example, imagine you want to build a model to estimate the time it will take for a given taxi ride. You retrain this model every week, and you want to use the data from the last six days to predict the ETAs (estimated time of arrival) for today. One of the features is DAY_OF_THE_WEEK, which you think is useful because the traffic on weekdays is usually worse than on the weekend. This feature coverage is 100%, because it's present in every feature. However, in the train split, the values for this feature are Monday to Saturday, whereas in the test split, the value for this feature is Sunday. If you include this feature in your model without a clever scheme to encode the days, it won't generalize to the test split, and might harm your model's performance.

On the other hand, HOUR_OF_THE_DAY is a great feature, because the time in the day affects the traffic too, and the range of values for this feature in the train split overlaps with the test split 100%.

When considering a feature's generalization, there's a trade-off between generalization and specificity. You might realize that the traffic during an hour only changes depending on whether that hour is the rush hour. So you generate the feature IS_RUSH_HOUR and set it to 1 if the hour is between 7 a.m. and 9 a.m. or between 4 p.m. and 6 p.m. IS_RUSH_HOUR is more generalizable but less specific than HOUR_OF_THE_DAY. Using IS_RUSH_HOUR without HOUR_OF_THE_DAY might cause models to lose important information about the hour.

Summary

Because the success of today's ML systems still depends on their features, it's important for organizations interested in using ML in production to invest time and effort into feature engineering.

How to engineer good features is a complex question with no foolproof answers. The best way to learn is through experience: trying out different features and observing how they affect your models' performance. It's also possible to learn from experts. I find it extremely useful to read about how the winning teams of Kaggle competitions engineer their features to learn more about their techniques and the considerations they went through.

Feature engineering often involves subject matter expertise, and subject matter experts might not always be engineers, so it's important to design your workflow in a way that allows nonengineers to contribute to the process.

Here is a summary of best practices for feature engineering:

- Split data by time into train/valid/test splits instead of doing it randomly.
- If you oversample your data, do it after splitting.
- Scale and normalize your data after splitting to avoid data leakage.
- Use statistics from only the train split, instead of the entire data, to scale your features and handle missing values.

- Understand how your data is generated, collected, and processed. Involve domain experts if possible.
- Keep track of your data's lineage.
- Understand feature importance to your model.
- Use features that generalize well.
- Remove no longer useful features from your models.

With a set of good features, we'll move to the next part of the workflow: training ML models. Before we move on, I just want to reiterate that moving to modeling doesn't mean we're done with handling data or feature engineering. We are never done with data and features. In most real-world ML projects, the process of collecting data and feature engineering goes on as long as your models are in production. We need to use new, incoming data to continually improve models, which we'll cover in Chapter 9.

CHAPTER 6

Model Development and Offline Evaluation

In Chapter 4, we discussed how to create training data for your model, and in Chapter 5, we discussed how to engineer features from that training data. With the initial set of features, we'll move to the ML algorithm part of ML systems. For me, this has always been the most fun step, as it allows me to play around with different algorithms and techniques, even the latest ones. This is also the first step where I can see all the hard work I've put into data and feature engineering transformed into a system whose outputs (predictions) I can use to evaluate the success of my effort.

To build an ML model, we first need to select the ML model to build. There are so many ML algorithms out there, with more actively being developed. This chapter starts with six tips for selecting the best algorithms for your task.

The section that follows discusses different aspects of model development, such as debugging, experiment tracking and versioning, distributed training, and AutoML.

Model development is an iterative process. After each iteration, you'll want to compare your model's performance against its performance in previous iterations and evaluate how suitable this iteration is for production. The last section of this chapter is dedicated to how to evaluate your model before deploying it to production, covering a range of evaluation techniques including perturbation tests, invariance tests, model calibration, and slide-based evaluation.

I expect that most readers already have an understanding of common ML algorithms such as linear models, decision trees, *k*-nearest neighbors, and different types of neural networks. This chapter will discuss techniques surrounding these algorithms but won't go into details of how they work. Because this chapter deals with ML algorithms, it requires a lot more ML knowledge than other chapters. If you're not familiar with them, I recommend taking an online course or reading a book on ML algorithms before reading this chapter. Readers wanting a quick refresh on basic ML concepts might find helpful the section "Basic ML Reviews" in the book's GitHub repository (*https://oreil.ly/designing-machine-learning-systems-code*).

Model Development and Training

In this section, we'll discuss necessary aspects to help you develop and train your model, including how to evaluate different ML models for your problem, creating ensembles of models, experiment tracking and versioning, and distributed training, which is necessary for the scale at which models today are usually trained at. We'll end this section with the more advanced topic of AutoML—using ML to automatically choose a model best for your problem.

Evaluating ML Models

There are many possible solutions to any given problem. Given a task that can leverage ML in its solution, you might wonder what ML algorithm you should use for it. For example, should you start with logistic regression, an algorithm that you're already familiar with? Or should you try out a new fancy model that is supposed to be the new state of the art for your problem? A more senior colleague mentioned that gradient-boosted trees have always worked for her for this task in the past—should you listen to her advice?

If you had unlimited time and compute power, the rational thing to do would be to try all possible solutions and see what is best for you. However, time and compute power are limited resources, and you have to be strategic about what models you select.

When talking about ML algorithms, many people think in terms of classical ML algorithms versus neural networks. There are a lot of interests and media coverage for neural networks, especially deep learning, which is understandable given that most of the AI progress in the last decade happened due to neural networks getting bigger and deeper.

These interests and coverage might give off the impression that deep learning is replacing classical ML algorithms. However, even though deep learning is finding more use cases in production, classical ML algorithms are not going away. Many recommender systems still rely on collaborative filtering and matrix factorization. Tree-based algorithms, including gradient-boosted trees, still power many classification tasks with strict latency requirements.

Even in applications where neural networks are deployed, classic ML algorithms are still being used in tandem. For example, neural networks and decision trees might be used together in an ensemble. A k-means clustering model might be used to extract features to input into a neural network. Vice versa, a pretrained neural network (like BERT or GPT-3) might be used to generate embeddings to input into a logistic regression model.

When selecting a model for your problem, you don't choose from every possible model out there, but usually focus on a set of models suitable for your problem. For example, if your boss tells you to build a system to detect toxic tweets, you know that this is a text classification problem—given a piece of text, classify whether it's toxic or not—and common models for text classification include naive Bayes, logistic regression, recurrent neural networks, and transformer-based models such as BERT, GPT, and their variants.

If your client wants you to build a system to detect fraudulent transactions, you know that this is the classic abnormality detection problem—fraudulent transactions are abnormalities that you want to detect—and common algorithms for this problem are many, including *k*-nearest neighbors, isolation forest, clustering, and neural networks.

Knowledge of common ML tasks and the typical approaches to solve them is essential in this process.

Different types of algorithms require different numbers of labels as well as different amounts of compute power. Some take longer to train than others, whereas some take longer to make predictions. Non-neural network algorithms tend to be more explainable (e.g., what features contributed the most to an email being classified as spam) than neural networks.

When considering what model to use, it's important to consider not only the model's performance, measured by metrics such as accuracy, F1 score, and log loss, but also its other properties, such as how much data, compute, and time it needs to train, what's its inference latency, and interpretability. For example, a simple logistic regression model might have lower accuracy than a complex neural network, but it requires less labeled data to start, it's much faster to train, it's much easier to deploy, and it's also much easier to explain why it's making certain predictions.

Comparing ML algorithms is out of the scope of this book. No matter how good a comparison is, it will be outdated as soon as new algorithms come out. Back in 2016, LSTM-RNNs were all the rage and the backbone of the architecture seq2seq (Sequence-to-Sequence) that powered many NLP tasks from machine translation to text summarization to text classification. However, just two years later, recurrent architectures were largely replaced by transformer architectures for NLP tasks.

To understand different algorithms, the best way is to equip yourself with basic ML knowledge and run experiments with the algorithms you're interested in. To keep up to date with so many new ML techniques and models, I find it helpful to monitor trends at major ML conferences such as NeurIPS, ICLR, and ICML, as well as following researchers whose work has a high signal-to-noise ratio on Twitter.

Six tips for model selection

Without getting into specifics of different algorithms, here are six tips that might help you decide what ML algorithms to work on next.

Avoid the state-of-the-art trap. While helping companies as well as recent graduates get started in ML, I usually have to spend a nontrivial amount of time steering them away from jumping straight into state-of-the-art models. I can see why people want state-of-the-art models. Many believe that these models would be the best solutions for their problems—why try an old solution if you believe that a newer and superior solution exists? Many business leaders also want to use state-of-the-art models because they want to make their businesses appear cutting edge. Developers might also be more excited getting their hands on new models than getting stuck into the same old things over and over again.

Researchers often only evaluate models in academic settings, which means that a model being state of the art often means that *it performs better than existing models on some static datasets*. It doesn't mean that this model will be fast enough or cheap enough for *you* to implement. It doesn't even mean that this model will perform better than other models on *your* data.

While it's essential to stay up to date with new technologies and beneficial to evaluate them for your business, the most important thing to do when solving a problem is finding solutions that can solve that problem. If there's a solution that can solve your problem that is much cheaper and simpler than state-of-the-art models, use the simpler solution.

Start with the simplest models. Zen of Python states that "simple is better than complex," and this principle is applicable to ML as well. Simplicity serves three purposes. First, simpler models are easier to deploy, and deploying your model early allows you to validate that your prediction pipeline is consistent with your training pipeline. Second, starting with something simple and adding more complex components step-by-step makes it easier to understand your model and debug it. Third, the simplest model serves as a baseline to which you can compare your more complex models.

Simplest models are not always the same as models with the least effort. For example, pretrained BERT models are complex, but they require little effort to get started with, especially if you use a ready-made implementation like the one in Hugging Face's Transformer. In this case, it's not a bad idea to use the complex solution, given that the community around this solution is well developed enough to help you get through any problems you might encounter. However, you might still want to experiment with simpler solutions to ensure that pretrained BERT is indeed better than those simpler solutions for your problem. Pretrained BERT might be low effort to start with, but it can be quite high effort to improve upon. Whereas if you start with a simpler model, there'll be a lot of room for you to improve upon your model.

Avoid human biases in selecting models. Imagine an engineer on your team is assigned the task of evaluating which model is better for your problem: a gradient-boosted tree or a pretrained BERT model. After two weeks, this engineer announces that the best BERT model outperforms the best gradient-boosted tree by 5%. Your team decides to go with the pretrained BERT model.

A few months later, however, a seasoned engineer joins your team. She decides to look into gradient-boosted trees again and finds out that this time, the best gradient-boosted tree outperforms the pretrained BERT model you currently have in production. What happened?

There are a lot of human biases in evaluating models. Part of the process of evaluating an ML architecture is to experiment with different features and different sets of hyperparameters to find the best model of that architecture. If an engineer is more excited about an architecture, they will likely spend a lot more time experimenting with it, which might result in better-performing models for that architecture.

When comparing different architectures, it's important to compare them under comparable setups. If you run 100 experiments for an architecture, it's not fair to only run a couple of experiments for the architecture you're evaluating it against. You might need to run 100 experiments for the other architecture too.

Because the performance of a model architecture depends heavily on the context it's evaluated in—e.g., the task, the training data, the test data, the hyperparameters, etc.—it's extremely difficult to make claims that a model architecture is better than another architecture. The claim might be true in a context, but unlikely true for all possible contexts.

Evaluate good performance now versus good performance later. The best model now does not always mean the best model two months from now. For example, a tree-based model might work better now because you don't have a ton of data yet, but two months from now, you might be able to double your amount of training data, and your neural network might perform much better.[1]

A simple way to estimate how your model's performance might change with more data is to use learning curves (*https://oreil.ly/9QZLa*). A learning curve of a model is a plot of its performance—e.g., training loss, training accuracy, validation accuracy—against the number of training samples it uses, as shown in Figure 6-1. The learning curve won't help you estimate exactly how much performance gain you can get from

1 Andrew Ng has a great lecture (*https://oreil.ly/o6tGK*) where he explains that if a learning algorithm suffers from high bias, getting more training data by itself won't help much. Whereas if a learning algorithm suffers from high variance, getting more training data is likely to help.

having more training data, but it can give you a sense of whether you can expect any performance gain at all from more training data.

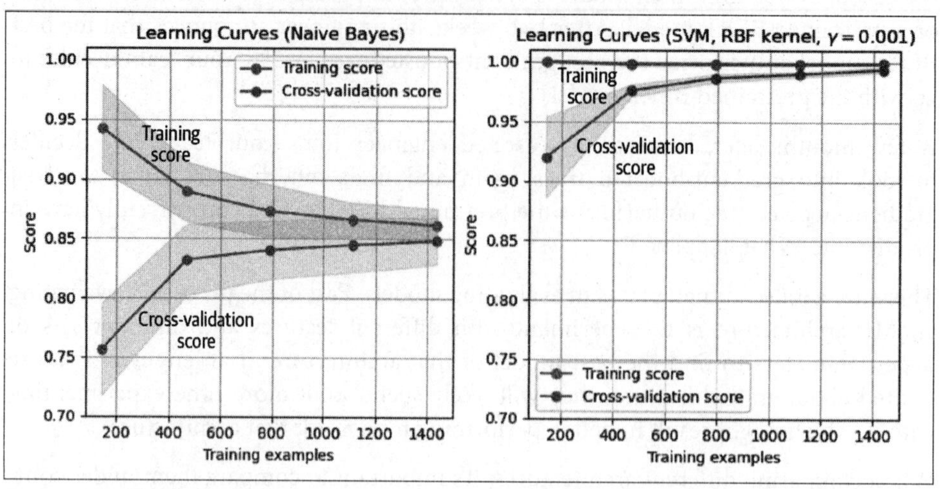

Figure 6-1. The learning curves of a naive Bayes model and an SVM model. Source: scikit-learn (https://oreil.ly/QA52c)

A situation that I've encountered is when a team evaluates a simple neural network against a collaborative filtering model for making recommendations. When evaluating both models offline, the collaborative filtering model outperformed. However, the simple neural network can update itself with each incoming example, whereas the collaborative filtering has to look at all the data to update its underlying matrix. The team decided to deploy both the collaborative filtering model and the simple neural network. They used the collaborative filtering model to make predictions for users, and continually trained the simple neural network in production with new, incoming data. After two weeks, the simple neural network was able to outperform the collaborative filtering model.

While evaluating models, you might want to take into account their potential for improvements in the near future, and how easy/difficult it is to achieve those improvements.

Evaluate trade-offs. There are many trade-offs you have to make when selecting models. Understanding what's more important in the performance of your ML system will help you choose the most suitable model.

One classic example of trade-off is the false positives and false negatives trade-off. Reducing the number of false positives might increase the number of false negatives, and vice versa. In a task where false positives are more dangerous than false negatives, such as fingerprint unlocking (unauthorized people shouldn't be classified as authorized and given access), you might prefer a model that makes fewer false positives. Similarly, in a task where false negatives are more dangerous than false positives, such as COVID-19 screening (patients with COVID-19 shouldn't be classified as no COVID-19), you might prefer a model that makes fewer false negatives.

Another example of trade-off is compute requirement and accuracy—a more complex model might deliver higher accuracy but might require a more powerful machine, such as a GPU instead of a CPU, to generate predictions with acceptable inference latency. Many people also care about the interpretability and performance trade-off. A more complex model can give a better performance, but its results are less interpretable.

Understand your model's assumptions. The statistician George Box said in 1976 that "all models are wrong, but some are useful." The real world is intractably complex, and models can only approximate using assumptions. Every single model comes with its own assumptions. Understanding what assumptions a model makes and whether our data satisfies those assumptions can help you evaluate which model works best for your use case.

Following are some of the common assumptions. It's not meant to be an exhaustive list, but just a demonstration:

Prediction assumption
: Every model that aims to predict an output Y from an input X makes the assumption that it's possible to predict Y based on X.

IID
: Neural networks assume that the examples are independent and identically distributed (*https://oreil.ly/hXRr2*), which means that all the examples are independently drawn from the same joint distribution.

Smoothness
: Every supervised machine learning method assumes that there's a set of functions that can transform inputs into outputs such that similar inputs are transformed into similar outputs. If an input X produces an output Y, then an input close to X would produce an output proportionally close to Y.

Tractability
: Let X be the input and Z be the latent representation of X. Every generative model makes the assumption that it's tractable to compute the probability $P(Z|X)$.

Boundaries
: A linear classifier assumes that decision boundaries are linear.

Conditional independence
: A naive Bayes classifier assumes that the attribute values are independent of each other given the class.

Normally distributed
: Many statistical methods assume that data is normally distributed.

Ensembles

When considering an ML solution to your problem, you might want to start with a system that contains just one model (the process of selecting one model for your problem was discussed earlier in the chapter). After developing one single model, you might think about how to continue improving its performance. One method that has consistently given a performance boost is to use an ensemble of multiple models instead of just an individual model to make predictions. Each model in the ensemble is called a *base learner*. For example, for the task of predicting whether an email is SPAM or NOT SPAM, you might have three different models. The final prediction for each email is the majority vote of all three models. So if at least two base learners output SPAM, the email will be classified as SPAM.

Twenty out of 22 winning solutions on Kaggle competitions in 2021, as of August 2021, use ensembles.[2] As of January 2022, 20 top solutions on SQuAD 2.0 (*https://oreil.ly/odo12*), the Stanford Question Answering Dataset, are ensembles, as shown in Figure 6-2.

Ensembling methods are less favored in production because ensembles are more complex to deploy and harder to maintain. However, they are still common for tasks where a small performance boost can lead to a huge financial gain, such as predicting click-through rate for ads.

2 I went through the winning solutions listed on Farid Rashidi's "Kaggle Solutions" web page (*https://oreil.ly/vNrPx*). One solution used 33 models (Giba, "1st Place-Winner Solution-Gilberto Titericz and Stanislav Semenov," Kaggle, *https://oreil.ly/z5od8*).

Rank	Model	EM	F1
	Human Performance *Stanford University* (Rajpurkar & Jia et al. '18)	86.831	89.452
1 Jun 04, 2021	IE-Net (ensemble) *RICOH_SRCB_DML*	**90.939**	**93.214**
2 Feb 21, 2021	FPNet (ensemble) *Ant Service Intelligence Team*	90.871	93.183
3 May 16, 2021	IE-NetV2 (ensemble) *RICOH_SRCB_DML*	90.860	93.100
4 Apr 06, 2020	SA-Net on Albert (ensemble) *QIANXIN*	90.724	93.011
5 May 05, 2020	SA-Net-V2 (ensemble) *QIANXIN*	90.679	92.948
5 Apr 05, 2020	Retro-Reader (ensemble) *Shanghai Jiao Tong University* http://arxiv.org/abs/2001.09694	90.578	92.978
5 Feb 05, 2021	FPNet (ensemble) *YuYang*	90.600	92.899
6 Apr 18, 2021	TransNets + SFVerifier + SFEnsembler (ensemble) *Senseforth AI Research*	90.487	92.894

Figure 6-2. As of January 2022, the top 20 solutions on SQuAD 2.0 (https://oreil.ly/odo12) are all ensembles

We'll go over an example to give you the intuition of why ensembling works. Imagine you have three email spam classifiers, each with an accuracy of 70%. Assuming that each classifier has an equal probability of making a correct prediction for each email, and that these three classifiers are not correlated, we'll show that by taking the majority vote of these three classifiers, we can get an accuracy of 78.4%.

For each email, each classifier has a 70% chance of being correct. The ensemble will be correct if at least two classifiers are correct. Table 6-1 shows the probabilities of different possible outcomes of the ensemble given an email. This ensemble will have an accuracy of 0.343 + 0.441 = 0.784, or 78.4%.

Table 6-1. Possible outcomes of the ensemble that takes the majority vote from three classifiers

Outputs of three models	Probability	Ensemble's output
All three are correct	0.7 * 0.7 * 0.7 = 0.343	Correct
Only two are correct	(0.7 * 0.7 * 0.3) * 3 = 0.441	Correct
Only one is correct	(0.3 * 0.3 * 0.7) * 3 = 0.189	Wrong
None are correct	0.3 * 0.3 * 0.3 = 0.027	Wrong

This calculation only holds if the classifiers in an ensemble are uncorrelated. If all classifiers are perfectly correlated—all three of them make the same prediction for every email—the ensemble will have the same accuracy as each individual classifier. When creating an ensemble, the less correlation there is among base learners, the better the ensemble will be. Therefore, it's common to choose very different types of models for an ensemble. For example, you might create an ensemble that consists of one transformer model, one recurrent neural network, and one gradient-boosted tree.

There are three ways to create an ensemble: bagging, boosting, and stacking. In addition to helping boost performance, according to several survey papers, ensemble methods such as boosting and bagging, together with resampling, have shown to help with imbalanced datasets.[3] We'll go over each of these three methods, starting with bagging.

Bagging

Bagging, shortened from *bootstrap aggregating*, is designed to improve both the training stability and accuracy of ML algorithms.[4] It reduces variance and helps to avoid overfitting.

Given a dataset, instead of training one classifier on the entire dataset, you sample with replacement to create different datasets, called bootstraps, and train a classification or regression model on each of these bootstraps. Sampling with replacement ensures that each bootstrap is created independently from its peers. Figure 6-3 shows an illustration of bagging.

3 Mikel Galar, Alberto Fernandez, Edurne Barrenechea, Humberto Bustince, and Francisco Herrera, "A Review on Ensembles for the Class Imbalance Problem: Bagging-, Boosting-, and Hybrid-Based Approaches," *IEEE Transactions on Systems, Man, and Cybernetics, Part C (Applications and Reviews)* 42, no. 4 (July 2012): 463–84, *https://oreil.ly/ZBlgE*; G. Rekha, Amit Kumar Tyagi, and V. Krishna Reddy, "Solving Class Imbalance Problem Using Bagging, Boosting Techniques, With and Without Using Noise Filtering Method," *International Journal of Hybrid Intelligent Systems* 15, no. 2 (January 2019): 67–76, *https://oreil.ly/hchzU*.

4 Training stability here means less fluctuation in the training loss.

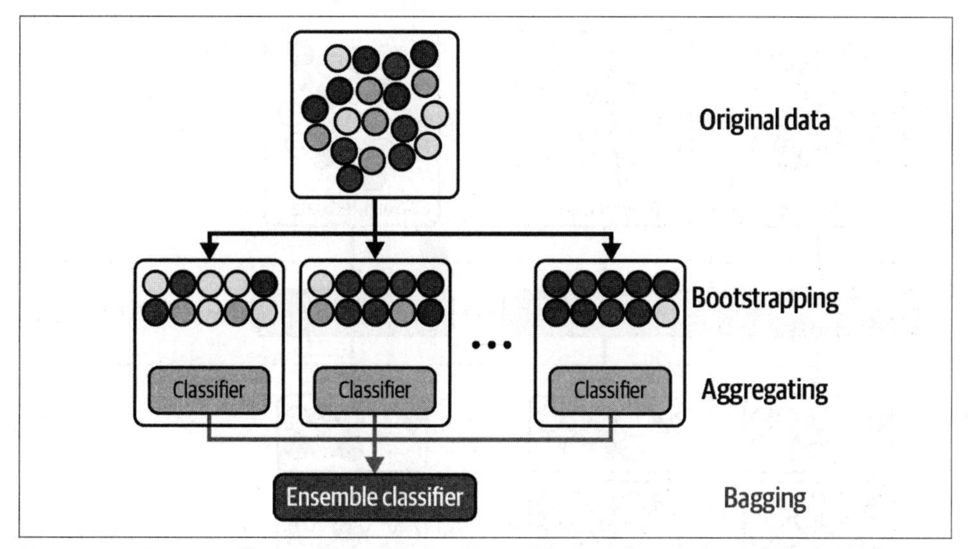

Figure 6-3. Bagging illustration. Source: Adapted from an image by Sirakorn (https://oreil.ly/KEAPl)

If the problem is classification, the final prediction is decided by the majority vote of all models. For example, if 10 classifiers vote SPAM and 6 models vote NOT SPAM, the final prediction is SPAM.

If the problem is regression, the final prediction is the average of all models' predictions.

Bagging generally improves unstable methods, such as neural networks, classification and regression trees, and subset selection in linear regression. However, it can mildly degrade the performance of stable methods such as *k*-nearest neighbors.[5]

A random forest is an example of bagging. A random forest is a collection of decision trees constructed by both bagging and feature randomness, where each tree can pick only from a random subset of features to use.

Boosting

Boosting is a family of iterative ensemble algorithms that convert weak learners to strong ones. Each learner in this ensemble is trained on the same set of samples, but the samples are weighted differently among iterations. As a result, future weak learners focus more on the examples that previous weak learners misclassified. Figure 6-4 shows an illustration of boosting, which involves the steps that follow.

5 Leo Breiman, "Bagging Predictors," *Machine Learning* 24 (1996): 123–40, *https://oreil.ly/adzJu*.

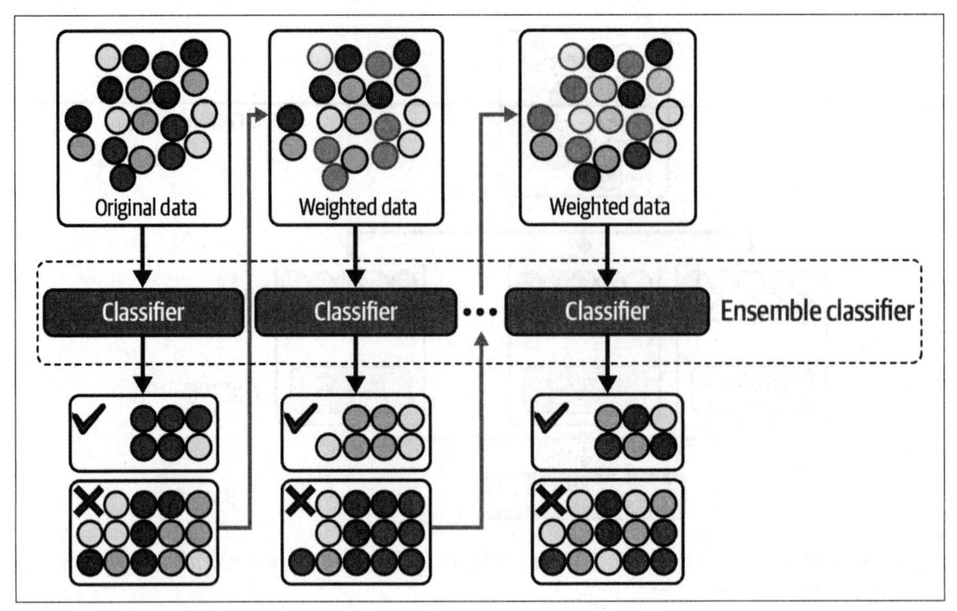

Figure 6-4. Boosting illustration. Source: Adapted from an image by Sirakorn (https://oreil.ly/h5cuS)

1. You start by training the first weak classifier on the original dataset.
2. Samples are reweighted based on how well the first classifier classifies them, e.g., misclassified samples are given higher weight.
3. Train the second classifier on this reweighted dataset. Your ensemble now consists of the first and the second classifiers.
4. Samples are weighted based on how well the ensemble classifies them.
5. Train the third classifier on this reweighted dataset. Add the third classifier to the ensemble.
6. Repeat for as many iterations as needed.
7. Form the final strong classifier as a weighted combination of the existing classifiers—classifiers with smaller training errors have higher weights.

An example of a boosting algorithm is a gradient boosting machine (GBM), which produces a prediction model typically from weak decision trees. It builds the model in a stage-wise fashion like other boosting methods do, and it generalizes them by allowing optimization of an arbitrary differentiable loss function.

XGBoost, a variant of GBM, used to be the algorithm of choice for many winning teams of ML competitions.[6] It's been used in a wide range of tasks from classification, ranking, to the discovery of the Higgs Boson.[7] However, many teams have been opting for LightGBM (*https://oreil.ly/1qyWf*), a distributed gradient boosting framework that allows parallel learning, which generally allows faster training on large datasets.

Stacking

Stacking means that you train base learners from the training data then create a meta-learner that combines the outputs of the base learners to output final predictions, as shown in Figure 6-5. The meta-learner can be as simple as a heuristic: you take the majority vote (for classification tasks) or the average vote (for regression tasks) from all base learners. It can be another model, such as a logistic regression model or a linear regression model.

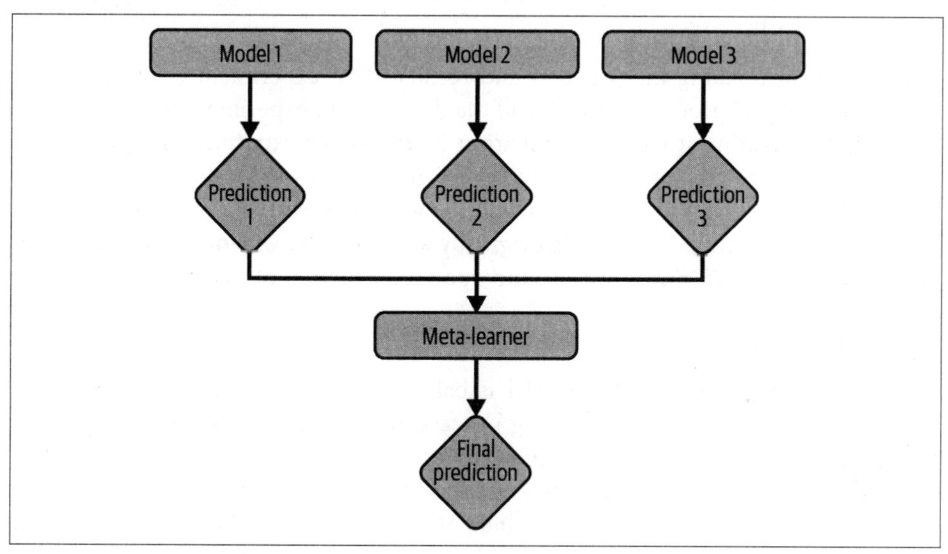

Figure 6-5. A visualization of a stacked ensemble from three base learners

For more great advice on how to create an ensemble, refer to the awesome ensemble guide (*https://oreil.ly/Nu6G6*) by one of Kaggle's legendary teams, MLWave.

6 "Machine Learning Challenge Winning Solutions," *https://oreil.ly/YjS8d*.

7 Tianqi Chen and Tong He, "Higgs Boson Discovery with Boosted Trees," *Proceedings of Machine Learning Research* 42 (2015): 69–80, *https://oreil.ly/ysBYO*.

Experiment Tracking and Versioning

During the model development process, you often have to experiment with many architectures and many different models to choose the best one for your problem. Some models might seem similar to each other and differ in only one hyperparameter—such as one model using a learning rate of 0.003 and another model using a learning rate of 0.002—and yet their performances are dramatically different. It's important to keep track of all the definitions needed to re-create an experiment and its relevant artifacts. An artifact is a file generated during an experiment—examples of artifacts can be files that show the loss curve, evaluation loss graph, logs, or intermediate results of a model throughout a training process. This enables you to compare different experiments and choose the one best suited for your needs. Comparing different experiments can also help you understand how small changes affect your model's performance, which, in turn, gives you more visibility into how your model works.

The process of tracking the progress and results of an experiment is called experiment tracking. The process of logging all the details of an experiment for the purpose of possibly recreating it later or comparing it with other experiments is called versioning. These two go hand in hand with each other. Many tools originally set out to be experiment tracking tools, such as MLflow and Weights & Biases, have grown to incorporate versioning. Many tools originally set out to be versioning tools, such as DVC (*https://oreil.ly/f3sBp*), have also incorporated experiment tracking.

Experiment tracking

A large part of training an ML model is babysitting the learning processes. Many problems can arise during the training process, including loss not decreasing, overfitting, underfitting, fluctuating weight values, dead neurons, and running out of memory. It's important to track what's going on during training not only to detect and address these issues but also to evaluate whether your model is learning anything useful.

When I just started getting into ML, all I was told to track was loss and speed. Fast-forward several years, and people are tracking so many things that their experiment tracking boards look both beautiful and terrifying at the same time. Following is just a short list of things you might want to consider tracking for each experiment during its training process:

- The *loss curve* corresponding to the train split and each of the eval splits.
- The *model performance metrics* that you care about on all nontest splits, such as accuracy, F1, perplexity.
- The log of *corresponding sample, prediction, and ground truth label.* This comes in handy for ad hoc analytics and sanity check.

- The *speed* of your model, evaluated by the number of steps per second or, if your data is text, the number of tokens processed per second.
- *System performance metrics* such as memory usage and CPU/GPU utilization. They're important to identify bottlenecks and avoid wasting system resources.
- The values over time of any *parameter and hyperparameter* whose changes can affect your model's performance, such as the learning rate if you use a learning rate schedule; gradient norms (both globally and per layer), especially if you're clipping your gradient norms; and weight norm, especially if you're doing weight decay.

In theory, it's not a bad idea to track everything you can. Most of the time, you probably don't need to look at most of them. But when something does happen, one or more of them might give you clues to understand and/or debug your model. In general, tracking gives you observability into the state of your model.[8] However, in practice, due to the limitations of tooling today, it can be overwhelming to track too many things, and tracking less important things can distract you from tracking what is really important.

Experiment tracking enables comparison across experiments. By observing how a certain change in a component affects the model's performance, you gain some understanding into what that component does.

A simple way to track your experiments is to automatically make copies of all the code files needed for an experiment and log all outputs with their timestamps.[9] Using third-party experiment tracking tools, however, can give you nice dashboards and allow you to share your experiments with your coworkers.

Versioning

Imagine this scenario. You and your team spent the last few weeks tweaking your model, and one of the runs finally showed promising results. You wanted to use it for more extensive tests, so you tried to replicate it using the set of hyperparameters you'd noted down somewhere, only to find out that the results weren't quite the same. You remembered that you'd made some changes to the code between that run and the next, so you tried your best to undo the changes from memory because your reckless past self had decided that the change was too minimal to be committed. But you still couldn't replicate the promising result because there are just too many possible ways to make changes.

8 We'll cover observability in Chapter 8.

9 I'm still waiting for an experiment tracking tool that integrates with Git commits and DVC commits.

This problem could have been avoided if you versioned your ML experiments. ML systems are part code, part data, so you need to not only version your code but your data as well. Code versioning has more or less become a standard in the industry. However, at this point, data versioning is like flossing. Everyone agrees it's a good thing to do, but few do it.

There are a few reasons why data versioning is challenging. One reason is that because data is often much larger than code, we can't use the same strategy that people usually use to version code to version data.

For example, code versioning is done by keeping track of all the changes made to a codebase. A change is known as a diff, short for difference. Each change is measured by line-by-line comparison. A line of code is usually short enough for line-by-line comparison to make sense. However, a line of your data, especially if it's stored in a binary format, can be indefinitely long. Saying that this line of 1,000,000 characters is different from the other line of 1,000,000 characters isn't going to be that helpful.

Code versioning tools allow users to revert to a previous version of the codebase by keeping copies of all the old files. However, a dataset used might be so large that duplicating it multiple times might be unfeasible.

Code versioning tools allow for multiple people to work on the same codebase at the same time by duplicating the codebase on each person's local machine. However, a dataset might not fit into a local machine.

Second, there's still confusion in what exactly constitutes a diff when we version data. Would diffs mean changes in the content of any file in your data repository, only when a file is removed or added, or when the checksum of the whole repository has changed?

As of 2021, data versioning tools like DVC only register a diff if the checksum of the total directory has changed and if a file is removed or added.

Another confusion is in how to resolve merge conflicts: if developer 1 uses data version X to train model A and developer 2 uses data version Y to train model B, it doesn't make sense to merge data versions X and Y to create Z, since there's no model corresponding with Z.

Third, if you use user data to train your model, regulations like General Data Protection Regulation (GDPR) might make versioning this data complicated. For example, regulations might mandate that you delete user data if requested, making it legally impossible to recover older versions of your data.

Aggressive experiment tracking and versioning helps with reproducibility, but it doesn't ensure reproducibility. The frameworks and hardware you use might introduce nondeterminism to your experiment results,[10] making it impossible to replicate the result of an experiment without knowing everything about the environment your experiment runs in.

The way we have to run so many experiments right now to find the best possible model is the result of us treating ML as a black box. Because we can't predict which configuration will work best, we have to experiment with multiple configurations. However, I hope that as the field progresses, we'll gain more understanding into different models and can reason about what model will work best instead of running hundreds or thousands of experiments.

Debugging ML Models

Debugging is an inherent part of developing any piece of software. ML models aren't an exception. Debugging is never fun, and debugging ML models can be especially frustrating for the following three reasons.

First, ML models fail silently, a topic we'll cover in depth in Chapter 8. The code compiles. The loss decreases as it should. The correct functions are called. The predictions are made, but the predictions are wrong. The developers don't notice the errors. And worse, users don't either and use the predictions as if the application was functioning as it should.

Second, even when you think you've found the bug, it can be frustratingly slow to validate whether the bug has been fixed. When debugging a traditional software program, you might be able to make changes to the buggy code and see the result immediately. However, when making changes to an ML model, you might have to retrain the model and wait until it converges to see whether the bug is fixed, which can take hours. In some cases, you can't even be sure whether the bugs are fixed until the model is deployed to the users.

Third, debugging ML models is hard because of their cross-functional complexity. There are many components in an ML system: data, labels, features, ML algorithms, code, infrastructure, etc. These different components might be owned by different teams. For example, data is managed by data engineers, labels by subject matter experts, ML algorithms by data scientists, and infrastructure by ML engineers or the ML platform team. When an error occurs, it could be because of any of these components or a combination of them, making it hard to know where to look or who should be looking into it.

10 Notable examples include atomic operations in CUDA where nondeterministic orders of operations lead to different floating point rounding errors between runs.

Here are some of the things that might cause an ML model to fail:

Theoretical constraints

As discussed previously, each model comes with its own assumptions about the data and the features it uses. A model might fail because the data it learns from doesn't conform to its assumptions. For example, you use a linear model for the data whose decision boundaries aren't linear.

Poor implementation of model

The model might be a good fit for the data, but the bugs are in the implementation of the model. For example, if you use PyTorch, you might have forgotten to stop gradient updates during evaluation when you should. The more components a model has, the more things that can go wrong, and the harder it is to figure out which goes wrong. However, with models being increasingly commoditized and more and more companies using off-the-shelf models, this is becoming less of a problem.

Poor choice of hyperparameters

With the same model, one set of hyperparameters can give you the state-of-the-art result but another set of hyperparameters might cause the model to never converge. The model is a great fit for your data, and its implementation is correct, but a poor set of hyperparameters might render your model useless.

Data problems

There are many things that could go wrong in data collection and preprocessing that might cause your models to perform poorly, such as data samples and labels being incorrectly paired, noisy labels, features normalized using outdated statistics, and more.

Poor choice of features

There might be many possible features for your models to learn from. Too many features might cause your models to overfit to the training data or cause data leakage. Too few features might lack predictive power to allow your models to make good predictions.

Debugging should be both preventive and curative. You should have healthy practices to minimize the opportunities for bugs to proliferate as well as a procedure for detecting, locating, and fixing bugs. Having the discipline to follow both the best practices and the debugging procedure is crucial in developing, implementing, and deploying ML models.

There is, unfortunately, still no scientific approach to debugging in ML. However, there have been a number of tried-and-true debugging techniques published by experienced ML engineers and researchers. The following are three of them. Readers interested in learning more might want to check out Andrej Karpathy's awesome post "A Recipe for Training Neural Networks" (*https://oreil.ly/8fJ08*).

Start simple and gradually add more components

Start with the simplest model and then slowly add more components to see if it helps or hurts the performance. For example, if you want to build a recurrent neural network (RNN), start with just one level of RNN cell before stacking multiple together or adding more regularization. If you want to use a BERT-like model (Devlin et al. 2018), which uses both a masked language model (MLM) and next sentence prediction (NSP) loss, you might want to use only the MLM loss before adding NSP loss.

Currently, many people start out by cloning an open source implementation of a state-of-the-art model and plugging in their own data. On the off-chance that it works, it's great. But if it doesn't, it's very hard to debug the system because the problem could have been caused by any of the many components in the model.

Overfit a single batch

After you have a simple implementation of your model, try to overfit a small amount of training data and run evaluation on the same data to make sure that it gets to the smallest possible loss. If it's for image recognition, overfit on 10 images and see if you can get the accuracy to be 100%, or if it's for machine translation, overfit on 100 sentence pairs and see if you can get to a BLEU score of near 100. If it can't overfit a small amount of data, there might be something wrong with your implementation.

Set a random seed

There are so many factors that contribute to the randomness of your model: weight initialization, dropout, data shuffling, etc. Randomness makes it hard to compare results across different experiments—you have no idea if the change in performance is due to a change in the model or a different random seed. Setting a random seed ensures consistency between different runs. It also allows you to reproduce errors and other people to reproduce your results.

Distributed Training

As models are getting bigger and more resource-intensive, companies care a lot more about training at scale.[11] Expertise in scalability is hard to acquire because it requires having regular access to massive compute resources. Scalability is a topic that merits a series of books. This section covers some notable issues to highlight the challenges of doing ML at scale and provide a scaffold to help you plan the resources for your project accordingly.

It's common to train a model using data that doesn't fit into memory. It's especially common when dealing with medical data such as CT scans or genome sequences. It can also happen with text data if you work for teams that train large language models (cue OpenAI, Google, NVIDIA, Cohere).

When your data doesn't fit into memory, your algorithms for preprocessing (e.g., zero-centering, normalizing, whitening), shuffling, and batching data will need to run out of core and in parallel.[12] When a sample of your data is large, e.g., one machine can handle a few samples at a time, you might only be able to work with a small batch size, which leads to instability for gradient descent-based optimization.

In some cases, a data sample is so large it can't even fit into memory and you will have to use something like gradient checkpointing, a technique that leverages the memory footprint and compute trade-off to make your system do more computation with less memory. According to the authors of the open source package gradient-checkpointing, "For feed-forward models we were able to fit more than 10x larger models onto our GPU, at only a 20% increase in computation time."[13] Even when a sample fits into memory, using checkpointing can allow you to fit more samples into a batch, which might allow you to train your model faster.

Data parallelism

It's now the norm to train ML models on multiple machines. The most common parallelization method supported by modern ML frameworks is data parallelism: you split your data on multiple machines, train your model on all of them, and accumulate gradients. This gives rise to a couple of issues.

11 For products that serve a large number of users, you also have to care about scalability in serving a model, which is outside of the scope of an ML project so not covered in this book.

12 According to Wikipedia, "Out-of-core algorithms are algorithms that are designed to process data that are too large to fit into a computer's main memory at once" (s.v. "External memory algorithm," *https://oreil.ly/apv5m*).

13 Tim Salimans, Yaroslav Bulatov, and contributors, gradient-checkpointing repository, 2017, *https://oreil.ly/GTUgC*.

A challenging problem is how to accurately and effectively accumulate gradients from different machines. As each machine produces its own gradient, if your model waits for all of them to finish a run—synchronous stochastic gradient descent (SGD)—stragglers will cause the entire system to slow down, wasting time and resources.[14] The straggler problem grows with the number of machines, as the more workers, the more likely that at least one worker will run unusually slowly in a given iteration. However, there have been many algorithms that effectively address this problem.[15]

If your model updates the weight using the gradient from each machine separately—asynchronous SGD—gradient staleness might become a problem because the gradients from one machine have caused the weights to change before the gradients from another machine have come in.[16]

The difference between synchronous SGD and asynchronous SGD is illustrated in Figure 6-6.

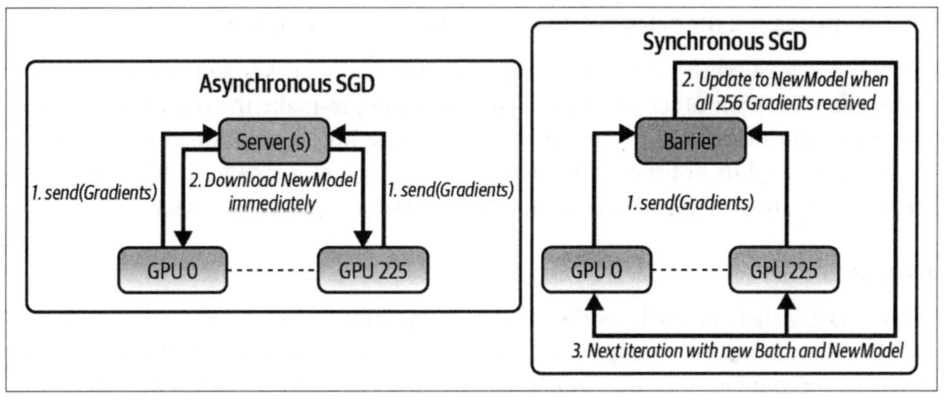

Figure 6-6. Synchronous SGD versus asynchronous SGD for data parallelism. Source: Adapted from an image by Jim Dowling[17]

14 Dipankar Das, Sasikanth Avancha, Dheevatsa Mudigere, Karthikeyan Vaidynathan, Srinivas Sridharan, Dhiraj Kalamkar, Bharat Kaul, and Pradeep Dubey, "Distributed Deep Learning Using Synchronous Stochastic Gradient Descent," *arXiv*, February 22, 2016, *https://oreil.ly/ma8Y6*.

15 Jianmin Chen, Xinghao Pan, Rajat Monga, Samy Bengio, and Rafal Jozefowicz, "Revisiting Distributed Synchronous SGD," ICLR 2017, *https://oreil.ly/dzVZ5*; Matei Zaharia, Andy Konwinski, Anthony D. Joseph, Randy Katz, and Ion Stoica, "Improving MapReduce Performance in Heterogeneous Environments," 8th USENIX Symposium on Operating Systems Design and Implementation, *https://oreil.ly/FWswd*; Aaron Harlap, Henggang Cui, Wei Dai, Jinliang Wei, Gregory R. Ganger, Phillip B. Gibbons, Garth A. Gibson, and Eric P. Xing, "Addressing the Straggler Problem for Iterative Convergent Parallel ML" (SoCC '16, Santa Clara, CA, October 5–7, 2016), *https://oreil.ly/wZgOO*.

16 Jeffrey Dean, Greg Corrado, Rajat Monga, Kai Chen, Matthieu Devin, Mark Mao, Marc'aurelio Ranzato, et al., "Large Scale Distributed Deep Networks," NIPS 2012, *https://oreil.ly/EWPun*.

17 Jim Dowling, "Distributed TensorFlow," O'Reilly Media, December 19, 2017, *https://oreil.ly/VYlOP*.

In theory, asynchronous SGD converges but requires more steps than synchronous SGD. However, in practice, when the number of weights is large, gradient updates tend to be sparse, meaning most gradient updates only modify small fractions of the parameters, and it's less likely that two gradient updates from different machines will modify the same weights. When gradient updates are sparse, gradient staleness becomes less of a problem and the model converges similarly for both synchronous and asynchronous SGD.[18]

Another problem is that spreading your model on multiple machines can cause your batch size to be very big. If a machine processes a batch size of 1,000, then 1,000 machines process a batch size of 1M (OpenAI's GPT-3 175B uses a batch size of 3.2M in 2020).[19] To oversimplify the calculation, if training an epoch on a machine takes 1M steps, training on 1,000 machines might take only 1,000 steps. An intuitive approach is to scale up the learning rate to account for more learning at each step, but we also can't make the learning rate too big as it will lead to unstable convergence. In practice, increasing the batch size past a certain point yields diminishing returns.[20]

Last but not least, with the same model setup, the main worker sometimes uses a lot more resources than other workers. If that's the case, to make the most use out of all machines, you need to figure out a way to balance out the workload among them. The easiest way, but not the most effective way, is to use a smaller batch size on the main worker and a larger batch size on other workers.

Model parallelism

With data parallelism, each worker has its own copy of the whole model and does all the computation necessary for its copy of the model. Model parallelism is when different components of your model are trained on different machines, as shown in Figure 6-7. For example, machine 0 handles the computation for the first two layers while machine 1 handles the next two layers, or some machines can handle the forward pass while several others handle the backward pass.

---

18 Feng Niu, Benjamin Recht, Christopher Ré, and Stephen J. Wright, "Hogwild!: A Lock-Free Approach to Parallelizing Stochastic Gradient Descent," 2011, *https://oreil.ly/sAEbv*.

19 Tom B. Brown, Benjamin Mann, Nick Ryder, Melanie Subbiah, Jared Kaplan, Prafulla Dhariwal, Arvind Neelakantan, et al., "Language Models Are Few-Shot Learners," *arXiv*, May 28, 2020, *https://oreil.ly/qjg2S*.

20 Sam McCandlish, Jared Kaplan, Dario Amodei, and OpenAI Dota Team, "An Empirical Model of Large-Batch Training," *arXiv*, December 14, 2018, *https://oreil.ly/mcjbV*; Christopher J. Shallue, Jaehoon Lee, Joseph Antognini, Jascha Sohl-Dickstein, Roy Frostig, and George E. Dahl, "Measuring the Effects of Data Parallelism on Neural Network Training," *Journal of Machine Learning Research* 20 (2019): 1–49, *https://oreil.ly/YAEOM*.

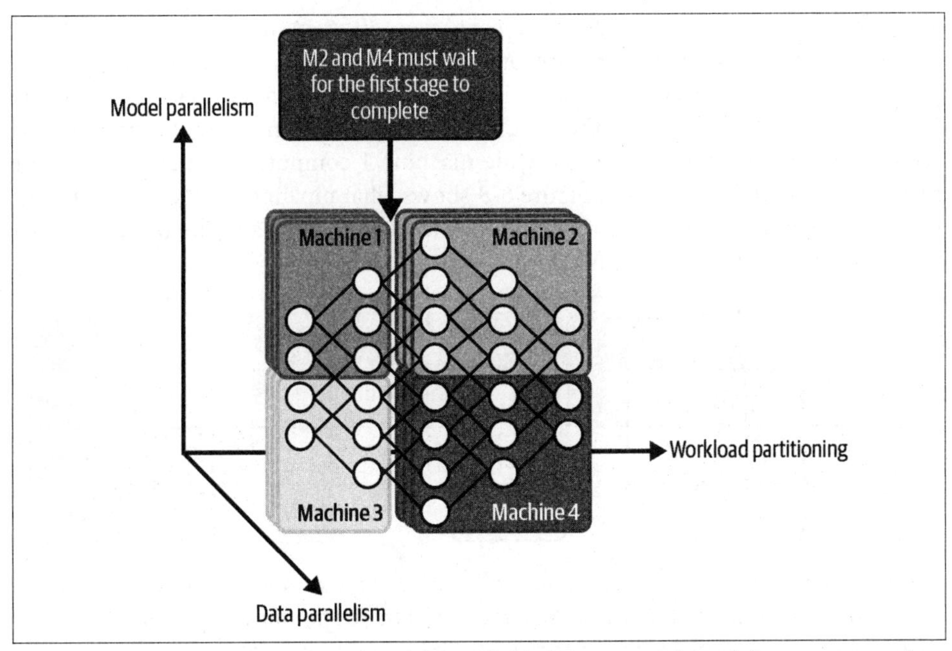

Figure 6-7. Data parallelism and model parallelism. Source: Adapted from an image by Jure Leskovec[21]

Model parallelism can be misleading because in some cases parallelism doesn't mean that different parts of the model in different machines are executed in parallel. For example, if your model is a massive matrix and the matrix is split into two halves on two machines, then these two halves might be executed in parallel. However, if your model is a neural network and you put the first layer on machine 1 and the second layer on machine 2, and layer 2 needs outputs from layer 1 to execute, then machine 2 has to wait for machine 1 to finish first to run.

Pipeline parallelism is a clever technique to make different components of a model on different machines run more in parallel. There are multiple variants to this, but the key idea is to break the computation of each machine into multiple parts. When machine 1 finishes the first part of its computation, it passes the result onto machine 2, then continues to the second part, and so on. Machine 2 now can execute its computation on the first part while machine 1 executes its computation on the second part.

21 Jure Leskovec, Mining Massive Datasets course, Stanford, lecture 13, 2020, *https://oreil.ly/gZcja*.

To make this concrete, consider you have four different machines and the first, second, third, and fourth layers are on machine 1, 2, 3, and 4 respectively. With pipeline parallelism, each mini-batch is broken into four micro-batches. Machine 1 computes the first layer on the first micro-batch, then machine 2 computes the second layer on machine 1's results while machine 1 computes the first layer on the second micro-batch, and so on. Figure 6-8 shows what pipeline parallelism looks like on four machines; each machine runs both the forward pass and the backward pass for one component of a neural network.

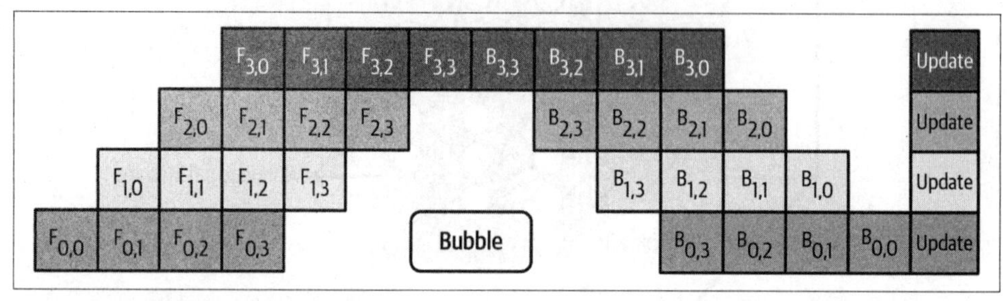

Figure 6-8. Pipeline parallelism for a neural network on four machines; each machine runs both the forward pass (F) and the backward pass (B) for one component of the neural network. Source: Adapted from an image by Huang et al.[22]

Model parallelism and data parallelism aren't mutually exclusive. Many companies use both methods for better utilization of their hardware, even though the setup to use both methods can require significant engineering effort.

AutoML

There's a joke that a good ML researcher is someone who will automate themselves out of job, designing an AI algorithm intelligent enough to design itself. It was funny until the TensorFlow Dev Summit 2018, where Jeff Dean took the stage and declared that Google intended on replacing ML expertise with 100 times more computational power, introducing AutoML to the excitement and horror of the community. Instead of paying a group of 100 ML researchers/engineers to fiddle with various models and eventually select a suboptimal one, why not use that money on compute to search for the optimal model? A screenshot from the recording of the event is shown in Figure 6-9.

22 Yanping Huang, Youlong Cheng, Ankur Bapna, Orhan Firat, Mia Xu Chen, Dehao Chen, HyoukJoong Lee, et al., "GPipe: Easy Scaling with Micro-Batch Pipeline Parallelism," *arXiv*, July 25, 2019, *https://oreil.ly/wehkx*.

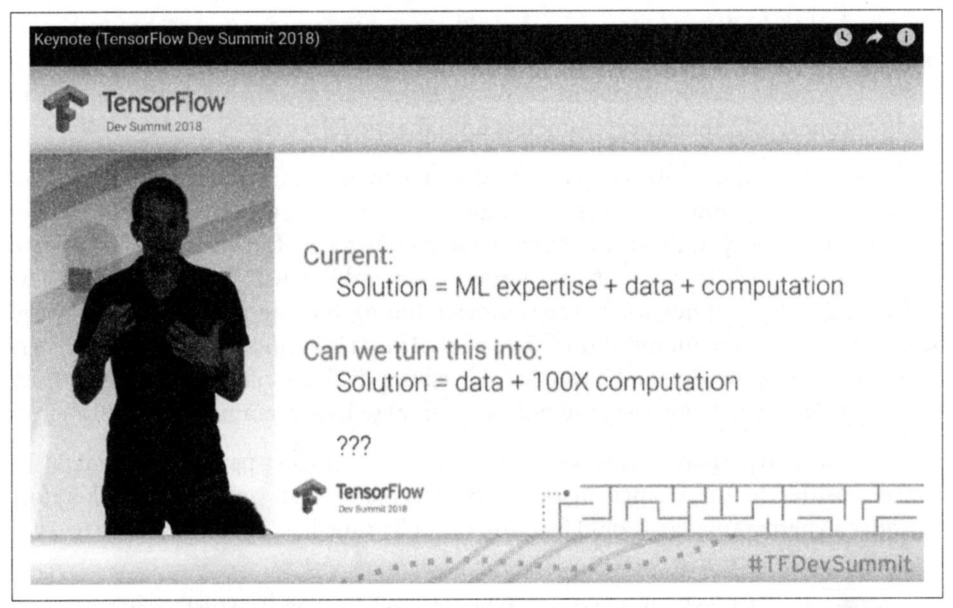

Figure 6-9. Jeff Dean unveiling Google's AutoML at TensorFlow Dev Summit 2018

Soft AutoML: Hyperparameter tuning

AutoML refers to automating the process of finding ML algorithms to solve real-world problems. One mild form, and the most popular form, of AutoML in production is hyperparameter tuning. A hyperparameter is a parameter supplied by users whose value is used to control the learning process, e.g., learning rate, batch size, number of hidden layers, number of hidden units, dropout probability, β_1 and β_2 in Adam optimizer, etc. Even quantization—e.g., whether to use 32 bits, 16 bits, or 8 bits to represent a number or a mixture of these representations—can be considered a hyperparameter to tune.[23]

With different sets of hyperparameters, the same model can give drastically different performances on the same dataset. Melis et al. showed in their 2018 paper "On the State of the Art of Evaluation in Neural Language Models" (*https://oreil.ly/AY2lF*) that weaker models with well-tuned hyperparameters can outperform stronger, fancier models. The goal of hyperparameter tuning is to find the optimal set of hyperparameters for a given model within a search space—the performance of each set evaluated on a validation set.

23 We'll cover quantization in Chapter 7.

Despite knowing its importance, many still ignore systematic approaches to hyperparameter tuning in favor of a manual, gut-feeling approach. The most popular is arguably graduate student descent (GSD), a technique in which a graduate student fiddles around with the hyperparameters until the model works.[24]

However, more and more people are adopting hyperparameter tuning as part of their standard pipelines. Popular ML frameworks either come with built-in utilities or have third-party utilities for hyperparameter tuning—for example, scikit-learn with auto-sklearn,[25] TensorFlow with Keras Tuner, and Ray with Tune (*https://oreil.ly/uulrC*). Popular methods for hyperparameter tuning include random search,[26] grid search, and Bayesian optimization.[27] The book *AutoML: Methods, Systems, Challenges* by the AutoML group at the University of Freiburg dedicates its first chapter (*https://oreil.ly/LfqJm*) (which you can read online for free) to hyperparameter optimization.

When tuning hyperparameters, keep in mind that a model's performance might be more sensitive to the change in one hyperparameter than another, and therefore sensitive hyperparameters should be more carefully tuned.

It's crucial to never use your test split to tune hyperparameters. Choose the best set of hyperparameters for a model based on its performance on a validation split, then report the model's final performance on the test split. If you use your test split to tune hyperparameters, you risk overfitting your model to the test split.

Hard AutoML: Architecture search and learned optimizer

Some teams take hyperparameter tuning to the next level: what if we treat other components of a model or the entire model as hyperparameters. The size of a convolution layer or whether or not to have a skip layer can be considered a hyperparameter. Instead of manually putting a pooling layer after a convolutional layer or ReLu (rectified linear unit) after linear, you give your algorithm these building blocks and let it figure out how to combine them. This area of research is known as architectural

---

24 GSD is a well-documented technique. See "How Do People Come Up With All These Crazy Deep Learning Architectures?," Reddit, *https://oreil.ly/5vEsH*; "Debate About Science at Organizations like Google Brain/FAIR/DeepMind," Reddit, *https://oreil.ly/2K77r*; "Grad Student Descent," *Science Dryad*, January 25, 2014, *https://oreil.ly/dIR9r*; and Guy Zyskind (@GuyZys), "Grad Student Descent: the preferred #nonlinear #optimization technique #machinelearning," Twitter, April 27, 2015, *https://oreil.ly/SW1or*.

25 auto-sklearn 2.0 also provides basic model selection capacity.

26 Our team at NVIDIA developed Milano (*https://oreil.ly/FYWaU*), a framework-agnostic tool for automatic hyperparameter tuning using random search.

27 A common practice I've observed is to start with coarse-to-fine random search, then experiment with Bayesian or grid search once the search space has been significantly reduced.

search, or neural architecture search (NAS) for neural networks, as it searches for the optimal model architecture.

A NAS setup consists of three components:

A search space
: Defines possible model architectures—i.e., building blocks to choose from and constraints on how they can be combined.

A performance estimation strategy
: To evaluate the performance of a candidate architecture without having to train each candidate architecture from scratch until convergence. When we have a large number of candidate architectures, say 1,000, training all of them until convergence can be costly.

A search strategy
: To explore the search space. A simple approach is random search—randomly choosing from all possible configurations—which is unpopular because it's prohibitively expensive even for NAS. Common approaches include reinforcement learning (rewarding the choices that improve the performance estimation) and evolution (adding mutations to an architecture, choosing the best-performing ones, adding mutations to them, and so on).[28]

For NAS, the search space is discrete—the final architecture uses only one of the available options for each layer/operation,[29] and you have to provide the set of building blocks. The common building blocks are various convolutions of different sizes, linear, various activations, pooling, identity, zero, etc. The set of building blocks varies based on the base architecture, e.g., convolutional neural networks or transformers.

In a typical ML training process, you have a model and then a learning procedure, an algorithm that helps your model find the set of parameters that minimize a given objective function for a given set of data. The most common learning procedure for neural networks today is gradient descent, which leverages an optimizer to specify how to update a model's weights given gradient updates.[30] Popular optimizers are, as you probably already know, Adam, Momentum, SGD, etc. In theory, you can include optimizers as building blocks in NAS and search for one that works best.

---

28 Barret Zoph and Quoc V. Le, "Neural Architecture Search with Reinforcement Learning," *arXiv*, November 5, 2016, *https://oreil.ly/FhsuQ*; Esteban Real, Alok Aggarwal, Yanping Huang, and Quoc V. Le, "Regularized Evolution for Image Classifier Architecture Search," AAAI 2019, *https://oreil.ly/FWYjn*.

29 You can make the search space continuous to allow differentiation, but the resulting architecture has to be converted into a discrete architecture. See "DARTS: Differentiable Architecture Search" (*https://oreil.ly/sms2H*) (Liu et al. 2018).

30 We cover learning procedures and optimizers in more detail in the section "Basic ML Reviews" in the book's GitHub repository (*https://oreil.ly/designing-machine-learning-systems-code*).

In practice, this is difficult to do, since optimizers are sensitive to the setting of their hyperparameters, and the default hyperparameters don't often work well across architectures.

This leads to an exciting research direction: what if we replace the functions that specify the update rule with a neural network? How much to update the model's weights will be calculated by this neural network. This approach results in learned optimizers, as opposed to hand-designed optimizers.

Since learned optimizers are neural networks, they need to be trained. You can train your learned optimizer on the same dataset you're training the rest of your neural network on, but this requires you to train an optimizer every time you have a task.

Another approach is to train a learned optimizer once on a set of existing tasks —using aggregated loss on those tasks as the loss function and existing designed optimizers as the learning rule—and use it for every new task after that. For example, Metz et al. constructed a set of thousands of tasks to train learned optimizers. Their learned optimizer was able to generalize to both new datasets and domains as well as new architectures.[31] And the beauty of this approach is that the learned optimizer can then be used to train a better-learned optimizer, an algorithm that improves on itself.

Whether it's architecture search or meta-learning learning rules, the up-front training cost is expensive enough that only a handful of companies in the world can afford to pursue them. However, it's important for people interested in ML in production to be aware of the progress in AutoML for two reasons. First, the resulting architectures and learned optimizers can allow ML algorithms to work off-the-shelf on multiple real-world tasks, saving production time and cost, during both training and inferencing. For example, EfficientNets, a family of models produced by Google's AutoML team, surpass state-of-the-art accuracy with up to 10x better efficiency.[32] Second, they might be able to solve many real-world tasks previously impossible with existing architectures and optimizers.

31 Luke Metz, Niru Maheswaranathan, C. Daniel Freeman, Ben Poole, and Jascha Sohl-Dickstein, "Tasks, Stability, Architecture, and Compute: Training More Effective Learned Optimizers, and Using Them to Train Themselves," *arXiv*, September 23, 2020, *https://oreil.ly/IH7eT*.

32 Mingxing Tan and Quoc V. Le, "EfficientNet: Improving Accuracy and Efficiency through AutoML and Model Scaling," *Google AI Blog*, May 29, 2019, *https://oreil.ly/gonEn*.

Four Phases of ML Model Development

Before we transition to model training, let's take a look at the four phases of ML model development. Once you've decided to explore ML, your strategy depends on which phase of ML adoption you are in. There are four phases of adopting ML. The solutions from a phase can be used as baselines to evaluate the solutions from the next phase:

Phase 1. Before machine learning

If this is your first time trying to make this type of prediction from this type of data, start with non-ML solutions. Your first stab at the problem can be the simplest heuristics. For example, to predict what letter users are going to type next in English, you can show the top three most common English letters, "e," "t," and "a," which might get your accuracy to be 30%.

Facebook newsfeed was introduced in 2006 without any intelligent algorithms—posts were shown in chronological order, as shown in Figure 6-10.[33] It wasn't until 2011 that Facebook started displaying news updates you were most interested in at the top of the feed.

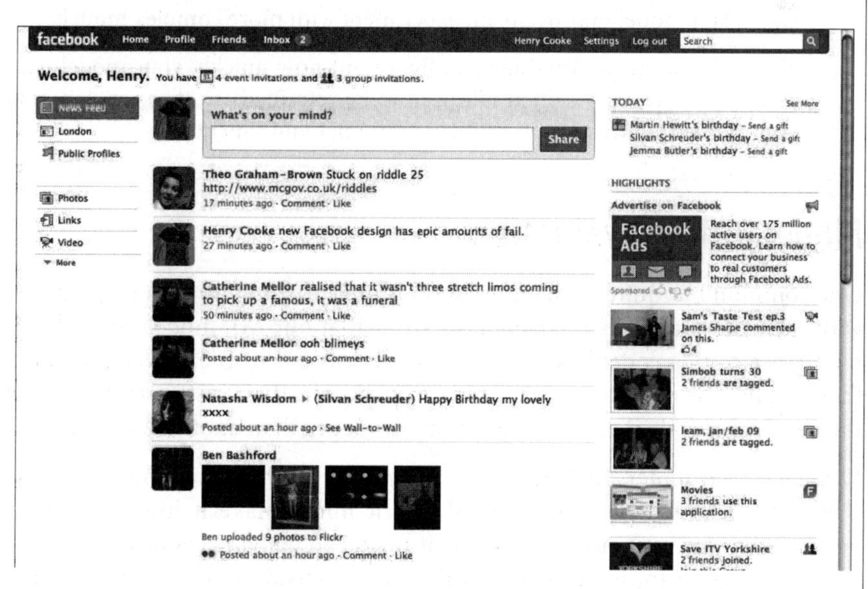

Figure 6-10. Facebook newsfeed circa 2006. Source: Iveta Ryšavá[34]

33 Samantha Murphy, "The Evolution of Facebook News Feed," *Mashable*, March 12, 2013, *https://oreil.ly/1HMXh*.

34 Iveta Ryšavá, "What Mark Zuckerberg's News Feed Looked Like in 2006," Newsfeed.org, January 14, 2016, *https://oreil.ly/XZT6Q*.

According to Martin Zinkevich in his magnificent "Rules of Machine Learning: Best Practices for ML Engineering": "If you think that machine learning will give you a 100% boost, then a heuristic will get you 50% of the way there."[35] You might even find that non-ML solutions work fine and you don't need ML yet.

Phase 2. Simplest machine learning models

For your first ML model, you want to start with a simple algorithm, something that gives you visibility into its working to allow you to validate the usefulness of your problem framing and your data. Logistic regression, gradient-boosted trees, *k*-nearest neighbors can be great for that. They are also easier to implement and deploy, which allows you to quickly build out a framework from data engineering to development to deployment that you can test out and gain confidence on.

Phase 3. Optimizing simple models

Once you have your ML framework in place, you can focus on optimizing the simple ML models with different objective functions, hyperparameter search, feature engineering, more data, and ensembles.

Phase 4. Complex models

Once you've reached the limit of your simple models and your use case demands significant model improvement, experiment with more complex models.

You'll also want to experiment to figure out how quickly your model decays in production (e.g., how often it'll need to be retrained) so that you can build out your infrastructure to support this retraining requirement.[36]

Model Offline Evaluation

One common but quite difficult question I often encounter when helping companies with their ML strategies is: "How do I know that our ML models are any good?" In one case, a company deployed ML to detect intrusions to 100 surveillance drones, but they had no way of measuring how many intrusions their system failed to detect, and they couldn't decide if one ML algorithm was better than another for their needs.

Lacking a clear understanding of how to evaluate your ML systems is not necessarily a reason for your ML project to fail, but it might make it impossible to find the best solution for your need, and make it harder to convince your managers to adopt ML. You might want to partner with the business team to develop metrics for model evaluation that are more relevant to your company's business.[37]

35 Martin Zinkevich, "Rules of Machine Learning: Best Practices for ML Engineering," Google, 2019, *https://oreil.ly/YtEsN*.

36 We'll go in depth about how often to update your models in Chapter 9.

37 See the section "Business and ML Objectives" on page 26.

Ideally, the evaluation methods should be the same during both development and production. But in many cases, the ideal is impossible because during development, you have ground truth labels, but in production, you don't.

For certain tasks, it's possible to infer or approximate labels in production based on users' feedback, as covered in the section "Natural Labels" on page 91. For example, for the recommendation task, it's possible to infer if a recommendation is good by whether users click on it. However, there are many biases associated with this.

For other tasks, you might not be able to evaluate your model's performance in production directly and might have to rely on extensive monitoring to detect changes and failures in your ML system's performance. We'll cover monitoring in Chapter 8.

Once your model is deployed, you'll need to continue monitoring and testing your model in production. In this section, we'll discuss methods to evaluate your model's performance before it's deployed. We'll start with the baselines against which we will evaluate our models. Then we'll cover some of the common methods to evaluate your model beyond overall accuracy metrics.

Baselines

Someone once told me that her new generative model achieved the FID score of 10.3 on ImageNet.[38] I had no idea what this number meant or whether her model would be useful for my problem.

Another time, I helped a company implement a classification model where the positive class appears 90% of the time. An ML engineer on the team told me, all excited, that their initial model achieved an F1 score of 0.90. I asked him how it was compared to random. He had no idea. It turned out that because for his task the POSITIVE class accounts for 90% of the labels, if his model randomly outputs the positive class 90% of the time, its F1 score would also be around 0.90.[39] His model might as well be making predictions at random.[40]

---

38 Fréchet inception distance, a common metric for measuring the quality of synthesized images. The smaller the value, the higher the quality is supposed to be.

39 The accuracy, in this case, would be around 0.80.

40 Revisit the section "Using the right evaluation metrics" on page 106 for a refresh on the asymmetry of F1.

Evaluation metrics, by themselves, mean little. When evaluating your model, it's essential to know the baseline you're evaluating it against. The exact baselines should vary from one use case to another, but here are the five baselines that might be useful across use cases:

Random baseline

If our model just predicts at random, what's the expected performance? The predictions are generated at random following a specific distribution, which can be the uniform distribution or the task's label distribution.

For example, consider the task that has two labels, NEGATIVE that appears 90% of the time and POSITIVE that appears 10% of the time. Table 6-2 shows the F1 and accuracy scores of baseline models making predictions at random. However, as an exercise to see how challenging it is for most people to have an intuition for these values, try to calculate these raw numbers in your head before looking at the table.

Table 6-2. F1 and accuracy scores of a baseline model predicting at random

Random distribution	Meaning	F1	Accuracy
Uniform random	Predicting each label with equal probability (50%)	0.167	0.5
Task's label distribution	Predicting NEGATIVE 90% of the time, and POSITIVE 10% of the time	0.1	0.82

Simple heuristic

Forget ML. If you just make predictions based on simple heuristics, what performance would you expect? For example, if you want to build a ranking system to rank items on a user's newsfeed with the goal of getting that user to spend more time on the newsfeed, how much time would a user spend if you just rank all the items in reverse chronological order, showing the latest one first?

Zero rule baseline

The zero rule baseline is a special case of the simple heuristic baseline when your baseline model always predicts the most common class.

For example, for the task of recommending the app a user is most likely to use next on their phone, the simplest model would be to recommend their most frequently used app. If this simple heuristic can predict the next app accurately 70% of the time, any model you build has to outperform it significantly to justify the added complexity.

Human baseline

In many cases, the goal of ML is to automate what would have been otherwise done by humans, so it's useful to know how your model performs compared to human experts. For example, if you work on a self-driving system, it's crucial to measure your system's progress compared to human drivers, because otherwise you might never be able to convince your users to trust this system. Even if your system isn't meant to replace human experts and only to aid them in improving their productivity, it's still important to know in what scenarios this system would be useful to humans.

Existing solutions
: In many cases, ML systems are designed to replace existing solutions, which might be business logic with a lot of if/else statements or third-party solutions. It's crucial to compare your new model to these existing solutions. Your ML model doesn't always have to be better than existing solutions to be useful. A model whose performance is a little bit inferior can still be useful if it's much easier or cheaper to use.

When evaluating a model, it's important to differentiate between "a good system" and "a useful system." A good system isn't necessarily useful, and a bad system isn't necessarily useless. A self-driving vehicle might be good if it's a significant improvement from previous self-driving systems, but it might not be useful if it doesn't perform at least as well as human drivers. In some cases, even if an ML system drives better than an average human, people might still not trust it, which renders it not useful. On the other hand, a system that predicts what word a user will type next on their phone might be considered bad if it's much worse than a native speaker. However, it might still be useful if its predictions can help users type faster some of the time.

Evaluation Methods

In academic settings, when evaluating ML models, people tend to fixate on their performance metrics. However, in production, we also want our models to be robust, fair, calibrated, and overall make sense. We'll introduce some evaluation methods that help with measuring these characteristics of a model.

Perturbation tests

A group of my students wanted to build an app to predict whether someone has COVID-19 through their cough. Their best model worked great on the training data, which consisted of two-second long cough segments collected by hospitals. However, when they deployed it to actual users, this model's predictions were close to random.

One of the reasons is that actual users' coughs contain a lot of noise compared to the coughs collected in hospitals. Users' recordings might contain background music or nearby chatter. The microphones they use are of varying quality. They might start recording their coughs as soon as recording is enabled or wait for a fraction of a second.

Ideally, the inputs used to develop your model should be similar to the inputs your model will have to work with in production, but it's not possible in many cases. This is especially true when data collection is expensive or difficult and the best available data you have access to for training is still very different from your real-world data. The inputs your models have to work with in production are often noisy compared to inputs in development.[41] The model that performs best on training data isn't necessarily the model that performs best on noisy data.

To get a sense of how well your model might perform with noisy data, you can make small changes to your test splits to see how these changes affect your model's performance. For the task of predicting whether someone has COVID-19 from their cough, you could randomly add some background noise or randomly clip the testing clips to simulate the variance in your users' recordings. You might want to choose the model that works best on the perturbed data instead of the one that works best on the clean data.

The more sensitive your model is to noise, the harder it will be to maintain it, since if your users' behaviors change just slightly, such as they change their phones, your model's performance might degrade. It also makes your model susceptible to adversarial attack.

Invariance tests

A Berkeley study found that between 2008 and 2015, 1.3 million creditworthy Black and Latino applicants had their mortgage applications rejected because of their races.[42] When the researchers used the income and credit scores of the rejected applications but deleted the race-identifying features, the applications were accepted.

Certain changes to the inputs shouldn't lead to changes in the output. In the preceding case, changes to race information shouldn't affect the mortgage outcome. Similarly, changes to applicants' names shouldn't affect their resume screening results nor should someone's gender affect how much they should be paid. If these happen, there are biases in your model, which might render it unusable no matter how good its performance is.

To avoid these biases, one solution is to do the same process that helped the Berkeley researchers discover the biases: keep the inputs the same but change the sensitive information to see if the outputs change. Better, you should exclude the sensitive information from the features used to train the model in the first place.[43]

---

41 Other examples of noisy data include images with different lighting or texts with accidental typos or intentional text modifications such as typing "long" as "loooooong."

42 Khristopher J. Brooks, "Disparity in Home Lending Costs Minorities Millions, Researchers Find," *CBS News*, November 15, 2019, *https://oreil.ly/TMPVl*.

43 It might also be mandated by law to exclude sensitive information from the model training process.

Directional expectation tests

Certain changes to the inputs should, however, cause predictable changes in outputs. For example, when developing a model to predict housing prices, keeping all the features the same but increasing the lot size shouldn't decrease the predicted price, and decreasing the square footage shouldn't increase it. If the outputs change in the opposite expected direction, your model might not be learning the right thing, and you need to investigate it further before deploying it.

Model calibration

Model calibration is a subtle but crucial concept to grasp. Imagine that someone makes a prediction that something will happen with a probability of 70%. What this prediction means is that out of all the times this prediction is made, the predicted outcome matches the actual outcome 70% of the time. If a model predicts that team A will beat team B with a 70% probability, and out of the 1,000 times these two teams play together, team A only wins 60% of the time, then we say that this model isn't calibrated. A calibrated model should predict that team A wins with a 60% probability.

Model calibration is often overlooked by ML practitioners, but it's one of the most important properties of any predictive system. To quote Nate Silver in his book *The Signal and the Noise*, calibration is "one of the most important tests of a forecast—I would argue that it is the single most important one."

We'll walk through two examples to show why model calibration is important. First, consider the task of building a recommender system to recommend what movies users will likely watch next. Suppose user A watches romance movies 80% of the time and comedy 20% of the time. If your recommender system shows exactly the movies A will most likely watch, the recommendations will consist of only romance movies because A is much more likely to watch romance than any other type of movies. You might want a more calibrated system whose recommendations are representative of users' actual watching habits. In this case, they should consist of 80% romance and 20% comedy.[44]

Second, consider the task of building a model to predict how likely it is that a user will click on an ad. For the sake of simplicity, imagine that there are only two ads, ad A and ad B. Your model predicts that this user will click on ad A with a 10% probability and on ad B with an 8% probability. You don't need your model to be calibrated to rank ad A above ad B. However, if you want to predict how many clicks your ads will get, you'll need your model to be calibrated. If your model predicts that a user will click on ad A with a 10% probability but in reality the ad is only clicked on

44 For more information on calibrated recommendations, check out the paper "Calibrated Recommendations" (*https://oreil.ly/yueHR*) by Harald Steck in 2018 based on his work at Netflix.

5% of the time, your estimated number of clicks will be way off. If you have another model that gives the same ranking but is better calibrated, you might want to consider the better calibrated one.

To measure a model's calibration, a simple method is counting: you count the number of times your model outputs the probability X and the frequency Y of that prediction coming true, and plot X against Y. The graph for a perfectly calibrated model will have X equal Y at all data points. In scikit-learn, you can plot the calibration curve of a binary classifier with the method `sklearn.calibration.calibration_curve`, as shown in Figure 6-11.

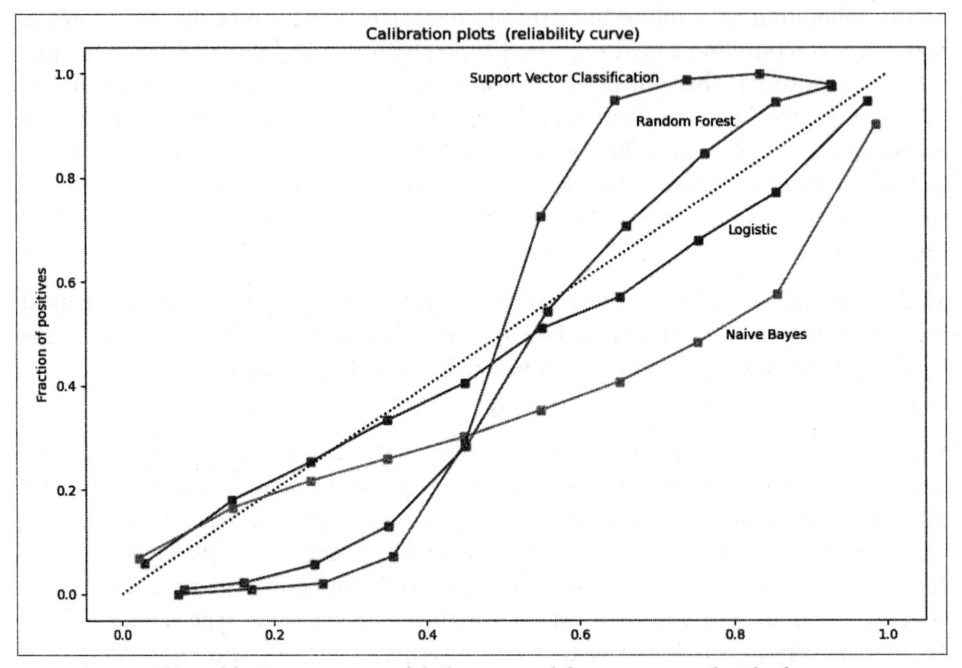

Figure 6-11. The calibration curves of different models on a toy task. The logistic regression model is the best calibrated model because it directly optimizes logistic loss. Source: scikit-learn (https://oreil.ly/Tnts7)

To calibrate your models, a common method is Platt scaling (*https://oreil.ly/pQ0TQ*), which is implemented in scikit-learn with `sklearn.calibration.CalibratedClassifierCV`. Another good open source implementation by Geoff Pleiss can be found on GitHub (*https://oreil.ly/e1Meh*). For readers who want to learn more about the importance of model calibration and how to calibrate neural networks, Lee Richardson and Taylor Pospisil have an excellent blog post (*https://oreil.ly/wPUkU*) based on their work at Google.

Confidence measurement

Confidence measurement can be considered a way to think about the usefulness threshold for each individual prediction. Indiscriminately showing all a model's predictions to users, even the predictions that the model is unsure about, can, at best, cause annoyance and make users lose trust in the system, such as an activity detection system on your smartwatch that thinks you're running even though you're just walking a bit fast. At worst, it can cause catastrophic consequences, such as a predictive policing algorithm that flags an innocent person as a potential criminal.

If you only want to show the predictions that your model is certain about, how do you measure that certainty? What is the certainty threshold at which the predictions should be shown? What do you want to do with predictions below that threshold—discard them, loop in humans, or ask for more information from users?

While most other metrics measure the system's performance on average, confidence measurement is a metric for each individual sample. System-level measurement is useful to get a sense of overall performance, but sample-level metrics are crucial when you care about your system's performance on every sample.

Slice-based evaluation

Slicing means to separate your data into subsets and look at your model's performance on each subset separately. A common mistake that I've seen in many companies is that they are focused too much on coarse-grained metrics like overall F1 or accuracy on the entire data and not enough on sliced-based metrics. This can lead to two problems.

One is that their model performs differently on different slices of data when the model should perform the same. For example, their data has two subgroups, one majority and one minority, and the majority subgroup accounts for 90% of the data:

- Model A achieves 98% accuracy on the majority subgroup but only 80% on the minority subgroup, which means its overall accuracy is 96.2%.
- Model B achieves 95% accuracy on the majority and 95% on the minority, which means its overall accuracy is 95%.

These two models are compared in Table 6-3. Which model would you choose?

Table 6-3. Two models' performance on the majority and minority subgroups

	Majority accuracy	Minority accuracy	Overall accuracy
Model A	98%	80%	96.2%
Model B	95%	95%	95%

If a company focuses only on overall metrics, they might go with model A. They might be very happy with this model's high accuracy until, one day, their end users discover that this model is biased against the minority subgroup because the minority subgroup happens to correspond to an underrepresented demographic group.[45] The focus on overall performance is harmful not only because of the potential public backlash, but also because it blinds the company to huge potential model improvements. If the company sees the two models' slice-based performance, they might follow different strategies. For example, they might decide to improve model A's performance on the minority subgroup, which leads to improving this model's performance overall. Or they might keep both models the same but now have more information to make a better-informed decision on which model to deploy.

Another problem is that their model performs the same on different slices of data when the model should perform differently. Some subsets of data are more critical. For example, when you build a model for user churn prediction (predicting when a user will cancel a subscription or a service), paid users are more critical than nonpaid users. Focusing on a model's overall performance might hurt its performance on these critical slices.

A fascinating and seemingly counterintuitive reason why slice-based evaluation is crucial is Simpson's paradox (*https://oreil.ly/clFB0*), a phenomenon in which a trend appears in several groups of data but disappears or reverses when the groups are combined. This means that model B can perform better than model A on all data together, but model A performs better than model B on each subgroup separately. Consider model A's and model B's performance on group A and group B as shown in Table 6-4. Model A outperforms model B for both group A and B, but when combined, model B outperforms model A.

Table 6-4. An example of Simpson's paradox[a]

	Group A	Group B	Overall
Model A	**93% (81/87)**	**73% (192/263)**	78% (273/350)
Model B	87% (234/270)	69% (55/80)	**83% (289/350)**

[a] Numbers from Charig et al.'s kidney stone treatment study in 1986: C. R. Charig, D. R. Webb, S. R. Payne, and J. E. Wickham, "Comparison of Treatment of Renal Calculi by Open Surgery, Percutaneous Nephrolithotomy, and Extracorporeal Shockwave Lithotripsy," *British Medical Journal* (Clinical Research Edition) 292, no. 6524 (March 1986): 879–82, *https://oreil.ly/X8oWr*.

45 Maggie Zhang, "Google Photos Tags Two African-Americans As Gorillas Through Facial Recognition Software," *Forbes*, July 1, 2015, *https://oreil.ly/VYG2j*.

Simpson's paradox is more common than you'd think. In 1973, Berkeley graduate statistics showed that the admission rate for men was much higher than for women, which caused people to suspect biases against women. However, a closer look into individual departments showed that the admission rates for women were actually higher than those for men in four out of six departments,[46] as shown in Table 6-5.

Table 6-5. Berkeley's 1973 graduate admission data[a]

	All		Men		Women	
Department	Applicants	Admitted	Applicants	Admitted	Applicants	Admitted
A	933	64%	***825***	62%	108	**82%**
B	585	63%	***560***	63%	25	**68 %**
C	918	35%	325	**37 %**	***593***	34%
D	792	34%	417	33%	375	**35 %**
E	584	25%	191	**28 %**	***393***	24%
F	714	6%	373	6%	341	**7 %**
Total	12,763	41%	8,442	**44%**	4,321	35%

[a] Data from Bickel et al. (1975)

Regardless of whether you'll actually encounter this paradox, the point here is that aggregation can conceal and contradict actual situations. To make informed decisions regarding what model to choose, we need to take into account its performance not only on the entire data, but also on individual slices. Slice-based evaluation can give you insights to improve your model's performance both overall and on critical data and help detect potential biases. It might also help reveal non-ML problems. Once, our team discovered that our model performed great overall but very poorly on traffic from mobile users. After investigating, we realized that it was because a button was half hidden on small screens (e.g., phone screens).

Even when you don't think slices matter, understanding how your model performs in a more fine-grained way can give you confidence in your model to convince other stakeholders, like your boss or your customers, to trust your ML models.

To track your model's performance on critical slices, you'd first need to know what your critical slices are. You might wonder how to discover critical slices in your data. Slicing is, unfortunately, still more of an art than a science, requiring intensive data exploration and analysis. Here are the three main approaches:

46 P. J. Bickel, E. A. Hammel, and J. W. O'Connell, "Sex Bias in Graduate Admissions: Data from Berkeley," *Science* 187 (1975): 398–404, *https://oreil.ly/TeR7E*.

Heuristics-based

Slice your data using domain knowledge you have of the data and the task at hand. For example, when working with web traffic, you might want to slice your data along dimensions like mobile versus desktop, browser type, and locations. Mobile users might behave very differently from desktop users. Similarly, internet users in different geographic locations might have different expectations on what a website should look like.[47]

Error analysis

Manually go through misclassified examples and find patterns among them. We discovered our model's problem with mobile users when we saw that most of the misclassified examples were from mobile users.

Slice finder

There has been research to systemize the process of finding slices, including Chung et al.'s "Slice Finder: Automated Data Slicing for Model Validation" (*https://oreil.ly/eypmq*) in 2019 and covered in Sumyea Helal's "Subgroup Discovery Algorithms: A Survey and Empirical Evaluation" (*https://oreil.ly/7yBJO*) (2016). The process generally starts with generating slice candidates with algorithms such as beam search, clustering, or decision, then pruning out clearly bad candidates for slices, and then ranking the candidates that are left.

Keep in mind that once you have discovered these critical slices, you will need sufficient, correctly labeled data for each of these slices for evaluation. The quality of your evaluation is only as good as the quality of your evaluation data.

Summary

In this chapter, we've covered the ML algorithm part of ML systems, which many ML practitioners consider to be the most fun part of an ML project lifecycle. With the initial models, we can bring to life (in the form of predictions) all our hard work in data and feature engineering, and can finally evaluate our hypothesis (i.e., we can predict the outputs given the inputs).

We started with how to select ML models best suited for our tasks. Instead of going into pros and cons of each individual model architecture—which is a fool's errand given the growing pools of existing models—the chapter outlined the aspects you need to consider to make an informed decision on which model is best for your objectives, constraints, and requirements.

47 For readers interested in learning more about UX design across cultures, Jenny Shen has a great post (*https://oreil.ly/MAJVB*).

We then continued to cover different aspects of model development. We covered not only individual models but also ensembles of models, a technique widely used in competitions and leaderboard-style research.

During the model development phase, you might experiment with many different models. Intensive tracking and versioning of your many experiments are generally agreed to be important, but many ML engineers still skip it because doing it might feel like a chore. Therefore, having tools and appropriate infrastructure to automate the tracking and versioning process is essential. We'll cover tools and infrastructure for ML production in Chapter 10.

As models today are getting bigger and consuming more data, distributed training is becoming an essential skill for ML model developers, and we discussed techniques for parallelism including data parallelism, model parallelism, and pipeline parallelism. Making your models work on a large distributed system, like the one that runs models with hundreds of millions, if not billions, of parameters, can be challenging and require specialized system engineering expertise.

We ended the chapter with how to evaluate your models to pick the best one to deploy. Evaluation metrics don't mean much unless you have a baseline to compare them to, and we covered different types of baselines you might want to consider for evaluation. We also covered a range of evaluation techniques necessary to sanity check your models before further evaluating your models in a production environment.

Often, no matter how good your offline evaluation of a model is, you still can't be sure of your model's performance in production until that model has been deployed. In the next chapter, we'll go over how to deploy a model.

CHAPTER 7
Model Deployment and Prediction Service

In Chapters 4 through 6, we have discussed the considerations for developing an ML model, from creating training data, extracting features, and developing the model to crafting metrics to evaluate this model. These considerations constitute the logic of the model—instructions on how to go from raw data into an ML model, as shown in Figure 7-1. Developing this logic requires both ML knowledge and subject matter expertise. In many companies, this is the part of the process that is done by the ML or data science teams.

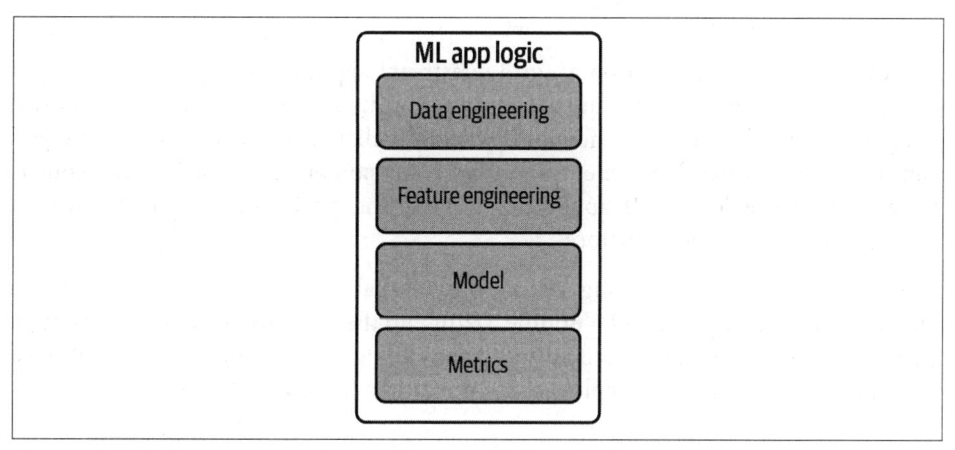

Figure 7-1. Different aspects that make up the ML model logic

In this chapter, we'll discuss another part in the iterative process: deploying your model. "Deploy" is a loose term that generally means making your model running and accessible. During model development, your model usually runs in a development environment.[1] To be deployed, your model will have to leave the development environment. Your model can be deployed to a staging environment for testing or to a production environment to be used by your end users. In this chapter, we focus on deploying models to production environments.

Before we move forward, I want to emphasize that production is a spectrum. For some teams, production means generating nice plots in notebooks to show to the business team. For other teams, production means keeping your models up and running for millions of users a day. If your work is in the first scenario, your production environment is similar to the development environment, and this chapter is less relevant for you. If your work is closer to the second scenario, read on.

I once read somewhere on the internet: deploying is easy if you ignore all the hard parts. If you want to deploy a model for your friends to play with, all you have to do is to wrap your predict function in a POST request endpoint using Flask or FastAPI, put the dependencies this predict function needs to run in a container,[2] and push your model and its associated container to a cloud service like AWS or GCP to expose the endpoint:

```
# Example of how to use FastAPI to turn your predict function
# into a POST endpoint
@app.route('/predict', methods=['POST'])
def predict():
    X = request.get_json()['X']
    y = MODEL.predict(X).tolist()
    return json.dumps({'y': y}), 200
```

You can use this exposed endpoint for downstream applications: e.g., when an application receives a prediction request from a user, this request is sent to the exposed endpoint, which returns a prediction. If you're familiar with the necessary tools, you can have a functional deployment in an hour. My students, after a 10-week course, were all able to deploy an ML application as their final projects even though few have had deployment experience before.[3]

The hard parts include making your model available to millions of users with a latency of milliseconds and 99% uptime, setting up the infrastructure so that the right person can be immediately notified when something goes wrong, figuring out what went wrong, and seamlessly deploying the updates to fix what's wrong.

1 We'll cover development environments in detail in Chapter 10.

2 We'll go more into containers in Chapter 9.

3 CS 329S: Machine Learning Systems Design (*https://oreil.ly/A6lFT*) at Stanford; you can see the project demos on YouTube (*https://oreil.ly/q4pjX*).

In many companies, the responsibility of deploying models falls into the hands of the same people who developed those models. In many other companies, once a model is ready to be deployed, it will be exported and handed off to another team to deploy it. However, this separation of responsibilities can cause high overhead communications across teams and make it slow to update your model. It also can make it hard to debug should something go wrong. We'll discuss more on team structures in Chapter 11.

Exporting a model means converting this model into a format that can be used by another application. Some people call this process "serialization."[4] There are two parts of a model that you can export: the model definition and the model's parameter values. The model definition defines the structure of your model, such as how many hidden layers it has and how many units in each layer. The parameter values provide the values for these units and layers. Usually, these two parts are exported together.

In TensorFlow 2, you might use `tf.keras.Model.save()` to export your model into TensorFlow's SavedModel format. In PyTorch, you might use `torch.onnx.export()` to export your model into ONNX format.

Regardless of whether your job involves deploying ML models, being cognizant of how your models are used can give you an understanding of their constraints and help you tailor them to their purposes.

In this chapter, we'll start off with some common myths about ML deployment that I've often heard from people who haven't deployed ML models. We'll then discuss the two main ways a model generates and serves its predictions to users: online prediction and batch prediction. The process of generating predictions is called *inference*.

We'll continue with where the computation for generating predictions should be done: on the device (also referred to as the edge) and the cloud. How a model serves and computes the predictions influences how it should be designed, the infrastructure it requires, and the behaviors that users encounter.

4 See the discussion on "data serialization" in the section "Data Formats" on page 53.

If you come from an academic background, some of the topics discussed in this chapter might be outside your comfort zone. If an unfamiliar term comes up, take a moment to look it up. If a section becomes too dense, feel free to skip it. This chapter is modular, so skipping a section shouldn't affect your understanding of another section.

Machine Learning Deployment Myths

As discussed in Chapter 1, deploying an ML model can be very different from deploying a traditional software program. This difference might cause people who have never deployed a model before to either dread the process or underestimate how much time and effort it will take. In this section, we'll debunk some of the common myths about the deployment process, which will, hopefully, put you in a good state of mind to begin the process. This section will be most helpful to people with little to no deploying experience.

Myth 1: You Only Deploy One or Two ML Models at a Time

When doing academic projects, I was advised to choose a small problem to focus on, which usually led to a single model. Many people from academic backgrounds I've talked to tend to also think of ML production in the context of a single model. Subsequently, the infrastructure they have in mind doesn't work for actual applications, because it can only support one or two models.

In reality, companies have many, many ML models. An application might have many different features, and each feature might require its own model. Consider a ride-sharing app like Uber. It needs a model to predict each of the following elements: ride demand, driver availability, estimated time of arrival, dynamic pricing, fraudulent transaction, customer churn, and more. Additionally, if this app operates in 20 countries, until you can have models that generalize across different user-profiles, cultures, and languages, each country would need its own set of models. So with 20 countries and 10 models for each country, you already have 200 models. Figure 7-2 shows a wide range of the tasks that leverage ML at Netflix.

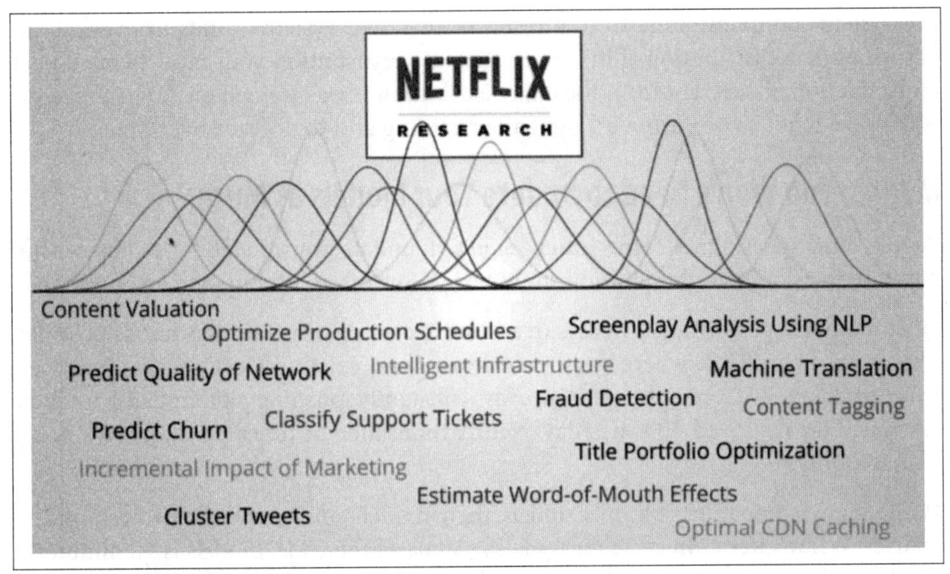

Figure 7-2. Different tasks that leverage ML at Netflix. Source: Ville Tuulos[5]

In fact, Uber has thousands of models in production.[6] At any given moment, Google has thousands of models training concurrently with hundreds of billions parameters in size.[7] Booking.com has 150+ models.[8] A 2021 study by Algorithmia shows that among organizations with over 25,000 employees, 41% have more than 100 models in production.[9]

Myth 2: If We Don't Do Anything, Model Performance Remains the Same

Software doesn't age like fine wine. It ages poorly. The phenomenon in which a software program degrades over time even if nothing seems to have changed is known as "software rot" or "bit rot."

5 Ville Tuulos, "Human-Centric Machine Learning Infrastructure @Netflix," InfoQ, 2018, video, 49:11, *https://oreil.ly/j4Hfx*.

6 Wayne Cunningham, "Science at Uber: Powering Machine Learning at Uber," *Uber Engineering Blog*, September 10, 2019, *https://oreil.ly/WfaCF*.

7 Daniel Papasian and Todd Underwood, "OpML '20—How ML Breaks: A Decade of Outages for One Large ML Pipeline," Google, 2020, video, 19:06, *https://oreil.ly/HjQm0*.

8 Lucas Bernardi, Themistoklis Mavridis, and Pablo Estevez, "150 Successful Machine Learning Models: 6 Lessons Learned at Booking.com," *KDD '19: Proceedings of the 25th ACM SIGKDD International Conference on Knowledge Discovery & Data Mining* (July 2019): 1743–51, *https://oreil.ly/Ea1Ke*.

9 "2021 Enterprise Trends in Machine Learning," Algorithmia, *https://oreil.ly/9kdcw*.

ML systems aren't immune to it. On top of that, ML systems suffer from what are known as data distribution shifts, when the data distribution your model encounters in production is different from the data distribution it was trained on.[10] Therefore, an ML model tends to perform best right after training and to degrade over time.

Myth 3: You Won't Need to Update Your Models as Much

People tend to ask me: "How often *should* I update my models?" It's the wrong question to ask. The right question should be: "How often *can* I update my models?"

Since a model's performance decays over time, we want to update it as fast as possible. This is an area of ML where we should learn from existing DevOps best practices. Even back in 2015, people were already constantly pushing out updates to their systems. Etsy deployed 50 times/day, Netflix thousands of times per day, AWS every 11.7 seconds.[11]

While many companies still only update their models once a month, or even once a quarter, Weibo's iteration cycle for updating some of their ML models is 10 minutes.[12] I've heard similar numbers at companies like Alibaba and ByteDance (the company behind TikTok).

In the words of Josh Wills, a former staff engineer at Google and director of data engineering at Slack, "We're always trying to bring new models into production just as fast as humanly possible."[13]

We'll discuss more on the frequency to retrain your models in Chapter 9.

Myth 4: Most ML Engineers Don't Need to Worry About Scale

What "scale" means varies from application to application, but examples include a system that serves hundreds of queries per second or millions of users a month.

You might argue that, if so, only a small number of companies need to worry about it. There is only one Google, one Facebook, one Amazon. That's true, but a small number of large companies employ the majority of the software engineering workforce. According to the Stack Overflow Developer Survey 2019, more than half of the respondents worked for a company of at least 100 employees (see Figure 7-3). This isn't a perfect correlation, but a company of 100 employees has a good chance of serving a reasonable number of users.

---

10 We'll discuss data distribution shifts further in Chapter 8.

11 Christopher Null, "10 Companies Killing It at DevOps," *TechBeacon*, 2015, *https://oreil.ly/JvNwu*.

12 Qian Yu, "Machine Learning with Flink in Weibo," QCon 2019, video, 17:57, *https://oreil.ly/RcTMv*.

13 Josh Wills, "Instrumentation, Observability and Monitoring of Machine Learning Models," InfoQ 2019, *https://oreil.ly/5Ot5m*.

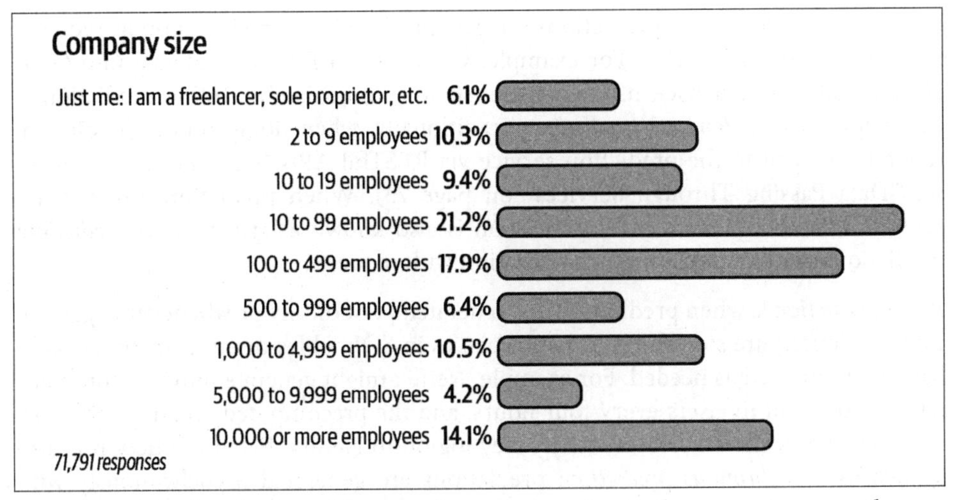

Figure 7-3. The distribution of the size of companies where software engineers work. Source: Adapted from an image by Stack Overflow[14]

I couldn't find a survey for ML-specific roles, so I asked on Twitter (*https://oreil.ly/e1fjn*) and found similar results. This means that if you're looking for an ML-related job in the industry, you'll likely work for a company of at least 100 employees, whose ML applications likely need to be scalable. Statistically speaking, an ML engineer should care about scale.

Batch Prediction Versus Online Prediction

One fundamental decision you'll have to make that will affect both your end users and developers working on your system is how it generates and serves its predictions to end users: online or batch. The terminologies surrounding batch and online prediction are still quite confusing due to the lack of standardized practices in the industry. I'll do my best to explain the nuances of each term in this section. If you find any of the terms mentioned here too confusing, feel free to ignore them for now. If you forget everything else, there are three main modes of prediction that I hope you'll remember:

- Batch prediction, which uses only batch features.
- Online prediction that uses only batch features (e.g., precomputed embeddings).
- Online prediction that uses both batch features and streaming features. This is also known as streaming prediction.

14 "Developer Survey Results," Stack Overflow, 2019, *https://oreil.ly/guYIq*.

Online prediction is when predictions are generated and returned as soon as requests for these predictions arrive. For example, you enter an English sentence into Google Translate and get back its French translation immediately. Online prediction is also known as *on-demand prediction*. Traditionally, when doing online prediction, requests are sent to the prediction service via RESTful APIs (e.g., HTTP requests—see "Data Passing Through Services" on page 73). When prediction requests are sent via HTTP requests, online prediction is also known as *synchronous prediction*: predictions are generated in synchronization with requests.

Batch prediction is when predictions are generated periodically or whenever triggered. The predictions are stored somewhere, such as in SQL tables or an in-memory database, and retrieved as needed. For example, Netflix might generate movie recommendations for all of its users every four hours, and the precomputed recommendations are fetched and shown to users when they log on to Netflix. Batch prediction is also known as *asynchronous prediction*: predictions are generated asynchronously with requests.

Terminology Confusion

The terms "online prediction" and "batch prediction" can be confusing. Both can make predictions for multiple samples (in batch) or one sample at a time. To avoid this confusion, people sometimes prefer the terms "synchronous prediction" and "asynchronous prediction." However, this distinction isn't perfect either, because when online prediction leverages a real-time transport to send prediction requests to your model, the requests and predictions technically are asynchronous.

Figure 7-4 shows a simplified architecture for batch prediction, and Figure 7-5 shows a simplified version of online prediction using only batch features. We'll go over what it means to use only batch features next.

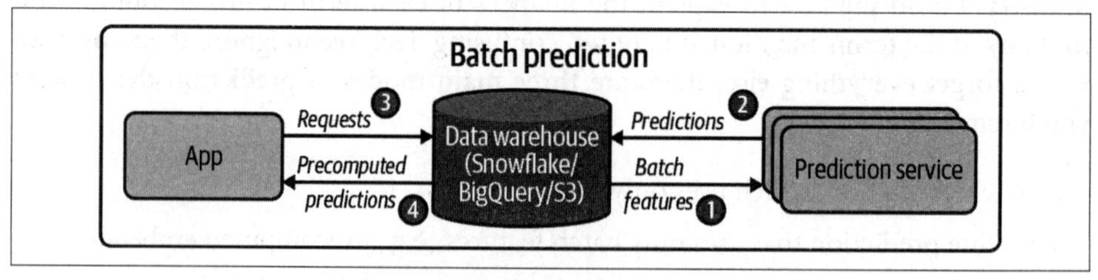

Figure 7-4. A simplified architecture for batch prediction

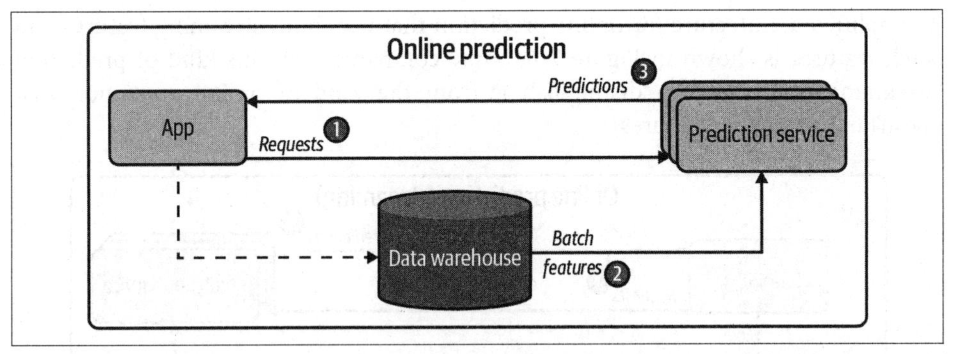

Figure 7-5. A simplified architecture for online prediction that uses only batch features

As discussed in Chapter 3, features computed from historical data, such as data in databases and data warehouses, are *batch features*. Features computed from streaming data—data in real-time transports—are *streaming features*. In batch prediction, only batch features are used. In online prediction, however, it's possible to use both batch features and streaming features. For example, after a user puts in an order on DoorDash, they might need the following features to estimate the delivery time:

Batch features
: The mean preparation time of this restaurant in the past

Streaming features
: In the last 10 minutes, how many other orders they have, and how many delivery people are available

Streaming Features Versus Online Features

I've heard the terms "streaming features" and "online features" used interchangeably. They are actually different. Online features are more general, as they refer to any feature used for online prediction, including batch features stored in memory.

A very common type of batch feature used for online prediction, especially session-based recommendations, is item embeddings. Item embeddings are usually precomputed in batch and fetched whenever they are needed for online prediction. In this case, embeddings can be considered online features but not streaming features.

Streaming features refer exclusively to features computed from streaming data.

A simplified architecture for online prediction that uses both streaming features and batch features is shown in Figure 7-6. Some companies call this kind of prediction "streaming prediction" to distinguish it from the kind of online prediction that doesn't use streaming features.

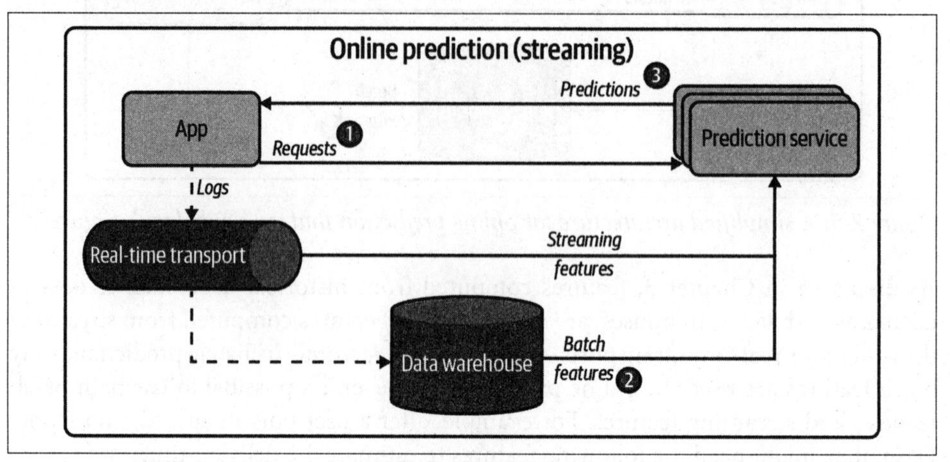

Figure 7-6. A simplified architecture for online prediction that uses both batch features and streaming features

However, online prediction and batch prediction don't have to be mutually exclusive. One hybrid solution is that you precompute predictions for popular queries, then generate predictions online for less popular queries. Table 7-1 summarizes the key points to consider for online prediction and batch prediction.

Table 7-1. Some key differences between batch prediction and online prediction

	Batch prediction (asynchronous)	Online prediction (synchronous)
Frequency	Periodical, such as every four hours	As soon as requests come
Useful for	Processing accumulated data when you don't need immediate results (such as recommender systems)	When predictions are needed as soon as a data sample is generated (such as fraud detection)
Optimized for	High throughput	Low latency

In many applications, online prediction and batch prediction are used side by side for different use cases. For example, food ordering apps like DoorDash and UberEats use batch prediction to generate restaurant recommendations—it'd take too long to generate these recommendations online because there are many restaurants. However, once you click on a restaurant, food item recommendations are generated using online prediction.

Many people believe that online prediction is less efficient, both in terms of cost and performance, than batch prediction because you might not be able to batch inputs together and leverage vectorization or other optimization techniques. This is not necessarily true, as we already discussed in the section "Batch Processing Versus Stream Processing" on page 78.

Also, with online prediction, you don't have to generate predictions for users who aren't visiting your site. Imagine you run an app where only 2% of your users log in daily—e.g., in 2020, Grubhub had 31 million users and 622,000 daily orders.[15] If you generate predictions for every user each day, the compute used to generate 98% of your predictions will be wasted.

From Batch Prediction to Online Prediction

To people coming to ML from an academic background, the more natural way to serve predictions is probably online. You give your model an input and it generates a prediction as soon as it receives that input. This is likely how most people interact with their models while prototyping. This is also likely easier to do for most companies when first deploying a model. You export your model, upload the exported model to Amazon SageMaker or Google App Engine, and get back an exposed endpoint.[16] Now, if you send a request that contains an input to that endpoint, it will send back a prediction generated on that input.

A problem with online prediction is that your model might take too long to generate predictions. Instead of generating predictions as soon as they arrive, what if you compute predictions in advance and store them in your database, and fetch them when requests arrive? This is exactly what batch prediction does. With this approach, you can generate predictions for multiple inputs at once, leveraging distributed techniques to process a high volume of samples efficiently.

15 David Curry, "Grubhub Revenue and Usage Statistics (2022)," Business of Apps, January 11, 2022, *https://oreil.ly/jX43M*; "Average Number of Grubhub Orders per Day Worldwide from 2011 to 2020," Statista, *https://oreil.ly/Tu9fm*.

16 The URL of the entry point for a service, which, in this case, is the prediction service of your ML model.

Because the predictions are precomputed, you don't have to worry about how long it'll take your models to generate predictions. For this reason, batch prediction can also be seen as a trick to reduce the inference latency of more complex models—the time it takes to retrieve a prediction is usually less than the time it takes to generate it.

Batch prediction is good for when you want to generate a lot of predictions and don't need the results immediately. You don't have to use all the predictions generated. For example, you can make predictions for all customers on how likely they are to buy a new product, and reach out to the top 10%.

However, the problem with batch prediction is that it makes your model less responsive to users' change preferences. This limitation can be seen even in more technologically progressive companies like Netflix. Say you've been watching a lot of horror movies lately, so when you first log in to Netflix, horror movies dominate recommendations. But you're feeling bright today, so you search "comedy" and start browsing the comedy category. Netflix should learn and show you more comedy in your list of their recommendations, right? As of writing this book, it can't update the list until the next batch of recommendations is generated, but I have no doubt that this limitation will be addressed in the near future.

Another problem with batch prediction is that you need to know what requests to generate predictions for in advance. In the case of recommending movies for users, you know in advance how many users to generate recommendations for.[17] However, for cases when you have unpredictable queries—if you have a system to translate from English to French, it might be impossible to anticipate every possible English text to be translated—you need to use online prediction to generate predictions as requests arrive.

In the Netflix example, batch prediction causes mild inconvenience (which is tightly coupled with user engagement and retention), not catastrophic failures. There are many applications where batch prediction would lead to catastrophic failures or just wouldn't work. Examples where online prediction is crucial include high-frequency trading, autonomous vehicles, voice assistants, unlocking your phone using face or fingerprints, fall detection for elderly care, and fraud detection. Being able to detect a fraudulent transaction that happened three hours ago is still better than not detecting it at all, but being able to detect it in real time can prevent the fraudulent transaction from going through.

17 If a new user joins, you can give them some generic recommendations.

Batch prediction is a workaround for when online prediction isn't cheap enough or isn't fast enough. Why generate one million predictions in advance and worry about storing and retrieving them if you can generate each prediction as needed at the exact same cost and same speed?

As hardware becomes more customized and powerful and better techniques are being developed to allow faster, cheaper online predictions, online prediction might become the default.

In recent years, companies have made significant investments to move from batch prediction to online prediction. To overcome the latency challenge of online prediction, two components are required:

- A (near) real-time pipeline that can work with incoming data, extract streaming features (if needed), input them into a model, and return a prediction in near real time. A streaming pipeline with real-time transport and a stream computation engine can help with that.
- A model that can generate predictions at a speed acceptable to its end users. For most consumer apps, this means milliseconds.

We've discussed stream processing in Chapter 3. We'll continue discussing the unification of the stream pipeline with the batch pipeline in the next section. Then we'll discuss how to speed up inference in the section "Model optimization" on page 216.

Unifying Batch Pipeline and Streaming Pipeline

Batch prediction is largely a product of legacy systems. In the last decade, big data processing has been dominated by batch systems like MapReduce and Spark, which allow us to periodically process a large amount of data very efficiently. When companies started with ML, they leveraged their existing batch systems to make predictions. When these companies want to use streaming features for their online prediction, they need to build a separate streaming pipeline. Let's go through an example to make this more concrete.

Imagine you want to build a model to predict arrival time for an application like Google Maps. The prediction is continually updated as a user's trip progresses. A feature you might want to use is the average speed of all the cars in your path in the last five minutes. For training, you might use data from the last month. To extract this feature from your training data, you might want to put all your data into a dataframe to compute this feature for multiple training samples at the same time. During inference, this feature will be continually computed on a sliding window. This means that in training this feature is computed in batch, whereas during inference this feature is computed in a streaming process.

Having two different pipelines to process your data is a common cause for bugs in ML production. One cause for bugs is when the changes in one pipeline aren't correctly replicated in the other, leading to two pipelines extracting two different sets of features. This is especially common if the two pipelines are maintained by two different teams, such as the ML team maintains the batch pipeline for training while the deployment team maintains the stream pipeline for inference, as shown in Figure 7-7.

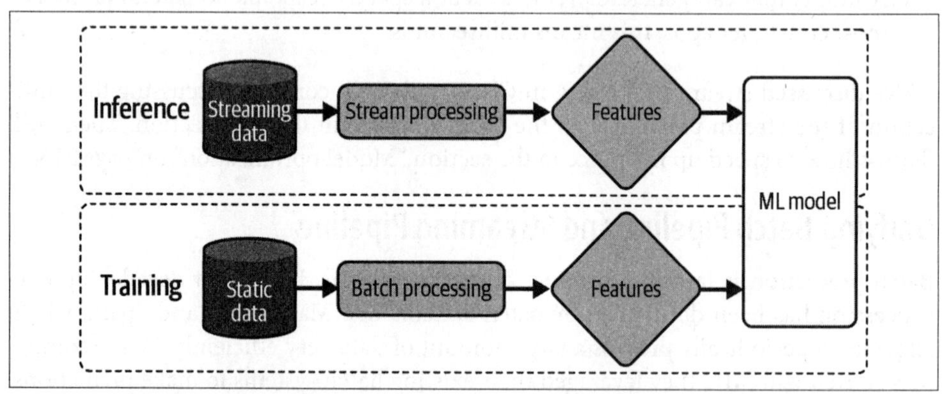

Figure 7-7. Having two different pipelines for training and inference is a common source for bugs for ML in production

Figure 7-8 shows a more detailed but also more complex feature of the data pipeline for ML systems that do online prediction. The boxed element labeled Research is what people are often exposed to in an academic environment.

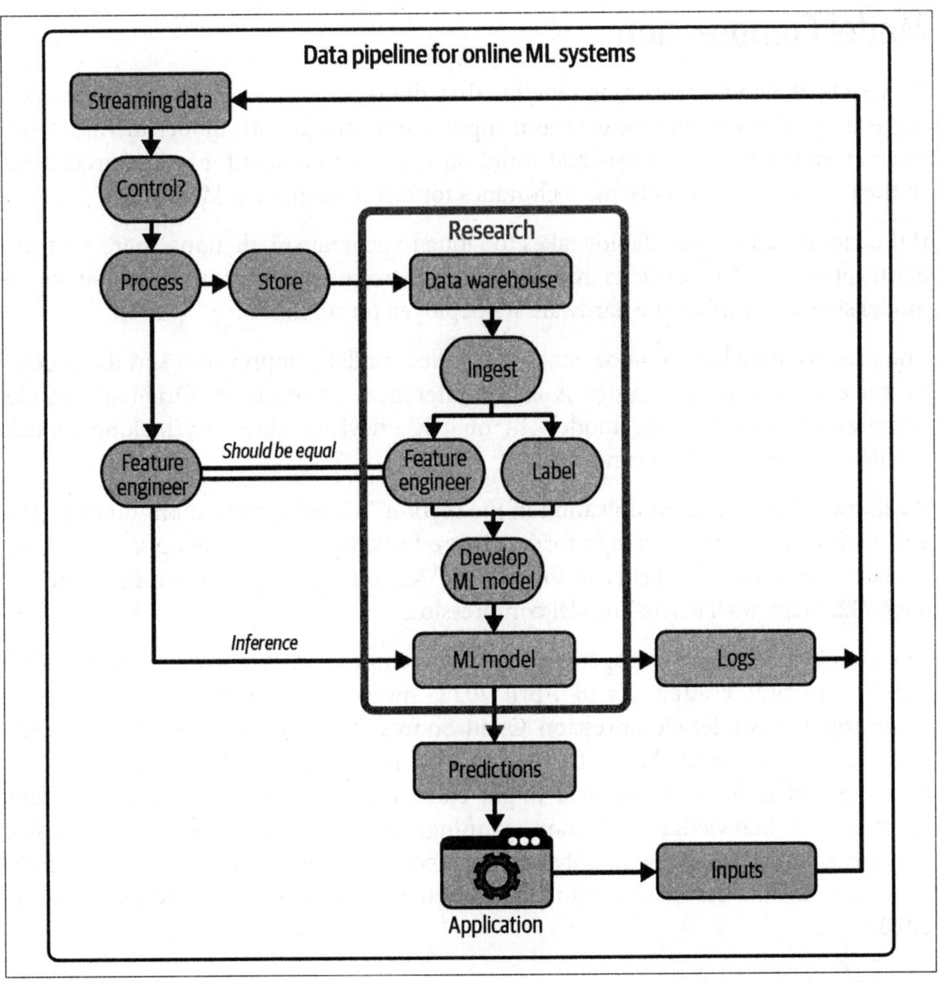

Figure 7-8. A data pipeline for ML systems that do online prediction

Building infrastructure to unify stream processing and batch processing has become a popular topic in recent years for the ML community. Companies including Uber and Weibo have made major infrastructure overhauls to unify their batch and stream processing pipelines by using a stream processor like Apache Flink.[18] Some companies use feature stores to ensure the consistency between the batch features used during training and the streaming features used in prediction. We'll discuss feature stores in Chapter 10.

18 Shuyi Chean and Fabian Hueske, "Streaming SQL to Unify Batch & Stream Processing w/ Apache Flink @Uber," *InfoQ*, *https://oreil.ly/XoaNu*; Yu, "Machine Learning with Flink in Weibo."

Model Compression

We've talked about a streaming pipeline that allows an ML system to extract streaming features from incoming data and input them into an ML model in (near) real time. However, having a near (real-time) pipeline isn't enough for online prediction. In the next section, we'll discuss techniques for fast inference for ML models.

If the model you want to deploy takes too long to generate predictions, there are three main approaches to reduce its inference latency: make it do inference faster, make the model smaller, or make the hardware it's deployed on run faster.

The process of making a model smaller is called model compression, and the process to make it do inference faster is called inference optimization. Originally, model compression was to make models fit on edge devices. However, making models smaller often makes them run faster.

We'll discuss inference optimization in the section "Model optimization" on page 216, and we'll discuss the landscape for hardware backends being developed specifically for running ML models faster in the section "ML on the Cloud and on the Edge" on page 212. Here, we'll discuss model compression.

The number of research papers on model compression is growing. Off-the-shelf utilities are proliferating. As of April 2022, Awesome Open Source has a list of "The Top 168 Model Compression Open Source Projects" (*https://oreil.ly/CYm82*), and that list is growing. While there are many new techniques being developed, the four types of techniques that you might come across the most often are low-rank optimization, knowledge distillation, pruning, and quantization. Readers interested in a comprehensive review might want to check out Cheng et al.'s "Survey of Model Compression and Acceleration for Deep Neural Networks," which was updated in 2020.[19]

Low-Rank Factorization

The key idea behind *low-rank factorization* is to replace high-dimensional tensors with lower-dimensional tensors.[20] One type of low-rank factorization is *compact convolutional filters*, where the over-parameterized (having too many parameters) convolution filters are replaced with compact blocks to both reduce the number of parameters and increase speed.

19 Yu Cheng, Duo Wang, Pan Zhou, and Tao Zhang, "A Survey of Model Compression and Acceleration for Deep Neural Networks," *arXiv*, June 14, 2020, *https://oreil.ly/1eMho*.

20 Max Jaderberg, Andrea Vedaldi, and Andrew Zisserman, "Speeding up Convolutional Neural Networks with Low Rank Expansions," *arXiv*, May 15, 2014, *https://oreil.ly/4Vf4s*.

For example, by using a number of strategies including replacing 3 × 3 convolution with 1 × 1 convolution, SqueezeNets achieves AlexNet-level accuracy on ImageNet with 50 times fewer parameters.[21]

Similarly, MobileNets decomposes the standard convolution of size $K \times K \times C$ into a depthwise convolution ($K \times K \times 1$) and a pointwise convolution ($1 \times 1 \times C$), with K being the kernel size and C being the number of channels. This means that each new convolution uses only $K^2 + C$ instead of K^2C parameters. If $K = 3$, this means an eight to nine times reduction in the number of parameters (see Figure 7-9).[22]

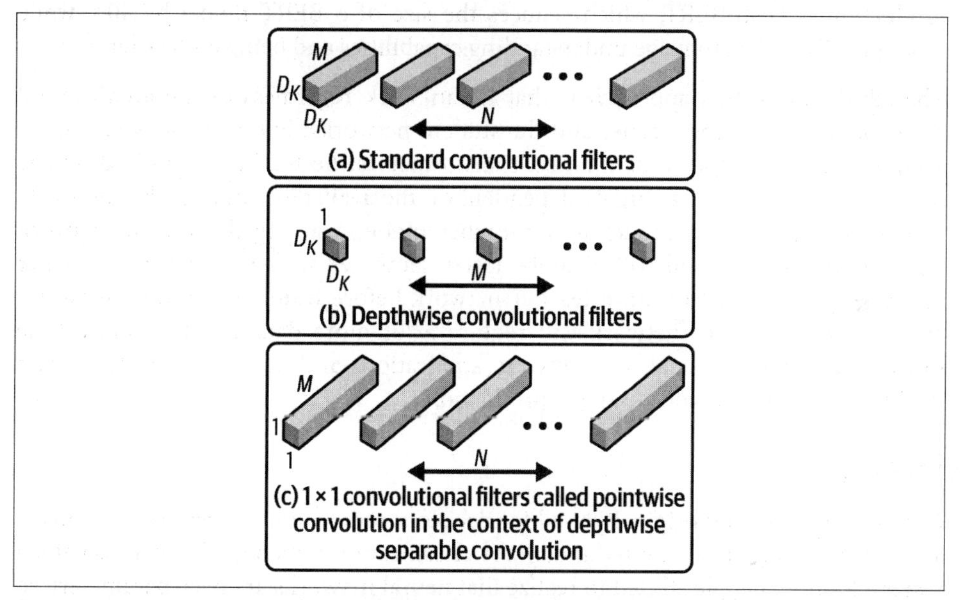

Figure 7-9. Compact convolutional filters in MobileNets. The standard convolutional filters in (a) are replaced by depthwise convolution in (b) and pointwise convolution in (c) to build a depthwise separable filter. Source: Adapted from an image by Howard et al.

This method has been used to develop smaller models with significant acceleration compared to standard models. However, it tends to be specific to certain types of models (e.g., compact convolutional filters are specific to convolutional neural

21 Forrest N. Iandola, Song Han, Matthew W. Moskewicz, Khalid Ashraf, William J. Dally, and Kurt Keutzer, "SqueezeNet: AlexNet-Level Accuracy with 50x Fewer Parameters and <0.5MB Model Size," *arXiv*, November 4, 2016, *https://oreil.ly/xs3mi*.

22 Andrew G. Howard, Menglong Zhu, Bo Chen, Dmitry Kalenichenko, Weijun Wang, Tobias Weyand, Marco Andreetto, and Hartwig Adam, "MobileNets: Efficient Convolutional Neural Networks for Mobile Vision Applications," *arXiv*, April 17, 2017, *https://oreil.ly/T84fD*.

networks) and requires a lot of architectural knowledge to design, so it's not widely applicable to many use cases yet.

Knowledge Distillation

Knowledge distillation is a method in which a small model (student) is trained to mimic a larger model or ensemble of models (teacher). The smaller model is what you'll deploy. Even though the student is often trained after a pretrained teacher, both may also be trained at the same time.[23] One example of a distilled network used in production is DistilBERT, which reduces the size of a BERT model by 40% while retaining 97% of its language understanding capabilities and being 60% faster.[24]

The advantage of this approach is that it can work regardless of the architectural differences between the teacher and the student networks. For example, you can get a random forest as the student and a transformer as the teacher. The disadvantage of this approach is that it's highly dependent on the availability of a teacher network. If you use a pretrained model as the teacher model, training the student network will require less data and will likely be faster. However, if you don't have a teacher available, you'll have to train a teacher network before training a student network, and training a teacher network will require a lot more data and take more time to train. This method is also sensitive to applications and model architectures, and therefore hasn't found wide usage in production.

Pruning

Pruning was a method originally used for decision trees where you remove sections of a tree that are uncritical and redundant for classification.[25] As neural networks gained wider adoption, people started to realize that neural networks are over-parameterized and began to find ways to reduce the workload caused by the extra parameters.

Pruning, in the context of neural networks, has two meanings. One is to remove entire nodes of a neural network, which means changing its architecture and reducing its number of parameters. The more common meaning is to find parameters least useful to predictions and set them to 0. In this case, pruning doesn't reduce the total number of parameters, only the number of nonzero parameters. The architecture of the neural network remains the same. This helps with reducing the size of a model because pruning makes a neural network more sparse, and sparse architecture tends to require less storage space than dense structure. Experiments show that pruning

23 Geoffrey Hinton, Oriol Vinyals, and Jeff Dean, "Distilling the Knowledge in a Neural Network," *arXiv*, March 9, 2015, *https://oreil.ly/OJEPW*.

24 Victor Sanh, Lysandre Debut, Julien Chaumond, and Thomas Wolf, "DistilBERT, a Distilled Version of BERT: Smaller, Faster, Cheaper and Lighter," *arXiv*, October 2, 2019, *https://oreil.ly/mQWBv*.

25 Hence the name "pruning."

techniques can reduce the nonzero parameter counts of trained networks by over 90%, decreasing storage requirements and improving computational performance of inference without compromising overall accuracy.[26] In Chapter 11, we'll discuss how pruning can introduce biases into your model.

While it's generally agreed that pruning works,[27] there have been many discussions on the actual value of pruning. Liu et al. argued that the main value of pruning isn't in the inherited "important weights" but in the pruned architecture itself.[28] In some cases, pruning can be useful as an architecture search paradigm, and the pruned architecture should be retrained from scratch as a dense model. However, Zhu et al. showed that the large sparse model after pruning outperformed the retrained dense counterpart.[29]

Quantization

Quantization is the most general and commonly used model compression method. It's straightforward to do and generalizes over tasks and architectures.

Quantization reduces a model's size by using fewer bits to represent its parameters. By default, most software packages use 32 bits to represent a float number (single precision floating point). If a model has 100M parameters and each requires 32 bits to store, it'll take up 400 MB. If we use 16 bits to represent a number, we'll reduce the memory footprint by half. Using 16 bits to represent a float is called half precision.

Instead of using floats, you can have a model entirely in integers; each integer takes only 8 bits to represent. This method is also known as "fixed point." In the extreme case, some have attempted the 1-bit representation of each weight (binary weight neural networks), e.g., BinaryConnect and XNOR-Net.[30] The authors of the XNOR-Net paper spun off Xnor.ai, a startup that focused on model compression. In early 2020, it was acquired by Apple for a reported $200M.[31]

---

26 Jonathan Frankle and Michael Carbin, "The Lottery Ticket Hypothesis: Finding Sparse, Trainable Neural Networks," ICLR 2019, *https://oreil.ly/ychdl*.

27 Davis Blalock, Jose Javier Gonzalez Ortiz, Jonathan Frankle, and John Guttag, "What Is the State of Neural Network Pruning?" *arXiv*, March 6, 2020, *https://oreil.ly/VQsC3*.

28 Zhuang Liu, Mingjie Sun, Tinghui Zhou, Gao Huang, and Trevor Darrell, "Rethinking the Value of Network Pruning," *arXiv*, March 5, 2019, *https://oreil.ly/mB4IZ*.

29 Michael Zhu and Suyog Gupta, "To Prune, or Not to Prune: Exploring the Efficacy of Pruning for Model Compression," *arXiv*, November 13, 2017, *https://oreil.ly/KBRjy*.

30 Matthieu Courbariaux, Yoshua Bengio, and Jean-Pierre David, "BinaryConnect: Training Deep Neural Networks with Binary Weights During Propagations," *arXiv*, November 2, 2015, *https://oreil.ly/Fwp2G*; Mohammad Rastegari, Vicente Ordonez, Joseph Redmon, and Ali Farhadi, "XNOR-Net: ImageNet Classification Using Binary Convolutional Neural Networks," *arXiv*, August 2, 2016, *https://oreil.ly/gr3Ay*.

31 Alan Boyle, Taylor Soper, and Todd Bishop, "Exclusive: Apple Acquires Xnor.ai, Edge AI Spin-out from Paul Allen's AI2, for Price in $200M Range," *GeekWire*, January 15, 2020, *https://oreil.ly/HgaxC*.

Quantization not only reduces memory footprint but also improves the computation speed. First, it allows us to increase our batch size. Second, less precision speeds up computation, which further reduces training time and inference latency. Consider the addition of two numbers. If we perform the addition bit by bit, and each takes x nanoseconds, it'll take $32x$ nanoseconds for 32-bit numbers but only $16x$ nanoseconds for 16-bit numbers.

There are downsides to quantization. Reducing the number of bits to represent your numbers means that you can represent a smaller range of values. For values outside that range, you'll have to round them up and/or scale them to be in range. Rounding numbers leads to rounding errors, and small rounding errors can lead to big performance changes. You also run the risk of rounding/scaling your numbers to under-/overflow and rendering it to 0. Efficient rounding and scaling is nontrivial to implement at a low level, but luckily, major frameworks have this built in.

Quantization can either happen during training (quantization aware training),[32] where models are trained in lower precision, or post-training, where models are trained in single-precision floating point and then quantized for inference. Using quantization during training means that you can use less memory for each parameter, which allows you to train larger models on the same hardware.

Recently, low-precision training has become increasingly popular, with support from most modern training hardware. NVIDIA introduced Tensor Cores, processing units that support mixed-precision training.[33] Google TPUs (tensor processing units) also support training with Bfloat16 (16-bit Brain Floating Point Format), which the company dubbed "the secret to high performance on Cloud TPUs."[34] Training in fixed-point is not yet as popular but has had a lot of promising results.[35]

Fixed-point inference has become a standard in the industry. Some edge devices only support fixed-point inference. Most popular frameworks for on-device ML inference—Google's TensorFlow Lite, Facebook's PyTorch Mobile, NVIDIA's TensorRT—offer post-training quantization for free with a few lines of code.

---

32 As of October 2020, TensorFlow's quantization aware training doesn't actually train models with weights in lower bits, but collects statistics to use for post-training quantization.

33 Chip Huyen, Igor Gitman, Oleksii Kuchaiev, Boris Ginsburg, Vitaly Lavrukhin, Jason Li, Vahid Noroozi, and Ravi Gadde, "Mixed Precision Training for NLP and Speech Recognition with OpenSeq2Seq," *NVIDIA Devblogs*, October 9, 2018, *https://oreil.ly/WDT1l*. It's my post!

34 Shibo Wang and Pankaj Kanwar, "BFloat16: The Secret to High Performance on Cloud TPUs," *Google Cloud Blog*, August 23, 2019, *https://oreil.ly/ZG5p0*.

35 Itay Hubara, Matthieu Courbariaux, Daniel Soudry, Ran El-Yaniv, and Yoshua Bengio, "Quantized Neural Networks: Training Neural Networks with Low Precision Weights and Activations," *Journal of Machine Learning Research* 18 (2018): 1–30; Benoit Jacob, Skirmantas Kligys, Bo Chen, Menglong Zhu, Matthew Tang, Andrew Howard, Hartwig Adam, and Dmitry Kalenichenko, "Quantization and Training of Neural Networks for Efficient Integer-Arithmetic-Only Inference," *arXiv*, December 15, 2017, *https://oreil.ly/sUuMT*.

Case Study

To get a better understanding of how to optimize models in production, consider a fascinating case study from Roblox on how they scaled BERT to serve 1+ billion daily requests on CPUs.[36] For many of their NLP services, they needed to handle over 25,000 inferences per second at a latency of under 20 ms, as shown in Figure 7-10. They started with a large BERT model with fixed shape input, then replaced BERT with DistilBERT and fixed shape input with dynamic shape input, and finally quantized it.

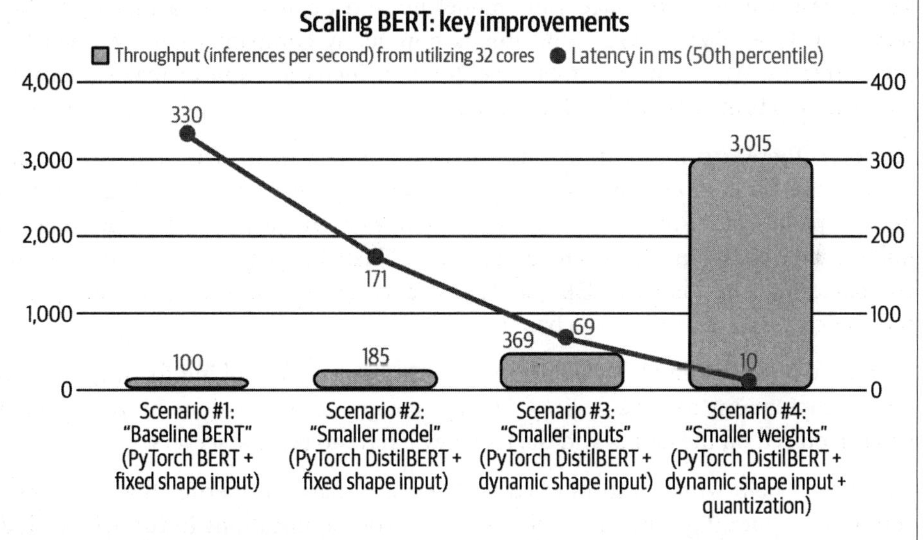

Figure 7-10. Latency improvement by various model compression methods. Source: Adapted from an image by Le and Kaehler

The biggest performance boost they got came from quantization. Converting 32-bit floating points to 8-bit integers reduces the latency 7 times and increases throughput 8 times.

The results here seem very promising to improve latency; however, they should be taken with a grain of salt since there's no mention of changes in output quality after each performance improvement.

36 Quoc Le and Kip Kaehler, "How We Scaled Bert To Serve 1+ Billion Daily Requests on CPUs," Roblox, May 27, 2020, *https://oreil.ly/U01Uj*.

ML on the Cloud and on the Edge

Another decision you'll want to consider is where your model's computation will happen: on the cloud or on the edge. On the cloud means a large chunk of computation is done on the cloud, either public clouds or private clouds. On the edge means a large chunk of computation is done on consumer devices—such as browsers, phones, laptops, smartwatches, cars, security cameras, robots, embedded devices, FPGAs (field programmable gate arrays), and ASICs (application-specific integrated circuits)—which are also known as edge devices.

The easiest way is to package your model up and deploy it via a managed cloud service such as AWS or GCP, and this is how many companies deploy when they get started in ML. Cloud services have done an incredible job to make it easy for companies to bring ML models into production.

However, there are many downsides to cloud deployment. The first is cost. ML models can be compute-intensive, and compute is expensive. Even back in 2018, big companies like Pinterest, Infor, and Intuit were already spending hundreds of millions of dollars on cloud bills every year.[37] That number for small and medium companies can be between $50K and $2M a year.[38] A mistake in handling cloud services can cause startups to go bankrupt.[39]

As their cloud bills climb, more and more companies are looking for ways to push their computations to edge devices. The more computation is done on the edge, the less is required on the cloud, and the less they'll have to pay for servers.

Other than help with controlling costs, there are many properties that make edge computing appealing. The first is that it allows your applications to run where cloud computing cannot. When your models are on public clouds, they rely on stable internet connections to send data to the cloud and back. Edge computing allows your models to work in situations where there are no internet connections or where the connections are unreliable, such as in rural areas or developing countries. I've worked with several companies and organizations that have strict no-internet policies, which means that whichever applications we wanted to sell them must not rely on internet connections.

---

37 Amir Efrati and Kevin McLaughlin, "As AWS Use Soars, Companies Surprised by Cloud Bills," *The Information*, February 25, 2019, *https://oreil.ly/H9ans*; Mats Bauer, "How Much Does Netflix Pay Amazon Web Services Each Month?" Quora, 2020, *https://oreil.ly/HtrBk*.

38 "2021 State of Cloud Cost Report," Anodot, *https://oreil.ly/5ZIJK*.

39 "Burnt $72K Testing Firebase and Cloud Run and Almost Went Bankrupt," Hacker News, December 10, 2020, *https://oreil.ly/vsHHC*; "How to Burn the Most Money with a Single Click in Azure," Hacker News, March 29, 2020, *https://oreil.ly/QvCiI*. We'll discuss in more detail how companies respond to high cloud bills in the section "Public Cloud Versus Private Data Centers" on page 300.

Second, when your models are already on consumers' devices, you can worry less about network latency. Requiring data transfer over the network (sending data to the model on the cloud to make predictions then sending predictions back to the users) might make some use cases impossible. In many cases, network latency is a bigger bottleneck than inference latency. For example, you might be able to reduce the inference latency of ResNet-50 from 30 ms to 20 ms, but the network latency can go up to seconds, depending on where you are and what services you're trying to use.

Putting your models on the edge is also appealing when handling sensitive user data. ML on the cloud means that your systems might have to send user data over networks, making it susceptible to being intercepted. Cloud computing also often means storing data of many users in the same place, which means a breach can affect many people. "Nearly 80% of companies experienced a cloud data breach in [the] past 18 months," according to *Security* magazine.[40]

Edge computing makes it easier to comply with regulations, like GDPR, about how user data can be transferred or stored. While edge computing might reduce privacy concerns, it doesn't eliminate them altogether. In some cases, edge computing might make it easier for attackers to steal user data, such as they can just take the device with them.

To move computation to the edge, the edge devices have to be powerful enough to handle the computation, have enough memory to store ML models and load them into memory, as well as have enough battery or be connected to an energy source to power the application for a reasonable amount of time. Running a full-sized BERT on your phone, if your phone is capable of running BERT, is a very quick way to kill its battery.

Because of the many benefits that edge computing has over cloud computing, companies are in a race to develop edge devices optimized for different ML use cases. Established companies including Google, Apple, and Tesla have all announced their plans to make their own chips. Meanwhile, ML hardware startups have raised billions of dollars to develop better AI chips.[41] It's projected that by 2025 the number of active edge devices worldwide will be over 30 billion.[42]

With so many new offerings for hardware to run ML models on, one question arises: how do we make our model run on arbitrary hardware efficiently? In the following section, we'll discuss how to compile and optimize a model to run it on a certain

40 "Nearly 80% of Companies Experienced a Cloud Data Breach in Past 18 Months," *Security*, June 5, 2020, *https://oreil.ly/gA1am*.

41 See slide #53, CS 329S's Lecture 8: Deployment - Prediction Service, 2022, *https://oreil.ly/cXTou*.

42 "Internet of Things (IoT) and Non-IoT Active Device Connections Worldwide from 2010 to 2025," Statista, https://oreil.ly/BChLN.

hardware backend. In the process, we'll introduce important concepts that you might encounter when handling models on the edge, including intermediate representations (IRs) and compilers.

Compiling and Optimizing Models for Edge Devices

For a model built with a certain framework, such as TensorFlow or PyTorch, to run on a hardware backend, that framework has to be supported by the hardware vendor. For example, even though TPUs were released publicly in February 2018, it wasn't until September 2020 that PyTorch was supported on TPUs. Before then, if you wanted to use a TPU, you'd have to use a framework that TPUs supported.

Providing support for a framework on a hardware backend is time-consuming and engineering-intensive. Mapping from ML workloads to a hardware backend requires understanding and taking advantage of that hardware's design, and different hardware backends have different memory layouts and compute primitives, as shown in Figure 7-11.

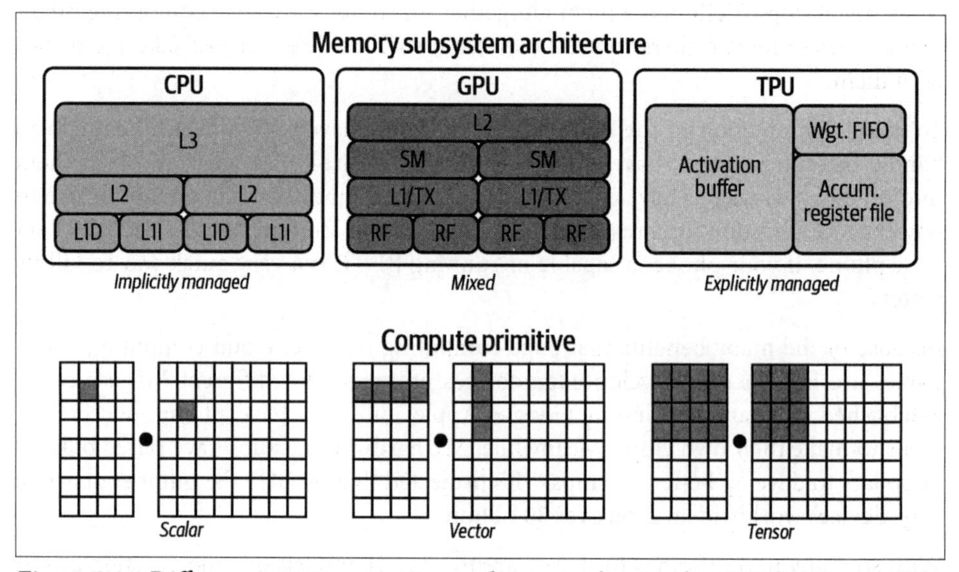

Figure 7-11. Different compute primitives and memory layouts for CPU, GPU, and TPU. Source: Adapted from an image by Chen et al.[43]

43 Tianqi Chen, Thierry Moreau, Ziheng Jiang, Lianmin Zheng, Eddie Yan, Meghan Cowan, Haichen Shen, et al., "TVM: An Automated End-to-End Optimizing Compiler for Deep Learning," *arXiv*, February 12, 2018, *https://oreil.ly/vGnkW*.

For example, the compute primitive of CPUs used to be a number (scalar) and the compute primitive of GPUs used to be a one-dimensional vector, whereas the compute primitive of TPUs is a two-dimensional vector (tensor).[44] Performing a convolution operator will be very different with one-dimensional vectors compared to two-dimensional vectors. Similarly, you'd need to take into account different L1, L2, and L3 layouts and buffer sizes to use them efficiently.

Because of this challenge, framework developers tend to focus on providing support to only a handful of server-class hardware, and hardware vendors tend to offer their own kernel libraries for a narrow range of frameworks. Deploying ML models to new hardware requires significant manual effort.

Instead of targeting new compilers and libraries for every new hardware backend, what if we create a middleman to bridge frameworks and platforms? Framework developers will no longer have to support every type of hardware; they will only need to translate their framework code into this middleman. Hardware vendors can then support one middleman instead of multiple frameworks.

This type of "middleman" is called an intermediate representation (IR). IRs lie at the core of how compilers work. From the original code for a model, compilers generate a series of high- and low-level IRs before generating the code native to a hardware backend so that it can run on that hardware backend, as shown in Figure 7-12.

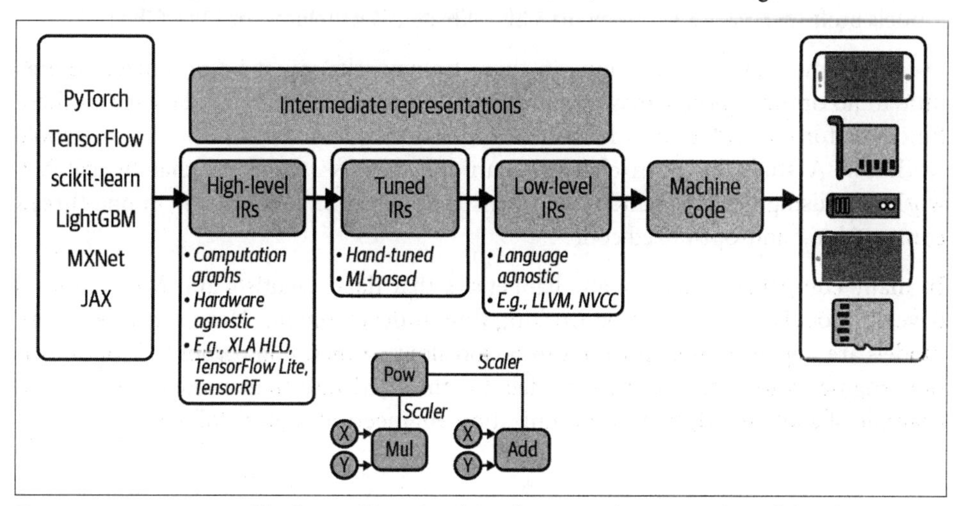

Figure 7-12. A series of high- and low-level IRs between the original model code to machine code that can run on a given hardware backend

44 Nowadays, many CPUs have vector instructions and some GPUs have tensor cores, which are two-dimensional.

This process is also called *lowering*, as in you "lower" your high-level framework code into low-level hardware-native code. It's not translating because there's no one-to-one mapping between them.

High-level IRs are usually computation graphs of your ML models. A computation graph is a graph that describes the order in which your computation is executed. Readers interested can read about computation graphs in PyTorch (*https://oreil.ly/who8P*) and TensorFlow (*https://oreil.ly/O8qR9*).

Model optimization

After you've "lowered" your code to run your models into the hardware of your choice, an issue you might run into is performance. The generated machine code might be able to run on a hardware backend, but it might not be able to do so efficiently. The generated code may not take advantage of data locality and hardware caches, or it may not leverage advanced features such as vector or parallel operations that could speed code up.

A typical ML workflow consists of many frameworks and libraries. For example, you might use pandas/dask/ray to extract features from your data. You might use NumPy to perform vectorization. You might use a pretrained model like Hugging Face's Transformers to generate features, then make predictions using an ensemble of models built with various frameworks like sklearn, TensorFlow, or LightGBM.

Even though individual functions in these frameworks might be optimized, there's little to no optimization across frameworks. A naive way of moving data across these functions for computation can cause an order of magnitude slowdown in the whole workflow. A study by researchers at Stanford DAWN lab found that typical ML workloads using NumPy, pandas, and TensorFlow run *23 times slower* in one thread compared to hand-optimized code.[45]

In many companies, what usually happens is that data scientists and ML engineers develop models that seem to be working fine in development. However, when these models are deployed, they turn out to be too slow, so their companies hire optimization engineers to optimize their models for the hardware their models run on. An example of a job description for optimization engineers at Mythic follows:

45 Shoumik Palkar, James Thomas, Deepak Narayanan, Pratiksha Thaker, Rahul Palamuttam, Parimajan Negi, Anil Shanbhag, et al., "Evaluating End-to-End Optimization for Data Analytics Applications in Weld," *Proceedings of the VLDB Endowment* 11, no. 9 (2018): 1002–15, *https://oreil.ly/ErUIo*.

> This vision comes together in the AI Engineering team, where our expertise is used to develop AI algorithms and models that are optimized for our hardware, as well as to provide guidance to Mythic's hardware and compiler teams.
>
> The AI Engineering team significantly impacts Mythic by:
>
> - Developing quantization and robustness AI retraining tools
> - Investigating new features for our compiler that leverage the adaptability of neural networks
> - Developing new neural networks that are optimized for our hardware products
> - Interfacing with internal and external customers to meet their development needs

Optimization engineers are hard to come by and expensive to hire because they need to have expertise in both ML and hardware architectures. Optimizing compilers (compilers that also optimize your code) are an alternative solution, as they can automate the process of optimizing models. In the process of lowering ML model code into machine code, compilers can look at the computation graph of your ML model and the operators it consists of—convolution, loops, cross-entropy—and find a way to speed it up.

There are two ways to optimize your ML models: locally and globally. Locally is when you optimize an operator or a set of operators of your model. Globally is when you optimize the entire computation graph end to end.

There are standard local optimization techniques that are known to speed up your model, most of them making things run in parallel or reducing memory access on chips. Here are four of the common techniques:

Vectorization
: Given a loop or a nested loop, instead of executing it one item at a time, execute multiple elements contiguous in memory at the same time to reduce latency caused by data I/O.

Parallelization
: Given an input array (or *n*-dimensional array), divide it into different, independent work chunks, and do the operation on each chunk individually.

Loop tiling[46]

Change the data accessing order in a loop to leverage hardware's memory layout and cache. This kind of optimization is hardware dependent. A good access pattern on CPUs is not a good access pattern on GPUs.

Operator fusion

Fuse multiple operators into one to avoid redundant memory access. For example, two operations on the same array require two loops over that array. In a fused case, it's just one loop. Figure 7-13 shows an example of operator fusion.

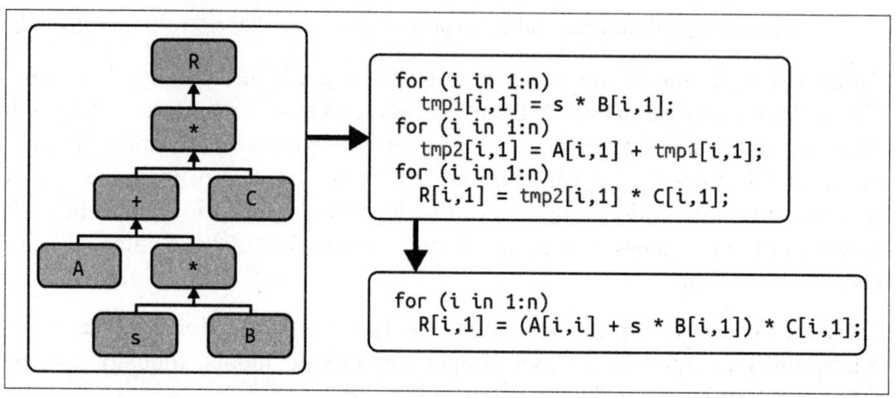

Figure 7-13. An example of an operator fusion. Source: Adapted from an image by Matthias Boehm[47]

To obtain a much bigger speedup, you'd need to leverage higher-level structures of your computation graph. For example, a convolution neural network with the computation graph can be fused vertically or horizontally to reduce memory access and speed up the model, as shown in Figure 7-14.

46 For a helpful visualization of loop tiling, see slide 33 from Colfax Research's presentation "Access to Caches and Memory" (*https://oreil.ly/7ipWQ*), session 10 of their Programming and Optimization for Intel Architecture: Hands-on Workshop series. The entire series is available at *https://oreil.ly/hT1g4*.

47 Matthias Boehm, "Architecture of ML Systems 04 Operator Fusion and Runtime Adaptation," Graz University of Technology, April 5, 2019, *https://oreil.ly/py43J*.

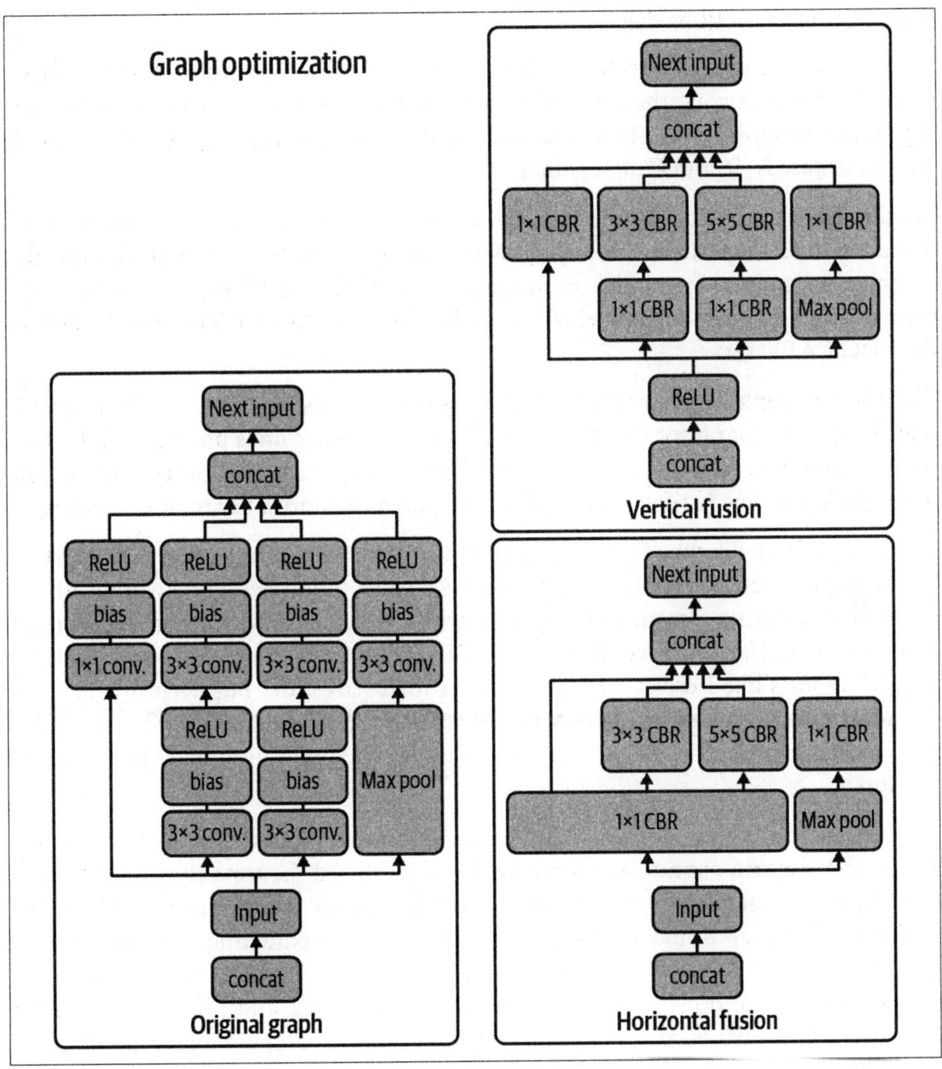

Figure 7-14. Vertical and horizontal fusion of the computation graph of a convolution neural network. Source: Adapted from an image by TensorRT team[48]

48 Shashank Prasanna, Prethvi Kashinkunti, and Fausto Milletari, "TensorRT 3: Faster TensorFlow Inference and Volta Support," NVIDIA Developer, December 4, 2017, *https://oreil.ly/d9h98*. CBR stands for "convolution, bias, and ReLU."

Using ML to optimize ML models

As hinted by the previous section with the vertical and horizontal fusion for a convolutional neural network, there are many possible ways to execute a given computation graph. For example, given three operators A, B, and C, you can fuse A with B, fuse B with C, or fuse A, B, and C all together.

Traditionally, framework and hardware vendors hire optimization engineers who, based on their experience, come up with heuristics on how to best execute the computation graph of a model. For example, NVIDIA might have an engineer or a team of engineers who focus exclusively on how to make ResNet-50 run really fast on their DGX A100 server.[49]

There are a couple of drawbacks to hand-designed heuristics. First, they're nonoptimal. There's no guarantee that the heuristics an engineer comes up with are the best possible solution. Second, they are nonadaptive. Repeating the process on a new framework or a new hardware architecture requires an enormous amount of effort.

This is complicated by the fact that model optimization is dependent on the operators its computation graph consists of. Optimizing a convolution neural network is different from optimizing a recurrent neural network, which is different from optimizing a transformer. Hardware vendors like NVIDIA and Google focus on optimizing popular models like ResNet-50 and BERT for their hardware. But what if you, as an ML researcher, come up with a new model architecture? You might need to optimize it yourself to show that it's fast first before it's adopted and optimized by hardware vendors.

If you don't have ideas for good heuristics, one possible solution might be to try all possible ways to execute a computation graph, record the time they need to run, then pick the best one. However, given a combinatorial number of possible paths, exploring them all would be intractable. Luckily, approximating the solutions to intractable problems is what ML is good at. What if we use ML to narrow down the search space so we don't have to explore that many paths, and predict how long a path will take so that we don't have to wait for the entire computation graph to finish executing?

To estimate how much time a path through a computation graph will take to run turns out to be difficult, as it requires making a lot of assumptions about that graph. It's much easier to focus on a small part of the graph.

---

49 This is also why you shouldn't read too much into benchmarking results, such as MLPerf's results (*https://oreil.ly/XrW2C*). A popular model running really fast on a type of hardware doesn't mean an arbitrary model will run really fast on that hardware. It might just be that this model is over-optimized.

If you use PyTorch on GPUs, you might have seen `torch.backends.cudnn.benchmark=True`. When this is set to True, *cuDNN autotune* will be enabled. cuDNN autotune searches over a predetermined set of options to execute a convolution operator and then chooses the fastest way. cuDNN autotune, despite its effectiveness, only works for convolution operators. A much more general solution is autoTVM (*https://oreil.ly/ZNgzH*), which is part of the open source compiler stack TVM. autoTVM works with subgraphs instead of just an operator, so the search spaces it works with are much more complex. The way autoTVM works is quite complicated, but in simple terms:

1. It first breaks your computation graph into subgraphs.
2. It predicts how big each subgraph is.
3. It allocates time to search for the best possible path for each subgraph.
4. It stitches the best possible way to run each subgraph together to execute the entire graph.

autoTVM measures the actual time it takes to run each path it goes down, which gives it ground truth data to train a cost model to predict how long a future path will take. The pro of this approach is that because the model is trained using the data generated during runtime, it can adapt to any type of hardware it runs on. The con is that it takes more time for the cost model to start improving. Figure 7-15 shows the performance gain that autoTVM gave compared to cuDNN for the model ResNet-50 on NVIDIA TITAN X.

While the results of ML-powered compilers are impressive, they come with a catch: they can be slow. You go through all the possible paths and find the most optimized ones. This process can take hours, even days for complex ML models. However, it's a one-time operation, and the results of your optimization search can be cached and used to both optimize existing models and provide a starting point for future tuning sessions. You optimize your model once for one hardware backend then run it on multiple devices of that same hardware type. This sort of optimization is ideal when you have a model ready for production and target hardware to run inference on.

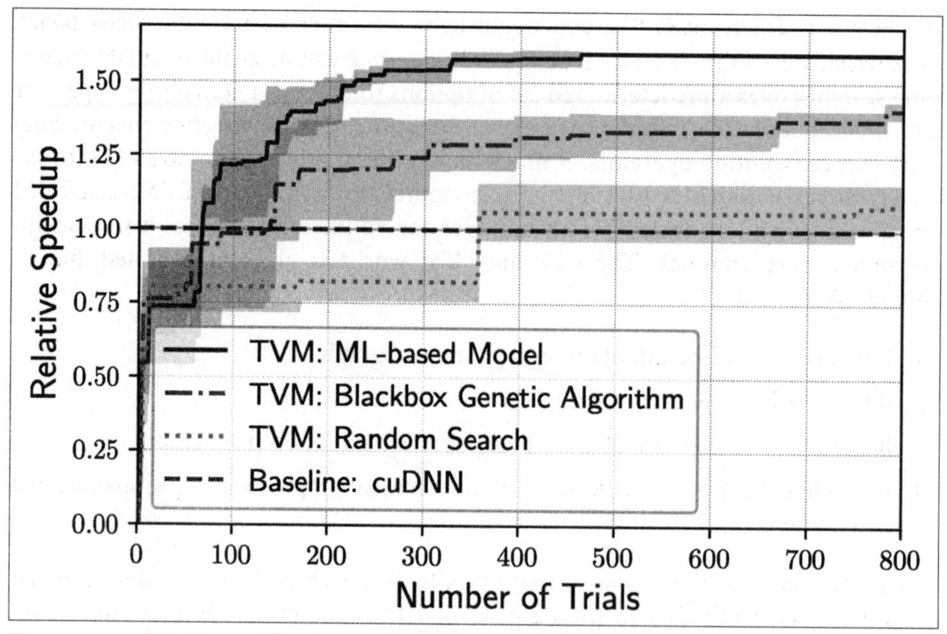

Figure 7-15. Speedup achieved by autoTVM over cuDNN for ResNet-50 on NVIDIA TITAN X. It takes ~70 trials for autoTVM to outperform cuDNN. Source: Chen et al.[50]

ML in Browsers

We've been talking about how compilers can help us generate machine-native code run models on certain hardware backends. It is, however, possible to generate code that can run on just any hardware backends by running that code in browsers. If you can run your model in a browser, you can run your model on any device that supports browsers: MacBooks, Chromebooks, iPhones, Android phones, and more. You wouldn't need to care what chips those devices use. If Apple decides to switch from Intel chips to ARM chips, it's not your problem.

When talking about browsers, many people think of JavaScript. There are tools that can help you compile your models into JavaScript, such as TensorFlow.js (*https://oreil.ly/3Afzv*), Synaptic (*https://oreil.ly/SYiLq*), and brain.js (*https://oreil.ly/83IIa*). However, JavaScript is slow, and its capacity as a programming language is limited for complex logics such as extracting features from data.

50 Chen et al., "TVM: An Automated End-to-End Optimizing Compiler for Deep Learning."

A more promising approach is WebAssembly (WASM). WASM is an open standard that allows you to run executable programs in browsers. After you've built your models in scikit-learn, PyTorch, TensorFlow, or whatever frameworks you've used, instead of compiling your models to run on specific hardware, you can compile your model to WASM. You get back an executable file that you can just use with JavaScript.

WASM is one of the most exciting technological trends I've seen in the last couple of years. It's performant, easy to use, and has an ecosystem that is growing like wildfire.[51] As of September 2021, it's supported by 93% of devices worldwide.[52]

The main drawback of WASM is that because WASM runs in browsers, it's slow. Even though WASM is already much faster than JavaScript, it's still slow compared to running code natively on devices (such as iOS or Android apps). A study by Jangda et al. showed that applications compiled to WASM run slower than native applications by an average of 45% (on Firefox) to 55% (on Chrome).[53]

Summary

Congratulations, you've finished possibly one of the most technical chapters in this book! The chapter is technical because deploying ML models is an engineering challenge, not an ML challenge.

We've discussed different ways to deploy a model, comparing online prediction with batch prediction, and ML on the edge with ML on the cloud. Each way has its own challenges. Online prediction makes your model more responsive to users' changing preferences, but you have to worry about inference latency. Batch prediction is a workaround for when your models take too long to generate predictions, but it makes your model less flexible.

Similarly, doing inference on the cloud is easy to set up, but it becomes impractical with network latency and cloud cost. Doing inference on the edge requires having edge devices with sufficient compute power, memory, and battery.

However, I believe that most of these challenges are due to the limitations of the hardware that ML models run on. As hardware becomes more powerful and optimized for ML, I believe that ML systems will transition to making online prediction on-device, illustrated in Figure 7-16.

---

51 Wasmer, *https://oreil.ly/dTRxr*; Awesome Wasm, *https://oreil.ly/hlIFb*.

52 Can I Use _____?, *https://oreil.ly/slI05*.

53 Abhinav Jangda, Bobby Powers, Emery D. Berger, and Arjun Guha, "Not So Fast: Analyzing the Performance of WebAssembly vs. Native Code," USENIX, *https://oreil.ly/uVzrX*.

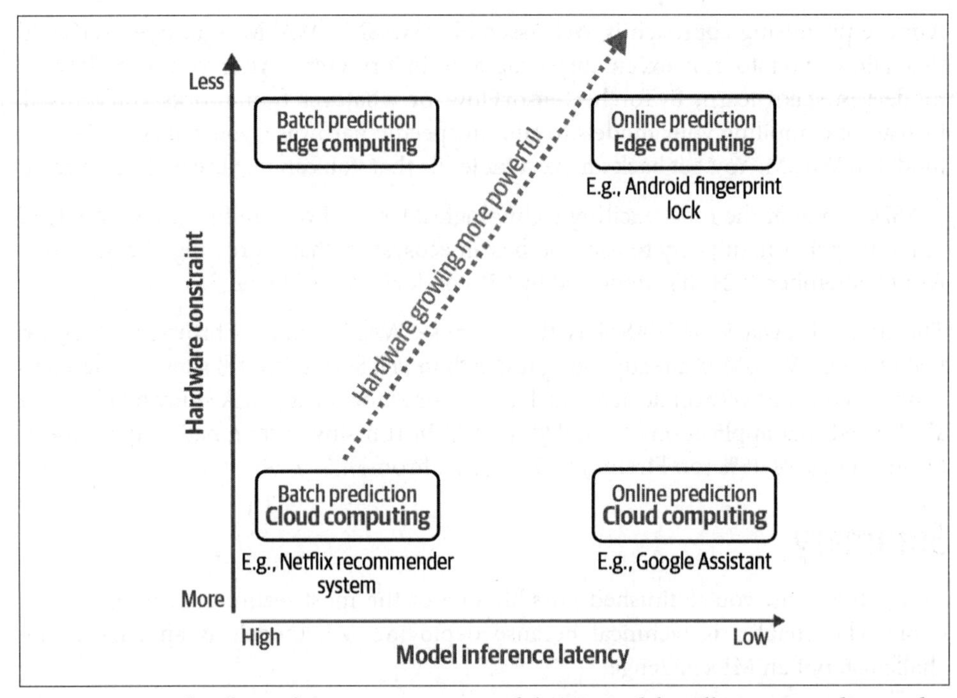

Figure 7-16. As hardware becomes more powerful, ML models will move to online and on the edge

I used to think that an ML project is done after the model is deployed, and I hope that I've made clear in this chapter that I was seriously mistaken. Moving the model from the development environment to the production environment creates a whole new host of problems. The first is how to keep that model in production. In the next chapter, we'll discuss how our models might fail in production, and how to continually monitor models to detect issues and address them as fast as possible.

CHAPTER 8

Data Distribution Shifts and Monitoring

Let's start the chapter with a story I was told by an executive that many readers might be able to relate to. About two years ago, his company hired a consulting firm to develop an ML model to help them predict how many of each grocery item they'd need next week, so they could restock the items accordingly. The consulting firm took six months to develop the model. When the consulting firm handed the model over, his company deployed it and was very happy with its performance. They could finally boast to their investors that they were an AI-powered company.

However, a year later, their numbers went down. The demand for some items was consistently being overestimated, which caused the extra items to expire. At the same time, the demand for some items was consistently being underestimated, leading to lost sales.[1] Initially, his inventory team manually changed the model's predictions to correct the patterns they noticed, but eventually, the model's predictions had become so bad that they could no longer use it. They had three options: pay the same consulting firm an obscene amount of money to update the model, pay another consulting firm even more money because this firm would need time to get up to speed, or hire an in-house team to maintain the model onward.

1 This seems to be a fairly common pattern for inventory prediction. Eugene Yan wrote about a similar story to illustrate the problem of degenerate feedback loops in his article "6 Little-Known Challenges After Deploying Machine Learning" (*https://oreil.ly/p1yCd*) (2021).

His company learned the hard way an important lesson that the rest of the industry is also discovering: deploying a model isn't the end of the process. A model's performance degrades over time in production. Once a model has been deployed, we still have to continually monitor its performance to detect issues as well as deploy updates to fix these issues.

In this chapter and the next, we'll cover the necessary topics to help you keep a model in production. We'll start by covering reasons why ML models that perform great during development fail in production. Then, we'll take a deep dive into one especially prevalent and thorny issue that affects almost all ML models in production: data distribution shifts. This occurs when the data distribution in production differs and diverges from the data distribution the model was exposed to during training. We'll continue with how to monitor for distribution shifts. In the next chapter, we'll cover how to continually update your models in production to adapt to shifts in data distributions.

Causes of ML System Failures

Before we identify the cause of ML system failures, let's briefly discuss what an ML system failure is. A failure happens when one or more expectations of the system is violated. In traditional software, we mostly care about a system's operational expectations: whether the system executes its logic within the expected operational metrics, e.g., latency and throughput.

For an ML system, we care about both its operational metrics and its ML performance metrics. For example, consider an English-French machine translation system. Its operational expectation might be that, given an English sentence, the system returns a French translation within a one-second latency. Its ML performance expectation is that the returned translation is an accurate translation of the original English sentence 99% of the time.

If you enter an English sentence into the system and don't get back a translation, the first expectation is violated, so this is a system failure.

If you get back a translation that isn't correct, it's not necessarily a system failure because the accuracy expectation allows some margin of error. However, if you keep entering different English sentences into the system and keep getting back wrong translations, the second expectation is violated, which makes it a system failure.

Operational expectation violations are easier to detect, as they're usually accompanied by an operational breakage such as a timeout, a 404 error on a webpage, an out-of-memory error, or a segmentation fault. However, ML performance expectation violations are harder to detect as doing so requires measuring and monitoring the performance of ML models in production. In the preceding example of the English-French machine translation system, detecting whether the returned translations are correct 99% of the time is difficult if we don't know what the correct translations are supposed to be. There are countless examples of Google Translate's painfully wrong translations being used by users because they aren't aware that these are wrong translations. For this reason, we say that ML systems often fail silently.

To effectively detect and fix ML system failures in production, it's useful to understand why a model, after proving to work well during development, would fail in production. We'll examine two types of failures: software system failures and ML-specific failures.

Software System Failures

Software system failures are failures that would have happened to non-ML systems. Here are some examples of software system failures:

Dependency failure
: A software package or a codebase that your system depends on breaks, which leads your system to break. This failure mode is common when the dependency is maintained by a third party, and especially common if the third party that maintains the dependency no longer exists.[2]

Deployment failure
: Failures caused by deployment errors, such as when you accidentally deploy the binaries of an older version of your model instead of the current version, or when your systems don't have the right permissions to read or write certain files.

2 This is one of the reasons why many companies are hesitant to use products by startups, and why many companies prefer to use open source software. When a product you use is no longer maintained by its creators, if that product is open source, at least you'll be able to access the codebase and maintain it yourself.

Hardware failures
When the hardware that you use to deploy your model, such as CPUs or GPUs, doesn't behave the way it should. For example, the CPUs you use might overheat and break down.[3]

Downtime or crashing
If a component of your system runs from a server somewhere, such as AWS or a hosted service, and that server is down, your system will also be down.

Just because some failures are not specific to ML doesn't mean they're not important for ML engineers to understand. In 2020, Daniel Papasian and Todd Underwood, two ML engineers at Google, looked at 96 cases where a large ML pipeline at Google broke. They reviewed data from over the previous 15 years to determine the causes and found out that 60 out of these 96 failures happened due to causes not directly related to ML.[4] Most of the issues are related to distributed systems, e.g., where the workflow scheduler or orchestrator makes a mistake, or related to the data pipeline, e.g., where data from multiple sources is joined incorrectly or the wrong data structures are being used.

Addressing software system failures requires not ML skills, but traditional software engineering skills, and addressing them is beyond the scope of this book. Because of the importance of traditional software engineering skills in deploying ML systems, ML engineering is mostly engineering, not ML.[5] For readers interested in learning how to make ML systems reliable from the software engineering perspective, I highly recommend the book *Reliable Machine Learning*, published by O'Reilly with Todd Underwood as one of the authors.

A reason for the prevalence of software system failures is that because ML adoption in the industry is still nascent, tooling around ML production is limited and best practices are not yet well developed or standardized. However, as toolings and best practices for ML production mature, there are reasons to believe that the proportion of software system failures will decrease and the proportion of ML-specific failures will increase.

---

3 Cosmic rays can cause your hardware to break down (Wikipedia, s.v. "Soft error," *https://oreil.ly/4cvNg*).

4 Daniel Papasian and Todd Underwood, "How ML Breaks: A Decade of Outages for One Large ML Pipeline," Google, July 17, 2020, video, 19:06, *https://oreil.ly/WGabN*. A non-ML failure might still be indirectly due to ML. For example, a server can crash for non-ML systems, but because ML systems tend to require more compute power, it might cause this server to crash more often.

5 The peak of my career: Elon Musk agreed with me (*https://oreil.ly/mBseG*).

ML-Specific Failures

ML-specific failures are failures specific to ML systems. Examples include data collection and processing problems, poor hyperparameters, changes in the training pipeline not correctly replicated in the inference pipeline and vice versa, data distribution shifts that cause a model's performance to deteriorate over time, edge cases, and degenerate feedback loops.

In this chapter, we'll focus on addressing ML-specific failures. Even though they account for a small portion of failures, they can be more dangerous than non-ML failures as they're hard to detect and fix, and they can prevent ML systems from being used altogether. We've covered data problems in great detail in Chapter 4, hyperparameter tuning in Chapter 6, and the danger of having two separate pipelines for training and inference in Chapter 7. In this chapter, we'll discuss three new but very common problems that arise after a model has been deployed: production data differing from training data, edge cases, and degenerate feedback loops.

Production data differing from training data

When we say that an ML model learns from the training data, it means that the model learns the underlying distribution of the training data with the goal of leveraging this learned distribution to generate accurate predictions for unseen data—data that it didn't see during training. We'll go into what this means mathematically in the section "Data Distribution Shifts" on page 237. When the model is able to generate accurate predictions for unseen data, we say that this model "generalizes to unseen data."[6] The test data that we use to evaluate a model during development is supposed to represent unseen data, and the model's performance on the test data is supposed to give us an idea of how well the model will generalize.

One of the first things I learned in ML courses is that it's essential for the training data and the unseen data to come from a similar distribution. The assumption is that the unseen data comes from a *stationary* distribution that is *the same* as the training data distribution. If the unseen data comes from a different distribution, the model might not generalize well.[7]

6 Back when in-person academic conferences were still a thing, I often heard researchers arguing about whose models can generalize better. "My model generalizes better than your model" is the ultimate flex.

7 Masashi Sugiyama and Motoaki Kawanabe, *Machine Learning in Non-stationary Environments: Introduction to Covariate Shift Adaptation* (Cambridge, MA: MIT Press, 2012).

This assumption is incorrect in most cases for two reasons. First, the underlying distribution of the real-world data is unlikely to be *the same* as the underlying distribution of the training data. Curating a training dataset that can accurately represent the data that a model will encounter in production turns out to be very difficult.[8] Real-world data is multifaceted and, in many cases, virtually infinite, whereas training data is finite and constrained by the time, compute, and human resources available during the dataset creation and processing. There are many different selection and sampling biases, as discussed in Chapter 4, that can happen and make real-world data diverge from training data. The divergence can be something as minor as real-world data using a different type of encoding of emojis. This type of divergence leads to a common failure mode known as *the train-serving skew*: a model that does great in development but performs poorly when deployed.

Second, the real world isn't *stationary*. Things change. Data distributions shift. In 2019, when people searched for Wuhan, they likely wanted to get travel information, but since COVID-19, when people search for Wuhan, they likely want to know about the place where COVID-19 originated. Another common failure mode is that a model does great when first deployed, but its performance degrades over time as the data distribution changes. This failure mode needs to be continually monitored and detected for as long as a model remains in production.

When I use COVID-19 as an example that causes data shifts, some people have the impression that data shifts only happen because of unusual events, which implies they don't happen often. Data shifts happen all the time, suddenly, gradually, or seasonally. They can happen suddenly because of a specific event, such as when your existing competitors change their pricing policies and you have to update your price predictions in response, or when you launch your product in a new region, or when a celebrity mentions your product, which causes a surge in new users, and so on. They can happen gradually because social norms, cultures, languages, trends, industries, etc. just change over time. They can also happen due to seasonal variations, such as people might be more likely to request rideshares in the winter when it's cold and snowy than in the spring.

Due to the complexity of ML systems and the poor practices in deploying them, a large percentage of what might look like data shifts on monitoring dashboards are caused by internal errors,[9] such as bugs in the data pipeline, missing values incorrectly inputted, inconsistencies between the features extracted during training and inference, features standardized using statistics from the wrong subset of data,

---

8 John Mcquaid, "Limits to Growth: Can AI's Voracious Appetite for Data Be Tamed?" *Undark*, October 18, 2021, *https://oreil.ly/LSjVD*.

9 The chief technology officer (CTO) of a monitoring service company told me that, in his estimate, 80% of the drifts captured by his service are caused by human errors.

wrong model version, or bugs in the app interface that force users to change their behaviors.

Since this is an error mode that affects almost all ML models, we'll cover this in detail in the section "Data Distribution Shifts" on page 237.

Edge cases

Imagine there existed a self-driving car that can drive you safely 99.99% of the time, but the other 0.01% of the time, it might get into a catastrophic accident that can leave you permanently injured or even dead.[10] Would you use that car?

If you're tempted to say no, you're not alone. An ML model that performs well on most cases but fails on a small number of cases might not be usable if these failures cause catastrophic consequences. For this reason, major self-driving car companies are focusing on making their systems work on edge cases.[11]

Edge cases are the data samples so extreme that they cause the model to make catastrophic mistakes. Even though edge cases generally refer to data samples drawn from the same distribution, if there is a sudden increase in the number of data samples in which your model doesn't perform well, it could be an indication that the underlying data distribution has shifted.

Autonomous vehicles are often used to illustrate how edge cases can prevent an ML system from being deployed. But this is also true for any safety-critical application such as medical diagnosis, traffic control, e-discovery,[12] etc. It can also be true for non-safety-critical applications. Imagine a customer service chatbot that gives reasonable responses to most of the requests, but sometimes, it spits out outrageously racist or sexist content. This chatbot will be a brand risk for any company that wants to use it, thus rendering it unusable.

---

10 This means the self-driving car is a bit safer than an average human driver. As of 2019, the ratio of traffic-related fatalities per 100,000 licensed drivers was 15.8, or 0.0158% ("Fatality Rate per 100,000 Licensed Drivers in the U.S. from 1990 to 2019," Statista, 2021, *https://oreil.ly/w3wYh*).

11 Rodney Brooks, "Edge Cases for Self Driving Cars," *Robots, AI, and Other Stuff*, June 17, 2017, *https://oreil.ly/Nyp4F*; Lance Eliot, "Whether Those Endless Edge or Corner Cases Are the Long-Tail Doom for AI Self-Driving Cars," *Forbes*, July 13, 2021, *https://oreil.ly/L2Sbp*; Kevin McAllister, "Self-Driving Cars Will Be Shaped by Simulated, Location Data," *Protocol*, March 25, 2021, *https://oreil.ly/tu8hs*.

12 e-discovery (*https://oreil.ly/KCets*), or electronic discovery, refers to discovery in legal proceedings, such as litigation, government investigations, or Freedom of Information Act requests, where the information sought is in electronic format.

Edge Cases and Outliers

You might wonder about the differences between an outlier and an edge case. The definition of what makes an edge case varies by discipline. In ML, because of its recent adoption in production, edge cases are still being discovered, which makes their definition contentious.

In this book, outliers refer to data: an example that differs significantly from other examples. Edge cases refer to performance: an example where a model performs significantly worse than other examples. An outlier can cause a model to perform unusually poorly, which makes it an edge case. However, not all outliers are edge cases. For example, a person jaywalking on a highway is an outlier, but it's not an edge case if your self-driving car can accurately detect that person and decide on a motion response appropriately.

During model development, outliers can negatively affect your model's performance, as shown in Figure 8-1. In many cases, it might be beneficial to remove outliers as it helps your model to learn better decision boundaries and generalize better to unseen data. However, during inference, you don't usually have the option to remove or ignore the queries that differ significantly from other queries. You can choose to transform it—for example, when you enter "mechin learnin" into Google Search, Google might ask if you mean "machine learning." But most likely you'll want to develop a model so that it can perform well even on unexpected inputs.

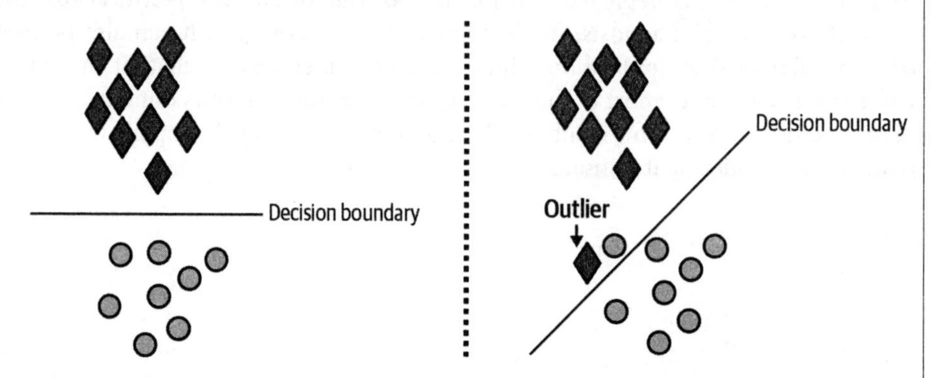

Figure 8-1. The image on the left shows the decision boundary when there's no outlier. The image on the right shows the decision boundary when there's one outlier, which is very different from the decision boundary in the first case, and probably less accurate.

Degenerate feedback loops

In the section "Natural Labels" on page 91, we discussed a feedback loop as the time it takes from when a prediction is shown until the time feedback on the prediction is provided. The feedback can be used to extract natural labels to evaluate the model's performance and train the next iteration of the model.

A *degenerate feedback loop* can happen when the predictions themselves influence the feedback, which, in turn, influences the next iteration of the model. More formally, a degenerate feedback loop is created when a system's outputs are used to generate the system's future inputs, which, in turn, influence the system's future outputs. In ML, a system's predictions can influence how users interact with the system, and because users' interactions with the system are sometimes used as training data to the same system, degenerate feedback loops can occur and cause unintended consequences. Degenerate feedback loops are especially common in tasks with natural labels from users, such as recommender systems and ads click-through-rate prediction.

To make this concrete, imagine you build a system to recommend to users songs that they might like. The songs that are ranked high by the system are shown first to users. Because they are shown first, users click on them more, which makes the system more confident that these recommendations are good. In the beginning, the rankings of two songs, A and B, might be only marginally different, but because A was originally ranked a bit higher, it showed up higher in the recommendation list, making users click on A more, which made the system rank A even higher. After a while, A's ranking became much higher than B's.[13] Degenerate feedback loops are one reason why popular movies, books, or songs keep getting more popular, which makes it hard for new items to break into popular lists. This type of scenario is incredibly common in production, and it's heavily researched. It goes by many different names, including "exposure bias," "popularity bias," "filter bubbles," and sometimes "echo chambers."

Here's another example to drive the danger of degenerative feedback loops home. Imagine building a resume-screening model to predict whether someone with a certain resume is qualified for the job. The model finds that feature X accurately predicts whether someone is qualified, so it recommends resumes with feature X. You can replace X with features like "went to Stanford," "worked at Google," or "identifies as male." Recruiters only interview people whose resumes are recommended by the model, which means they only interview candidates with feature X, which means the company only hires candidates with feature X. This, in turn, makes the model put even more weight on feature X.[14] Having visibility into how your model makes

13 Ray Jiang, Silvia Chiappa, Tor Lattimore, András György, and Pushmeet Kohli, "Degenerate Feedback Loops in Recommender Systems," *arXiv*, February 27, 2019, *https://oreil.ly/b9G7o*.

14 This is related to "survivorship bias."

predictions—such as measuring the importance of each feature for the model, as discussed in Chapter 5—can help detect the bias toward feature X in this case.

Left unattended, degenerate feedback loops can cause your model to perform suboptimally at best. At worst, they can perpetuate and magnify biases embedded in data, such as biasing against candidates without feature X.

Detecting degenerate feedback loops. If degenerate feedback loops are so bad, how do we know if a feedback loop in a system is degenerate? When a system is offline, degenerate feedback loops are difficult to detect. Degenerate loops result from user feedback, and a system won't have users until it's online (i.e., deployed to users).

For the task of recommender systems, it's possible to detect degenerate feedback loops by measuring the popularity diversity of a system's outputs even when the system is offline. An item's popularity can be measured based on how many times it has been interacted with (e.g., seen, liked, bought, etc.) in the past. The popularity of all the items will likely follow a long-tail distribution: a small number of items are interacted with a lot, while most items are rarely interacted with at all. Various metrics such as *aggregate diversity* and *average coverage of long-tail items* proposed by Brynjolfsson et al. (*https://oreil.ly/8EKPf*) (2011), Fleder and Hosanagar (*https://oreil.ly/PmNQm*) (2009), and Abdollahpouri et al. (*https://oreil.ly/EkiFw*) (2019) can help you measure the diversity of the outputs of a recommender system.[15] Low scores mean that the outputs of your system are homogeneous, which might be caused by popularity bias.

In 2021, Chia et al. went a step further and proposed the measurement of hit rate against popularity. They first divided items into buckets based on their popularity—e.g., bucket 1 consists of items that have been interacted with less than 100 times, bucket 2 consists of items that have been interacted with more than 100 times but less than 1,000 times, etc. Then they measured the prediction accuracy of a recommender system for each of these buckets. If a recommender system is much better at recommending popular items than recommending less popular items, it likely suffers from popularity bias.[16] Once your system is in production and you notice that its predictions become more homogeneous over time, it likely suffers from degenerate feedback loops.

15 Erik Brynjolfsson, Yu (Jeffrey) Hu, and Duncan Simester, "Goodbye Pareto Principle, Hello Long Tail: The Effect of Search Costs on the Concentration of Product Sales," *Management Science* 57, no. 8 (2011): 1373–86, *https://oreil.ly/tGhHi*; Daniel Fleder and Kartik Hosanagar, "Blockbuster Culture's Next Rise or Fall: The Impact of Recommender Systems on Sales Diversity," *Management Science* 55, no. 5 (2009), *https://oreil.ly/Zwkh8*; Himan Abdollahpouri, Robin Burke, and Bamshad Mobasher, "Managing Popularity Bias in Recommender Systems with Personalized Re-ranking," *arXiv*, January 22, 2019, *https://oreil.ly/jgYLr*.

16 Patrick John Chia, Jacopo Tagliabue, Federico Bianchi, Chloe He, and Brian Ko, "Beyond NDCG: Behavioral Testing of Recommender Systems with RecList," *arXiv*, November 18, 2021, *https://oreil.ly/7GfHk*.

Correcting degenerate feedback loops. Because degenerate feedback loops are a common problem, there are many proposed methods on how to correct them. In this chapter, we'll discuss two methods. The first one is to use randomization, and the second one is to use positional features.

We've discussed that degenerate feedback loops can cause a system's outputs to be more homogeneous over time. Introducing randomization in the predictions can reduce their homogeneity. In the case of recommender systems, instead of showing the users only the items that the system ranks highly for them, we show users random items and use their feedback to determine the true quality of these items. This is the approach that TikTok follows. Each new video is randomly assigned an initial pool of traffic (which can be up to hundreds of impressions). This pool of traffic is used to evaluate each video's unbiased quality to determine whether it should be moved to a bigger pool of traffic or be marked as irrelevant.[17]

Randomization has been shown to improve diversity, but at the cost of user experience.[18] Showing our users completely random items might cause users to lose interest in our product. An intelligent exploration strategy, such as those discussed in the section "Contextual bandits as an exploration strategy" on page 289, can help increase item diversity with acceptable prediction accuracy loss. Schnabel et al. use a small amount of randomization and causal inference techniques to estimate the unbiased value of each song.[19] They were able to show that this algorithm was able to correct a recommender system to make recommendations fair to creators.

We've also discussed that degenerate feedback loops are caused by users' feedback on predictions, and users' feedback on a prediction is biased based on where it is shown. Consider the preceding recommender system example, where each time you recommend five songs to users. You realize that the top recommended song is much more likely to be clicked on compared to the other four songs. You are unsure whether your model is exceptionally good at picking the top song, or whether users click on any song as long as it's recommended on top.

---

17 Catherine Wang, "Why TikTok Made Its User So Obsessive? The AI Algorithm That Got You Hooked," *Towards Data Science*, June 7, 2020, *https://oreil.ly/J7nJ9*.

18 Gediminas Adomavicius and YoungOk Kwon, "Improving Aggregate Recommendation Diversity Using Ranking-Based Techniques," *IEEE Transactions on Knowledge and Data Engineering* 24, no. 5 (May 2012): 896–911, *https://oreil.ly/0JjUV*.

19 Tobias Schnabel, Adith Swaminathan, Ashudeep Singh, Navin Chandak, and Thorsten Joachims, "Recommendations as Treatments: Debiasing Learning and Evaluation," *arXiv*, February 17, 2016, *https://oreil.ly/oDPSK*.

If the position in which a prediction is shown affects its feedback in any way, you might want to encode the position information using *positional features*. Positional features can be numerical (e.g., positions are 1, 2, 3,...) or Boolean (e.g., whether a prediction is shown in the first position or not). Note that "positional features" are different from "positional embeddings" mentioned in Chapter 5.

Here is a naive example to show how to use positional features. During training, you add "whether a song is recommended first" as a feature to your training data, as shown in Table 8-1. This feature allows your model to learn how much being a top recommendation influences how likely a song is clicked on.

Table 8-1. Adding positional features to your training data to mitigate degenerate feedback loops

ID	Song	Genre	Year	Artist	User	1st Position	Click
1	Shallow	Pop	2020	Lady Gaga	listenr32	**False**	No
2	Good Vibe	Funk	2019	Funk Overlord	listenr32	**False**	No
3	Beat It	Rock	1989	Michael Jackson	fancypants	**False**	No
4	In Bloom	Rock	1991	Nirvana	fancypants	**True**	Yes
5	Shallow	Pop	2020	Lady Gaga	listenr32	**True**	Yes

During inference, you want to predict whether a user will click on a song regardless of where the song is recommended, so you might want to set the 1st Position feature to be False. Then you look at the model's predictions for various songs for each user and can choose the order in which to show each song.

This is a naive example because doing this alone might not be enough to combat degenerate feedback loops. A more sophisticated approach would be to use two different models. The first model predicts the probability that the user will see and consider a recommendation taking into account the position at which that recommendation will be shown. The second model then predicts the probability that the user will click on the item given that they saw and considered it. The second model doesn't concern positions at all.

Data Distribution Shifts

In the previous section, we discussed common causes for ML system failures. In this section, we'll zero in onto one especially sticky cause of failures: data distribution shifts, or data shifts for short. Data distribution shift refers to the phenomenon in supervised learning when the data a model works with changes over time, which causes this model's predictions to become less accurate as time passes. The distribution of the data the model is trained on is called the *source distribution*. The distribution of the data the model runs inference on is called the *target distribution*.

Even though discussions around data distribution shift have only become common in recent years with the growing adoption of ML in the industry, data distribution shift in systems that learned from data has been studied as early as in 1986.[20] There's also a book on dataset distribution shifts, *Dataset Shift in Machine Learning* by Quiñonero-Candela et al., published by MIT Press in 2008.

Types of Data Distribution Shifts

While data distribution shift is often used interchangeably with concept drift and covariate shift and occasionally label shift, these are three distinct subtypes of data shift. Note that this discussion on different types of data shifts is math-heavy and mostly useful from a research perspective: to develop efficient algorithms to detect and address data shifts requires understanding the causes of those shifts. In production, when encountering a distribution shift, data scientists don't usually stop to wonder what type of shift it is. They mostly care about what they can do to handle this shift. If you find this discussion dense, feel free to skip to the section "General Data Distribution Shifts" on page 241.

To understand what concept drift, covariate shift, and label shift mean, we first need to define a couple of mathematical notations. Let's call the inputs to a model X and its outputs Y. We know that in supervised learning, the training data can be viewed as a set of samples from the joint distribution $P(X, Y)$, and then ML usually models $P(Y|X)$. This joint distribution $P(X, Y)$ can be decomposed in two ways:

- $P(X, Y) = P(Y|X)P(X)$
- $P(X, Y) = P(X|Y)P(Y)$

20 Jeffrey C. Schlimmer and Richard H. Granger, Jr., "Incremental Learning from Noisy Data," *Machine Learning* 1 (1986): 317–54, *https://oreil.ly/FxFQi*.

$P(Y|X)$ denotes the conditional probability of an output given an input—for example, the probability of an email being spam given the content of the email. $P(X)$ denotes the probability density of the input. $P(Y)$ denotes the probability density of the output. Label shift, covariate shift, and concept drift are defined as follows:

Covariate shift
: When $P(X)$ changes but $P(Y|X)$ remains the same. This refers to the first decomposition of the joint distribution.

Label shift
: When $P(Y)$ changes but $P(X|Y)$ remains the same. This refers to the second decomposition of the joint distribution.

Concept drift
: When $P(Y|X)$ changes but $P(X)$ remains the same. This refers to the first decomposition of the joint distribution.[21]

If you find this confusing, don't panic. We'll go over examples in the following section to illustrate their differences.

Covariate shift

Covariate shift is one of the most widely studied forms of data distribution shift.[22] In statistics, a covariate is an independent variable that can influence the outcome of a given statistical trial but which is not of direct interest. Consider that you are running an experiment to determine how locations affect the housing prices. The housing price variable is your direct interest, but you know the square footage affects the price, so the square footage is a covariate. In supervised ML, the label is the variable of direct interest, and the input features are covariate variables.

Mathematically, covariate shift is when $P(X)$ changes, but $P(Y|X)$ remains the same, which means that the distribution of the input changes, but the conditional probability of an output given an input remains the same.

To make this concrete, consider the task of detecting breast cancer. You know that the risk of breast cancer is higher for women over the age of 40,[23] so you have a variable "age" as your input. You might have more women over the age of 40 in your training data than in your inference data, so the input distributions differ for your training and inference data. However, for an example with a given age, such as above 40, the

---

21 You might wonder what about the case when $P(X|Y)$ changes but $P(Y)$ remains the same, as in the second decomposition. I've never encountered any research in this setting. I asked a couple of researchers who specialize in data shifts about it, and they also told me that setting would be too difficult to study.

22 Wouter M. Kouw and Marco Loog, "An Introduction to Domain Adaptation and Transfer Learning," *arXiv*, December 31, 2018, *https://oreil.ly/VKSVP*.

23 "Breast Cancer Risk in American Women," National Cancer Institute, *https://oreil.ly/BFP3U*.

probability that this example has breast cancer is constant. So $P(Y|X)$, the probability of having breast cancer given age over 40, is the same.

During model development, covariate shifts can happen due to biases during the data selection process, which could result from difficulty in collecting examples for certain classes. For example, suppose that to study breast cancer, you get data from a clinic where women go to test for breast cancer. Because people over 40 are encouraged by their doctors to get checkups, your data is dominated by women over 40. For this reason, covariate shift is closely related to the sample selection bias problem.[24]

Covariate shifts can also happen because the training data is artificially altered to make it easier for your model to learn. As discussed in Chapter 4, it's hard for ML models to learn from imbalanced datasets, so you might want to collect more samples of the rare classes or oversample your data on the rare classes to make it easier for your model to learn the rare classes.

Covariate shift can also be caused by the model's learning process, especially through active learning. In Chapter 4, we defined active learning as follows: instead of randomly selecting samples to train a model on, we use the samples most helpful to that model according to some heuristics. This means that the training input distribution is altered by the learning process to differ from the real-world input distribution, and covariate shifts are a by-product.[25]

In production, covariate shift usually happens because of major changes in the environment or in the way your application is used. Imagine you have a model to predict how likely a free user will be to convert to a paid user. The income level of the user is a feature. Your company's marketing department recently launched a campaign that attracts users from a demographic more affluent than your current demographic. The input distribution into your model has changed, but the probability that a user with a given income level will convert remains the same.

If you know in advance how the real-world input distribution will differ from your training input distribution, you can leverage techniques such as *importance weighting* to train your model to work for the real-world data. Importance weighting consists of two steps: estimate the density ratio between the real-world input distribution and the training input distribution, then weight the training data according to this ratio and train an ML model on this weighted data.[26]

---

24 Arthur Gretton, Alex Smola, Jiayuan Huang, Marcel Schmittfull, Karsten Borgwardt, and Bernard Schölkopf, "Covariate Shift by Kernel Mean Matching," *Journal of Machine Learning Research* (2009), *https://oreil.ly/s49MI*.

25 Sugiyama and Kawanabe, *Machine Learning in Non-stationary Environments*.

26 Tongtong Fang, Nan Lu, Gang Niu, and Masashi Sugiyama, "Rethinking Importance Weighting for Deep Learning under Distribution Shift," *NeurIPS Proceedings* 2020, *https://oreil.ly/GzJ1r*; Gretton et al., "Covariate Shift by Kernel Mean Matching."

However, because we don't know in advance how the distribution will change in the real world, it's very difficult to preemptively train your models to make them robust to new, unknown distributions. There has been research that attempts to help models learn representations of latent variables that are invariant across data distributions,[27] but I'm not aware of their adoption in the industry.

Label shift

Label shift, also known as prior shift, prior probability shift, or target shift, is when $P(Y)$ changes but $P(X|Y)$ remains the same. You can think of this as the case when the output distribution changes but, *for a given output*, the input distribution stays the same.

Remember that covariate shift is when the input distribution changes. When the input distribution changes, the output distribution also changes, resulting in both covariate shift and label shift happening at the same time. Consider the preceding breast cancer example for covariate shift. Because there are more women over 40 in our training data than in our inference data, the percentage of POSITIVE labels is higher during training. However, if you randomly select person A with breast cancer from your training data and person B with breast cancer from your test data, A and B have the same probability of being over 40. This means that $P(X|Y)$, or probability of age over 40 given having breast cancer, is the same. So this is also a case of label shift.

However, not all covariate shifts result in label shifts. It's a subtle point, so we'll consider another example. Imagine that there is now a preventive drug that every woman takes that helps reduce their chance of getting breast cancer. The probability $P(Y|X)$ reduces for women of all ages, so it's no longer a case of covariate shift. However, given a person with breast cancer, the age distribution remains the same, so this is still a case of label shift.

Because label shift is closely related to covariate shift, methods for detecting and adapting models to label shifts are similar to covariate shift adaptation methods. We'll discuss them more later in this chapter.

27 Han Zhao, Remi Tachet Des Combes, Kun Zhang, and Geoffrey Gordon, "On Learning Invariant Representations for Domain Adaptation," *Proceedings of Machine Learning Research* 97 (2019): 7523–32, *https://oreil.ly/ZxYWD*.

Concept drift

Concept drift, also known as posterior shift, is when the input distribution remains the same but the conditional distribution of the output given an input changes. You can think of this as "same input, different output." Consider you're in charge of a model that predicts the price of a house based on its features. Before COVID-19, a three-bedroom apartment in San Francisco could cost $2,000,000. However, at the beginning of COVID-19, many people left San Francisco, so the same apartment would cost only $1,500,000. So even though the distribution of house features remains the same, the conditional distribution of the price of a house given its features has changed.

In many cases, concept drifts are cyclic or seasonal. For example, rideshare prices will fluctuate on weekdays versus weekends, and flight ticket prices rise during holiday seasons. Companies might have different models to deal with cyclic and seasonal drifts. For example, they might have one model to predict rideshare prices on weekdays and another model for weekends.

General Data Distribution Shifts

There are other types of changes in the real world that, even though not well studied in research, can still degrade your models' performance.

One is *feature change*, such as when new features are added, older features are removed, or the set of all possible values of a feature changes.[28] For example, your model was using years for the "age" feature, but now it uses months, so the range of this feature's values has drifted. One time, our team realized that our model's performance plummeted because a bug in our pipeline caused a feature to become NaNs (short for "not a number").

Label schema change is when the set of possible values for Y change. With label shift, $P(Y)$ changes but $P(X|Y)$ remains the same. With label schema change, both $P(Y)$ and $P(X|Y)$ change. A schema describes the structure of the data, so the label schema of a task describes the structure of the labels of that task. For example, a dictionary that maps from a class to an integer value, such as `{"POSITIVE": 0, "NEGATIVE": 1}`, is a schema.

28 You can think of this as the case where both $P(X)$ and $P(Y|X)$ change.

With regression tasks, label schema change could happen because of changes in the possible range of label values. Imagine you're building a model to predict someone's credit score. Originally, you used a credit score system that ranged from 300 to 850, but you switched to a new system that ranges from 250 to 900.

With classification tasks, label schema change could happen because you have new classes. For example, suppose you are building a model to diagnose diseases and there's a new disease to diagnose. Classes can also become outdated or more fine-grained. Imagine that you're in charge of a sentiment analysis model for tweets that mention your brand. Originally, your model predicted only three classes: POSITIVE, NEGATIVE, and NEUTRAL. However, your marketing department realized the most damaging tweets are the angry ones, so they wanted to break the NEGATIVE class into two classes: SAD and ANGRY. Instead of having three classes, your task now has four classes. When the number of classes changes, your model's structure might change,[29] and you might need to both relabel your data and retrain your model from scratch. Label schema change is especially common with high-cardinality tasks—tasks with a high number of classes—such as product or documentation categorization.

There's no rule that says that only one type of shift should happen at one time. A model might suffer from multiple types of drift, which makes handling them a lot more difficult.

Detecting Data Distribution Shifts

Data distribution shifts are only a problem if they cause your model's performance to degrade. So the first idea might be to monitor your model's accuracy-related metrics—accuracy, F1 score, recall, AUC-ROC, etc.—in production to see whether they have changed. "Change" here usually means "decrease," but if my model's accuracy suddenly goes up or fluctuates significantly for no reason that I'm aware of, I'd want to investigate.

Accuracy-related metrics work by comparing the model's predictions to ground truth labels.[30] During model development, you have access to labels, but in production, you don't always have access to labels, and even if you do, labels will be delayed, as discussed in the section "Natural Labels" on page 91. Having access to labels within a reasonable time window will vastly help with giving you visibility into your model's performance.

29 If you use a neural network using softmax as your last layer for your classification tax, the dimension of this softmax layer is [number_of_hidden_units × number_of_classes]. When the number of classes changes, the number of parameters in your softmax layer changes.

30 You don't need ground truth labels if you use an unsupervised learning method, but the vast majority of applications today are supervised.

When ground truth labels are unavailable or too delayed to be useful, we can monitor other distributions of interest instead. The distributions of interest are the input distribution $P(X)$, the label distribution $P(Y)$, and the conditional distributions $P(X|Y)$ and $P(Y|X)$.

While we don't need to know the ground truth labels Y to monitor the input distribution, monitoring the label distribution and both of the conditional distributions require knowing Y. In research, there have been efforts to understand and detect label shifts without labels from the target distribution. One such effort is Black Box Shift Estimation (*https://oreil.ly/4rKh7*) by Lipton et al. (2018). However, in the industry, most drift detection methods focus on detecting changes in the input distribution, especially the distributions of features, as we discuss in detail in this chapter.

Statistical methods

In industry, a simple method many companies use to detect whether the two distributions are the same is to compare their statistics like min, max, mean, median, variance, various quantiles (such as 5th, 25th, 75th, or 95th quantile), skewness, kurtosis, etc. For example, you can compute the median and variance of the values of a feature during inference and compare them to the metrics computed during training. As of October 2021, even TensorFlow Extended's built-in data validation tools (*https://oreil.ly/knwm0*) use only summary statistics to detect the skew between the training and serving data and shifts between different days of training data. This is a good start, but these metrics are far from sufficient.[31] Mean, median, and variance are only useful with the distributions for which the mean/median/variance are useful summaries. If those metrics differ significantly, the inference distribution might have shifted from the training distribution. However, if those metrics are similar, there's no guarantee that there's no shift.

A more sophisticated solution is to use a two-sample hypothesis test, shortened as two-sample test. It's a test to determine whether the difference between two populations (two sets of data) is statistically significant. If the difference is statistically significant, then the probability that the difference is a random fluctuation due to sampling variability is very low, and, therefore, the difference is caused by the fact that these two populations come from two distinct distributions. If you consider the data from yesterday to be the source population and the data from today to be the target population and they are statistically different, it's likely that the underlying data distribution has shifted between yesterday and today.

31 Hamel Husain gave a great lecture on why TensorFlow Extended's skew detection is so bad for CS 329S: Machine Learning Systems Design (*https://oreil.ly/Y9hAW*) (Stanford, 2022). You can find the video on YouTube (*https://oreil.ly/ivxbQ*).

A caveat is that just because the difference is statistically significant doesn't mean that it is practically important. However, a good heuristic is that if you are able to detect the difference from a relatively small sample, then it is probably a serious difference. If it takes a huge number of samples to detect, then the difference is probably not worth worrying about.

A basic two-sample test is the Kolmogorov–Smirnov test, also known as the K-S or KS test.[32] It's a nonparametric statistical test, which means it doesn't require any parameters of the underlying distribution to work. It doesn't make any assumption about the underlying distribution, which means it can work for any distribution. However, one major drawback of the KS test is that it can only be used for one-dimensional data. If your model's predictions and labels are one-dimensional (scalar numbers), then the KS test is useful to detect label or prediction shifts. However, it won't work for high-dimensional data, and features are usually high-dimensional.[33] KS tests can also be expensive and produce too many false positive alerts.[34]

Another test is Least-Squares Density Difference, an algorithm that is based on the least squares density-difference estimation method.[35] There is also MMD, Maximum Mean Discrepancy (*https://oreil.ly/KzUuw*) (Gretton et al. 2012), a kernel-based technique for multivariate two-sample testing and its variant Learned Kernel MMD (*https://oreil.ly/C5dXI*) (Liu et al. 2020). MMD is popular in research, but as of writing this book, I'm not aware of any company that is using it in the industry. Alibi Detect (*https://oreil.ly/162tf*) is a great open source package with the implementations of many drift detection algorithms, as shown in Figure 8-2.

Because two-sample tests often work better on low-dimensional data than on high-dimensional data, it's highly recommended that you reduce the dimensionality of your data before performing a two-sample test on it.[36]

---

32 I. M. Chakravarti, R. G. Laha, and J. Roy, *Handbook of Methods of Applied Statistics*, vol. 1, *Techniques of Computation, Descriptive Methods, and Statistical Inference* (New York: Wiley, 1967).

33 Eric Feigelson and G. Jogesh Babu, "Beware the Kolmogorov-Smirnov Test!" Center for Astrostatistics, Penn State University, *https://oreil.ly/7AHcT*.

34 Eric Breck, Marty Zinkevich, Neoklis Polyzotis, Steven Whang, and Sudip Roy, "Data Validation for Machine Learning," *Proceedings of SysML*, 2019, *https://oreil.ly/xoneh*.

35 Li Bu, Cesare Alippi, and Dongbin Zhao, "A pdf-Free Change Detection Test Based on Density Difference Estimation," *IEEE Transactions on Neural Networks and Learning Systems* 29, no. 2 (February 2018): 324–34, *https://oreil.ly/RD8Uy*. The authors claim that the method works on multidimensional inputs.

36 Stephan Rabanser, Stephan Günnemann, and Zachary C. Lipton, "Failing Loudly: An Empirical Study of Methods for Detecting Dataset Shift," *arXiv*, October 29, 2018, *https://oreil.ly/HxAwV*.

Detector	Tabular	Image	Time Series	Text	Categorical Features	Online	Feature Level
Kolmogorov-Smirnov	✓	✓		✓	✓		✓
Cramér-von Mises	✓	✓				✓	✓
Fisher's Exact Test	✓				✓	✓	✓
Maximum Mean Discrepancy (MMD)	✓	✓		✓	✓	✓	
Learned Kernel MMD	✓	✓		✓	✓		
Context-aware MMD	✓	✓	✓	✓	✓		
Least-Squares Density Difference	✓	✓		✓	✓	✓	
Chi-Squared	✓				✓		✓
Mixed-type tabular data	✓				✓		✓
Classifier	✓	✓	✓	✓	✓		
Spot-the-diff	✓	✓	✓	✓	✓		✓
Classifier Uncertainty	✓	✓	✓	✓	✓		
Regressor Uncertainty	✓	✓	✓	✓	✓		

Figure 8-2. Some drift detection algorithms implemented by Alibi Detect (https://oreil.ly/162tf). Source: Screenshot of the project's GitHub repository

Time scale windows for detecting shifts

Not all types of shifts are equal—some are harder to detect than others. For example, shifts happen at different rates, and abrupt changes are easier to detect than slow, gradual changes.[37] Shifts can also happen across two dimensions: spatial or temporal. Spatial shifts are shifts that happen across access points, such as your application gets a new group of users or your application is now served on a different type of device. Temporal shifts are shifts that happen over time. To detect temporal shifts, a common approach is to treat input data to ML applications as time-series data.[38]

37 Manuel Baena-García, José del Campo-Ávila, Raúl Fidalgo, Albert Bifet, Ricard Gavaldà, and Rafael Morales-Bueno, "Early Drift Detection Method," 2006, *https://oreil.ly/DnvOs.*

38 Nandini Ramanan, Rasool Tahmasbi, Marjorie Sayer, Deokwoo Jung, Shalini Hemachandran, and Claudionor Nunes Coelho Jr., "Real-time Drift Detection on Time-series Data," *arXiv*, October 12, 2021, *https://oreil.ly/xmdqW.*

When dealing with temporal shifts, the time scale window of the data we look at affects the shifts we can detect. If your data has a weekly cycle, then a time scale of less than a week won't detect the cycle. Consider the data in Figure 8-3. If we use data from day 9 to day 14 as the source distribution, then day 15 looks like a shift. However, if we use data from day 1 to day 14 as the source distribution, then all data points from day 15 are likely being generated by that same distribution. As illustrated by this example, detecting temporal shifts is hard when shifts are confounded by seasonal variation.

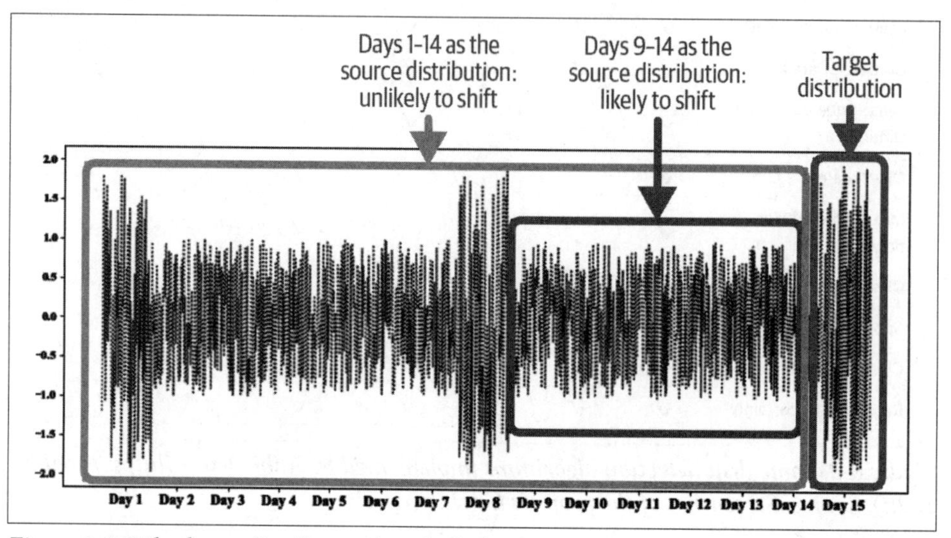

Figure 8-3. Whether a distribution has drifted over time depends on the time scale window specified

When computing running statistics over time, it's important to differentiate between *cumulative and sliding statistics*. Sliding statistics are computed within a single time scale window, e.g., an hour. Cumulative statistics are continually updated with more data. This means, for the beginning of each time scale window, the sliding accuracy is reset, whereas the cumulative sliding accuracy is not. Because cumulative statistics contain information from previous time windows, they might obscure what happens in a specific time window. Figure 8-4 shows an example of how cumulative accuracy can hide the sudden dip in accuracy between hours 16 and 18.

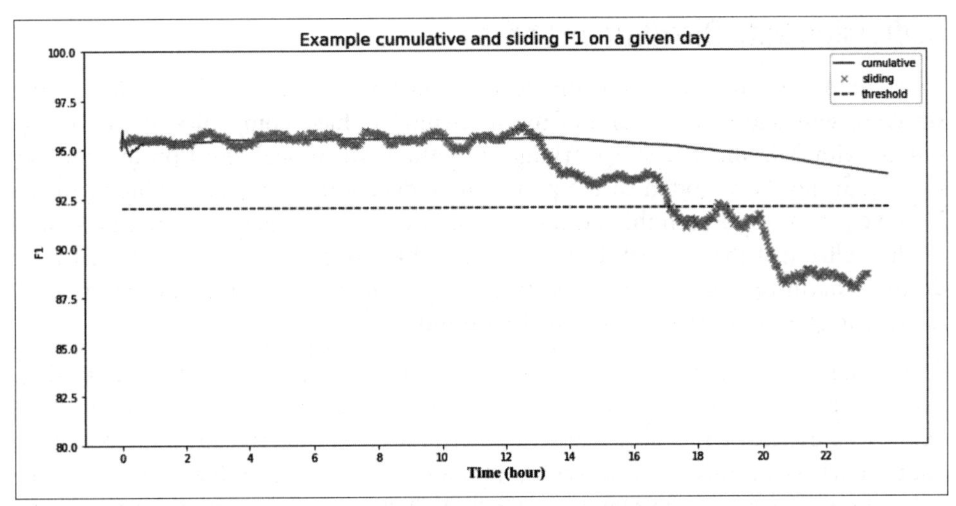

Figure 8-4. Cumulative accuracy hides the sudden dip in accuracy between hours 16 and 18. Source: Adapted from an image by MadeWithML (https://oreil.ly/viegx)

Working with data in the temporal space makes things so much more complicated, requiring knowledge of time-series analysis techniques such as time-series decompositions that are beyond the scope of this book. For readers interested in time-series decomposition, Lyft engineering has a great case study (*https://oreil.ly/zi1kk*) on how they decompose their time-series data to deal with the seasonality of the market.

As of today, many companies use the distribution of the training data as the base distribution and monitor the production data distribution at a certain granularity level, such as hourly and daily.[39] The shorter your time scale window, the faster you'll be able to detect changes in your data distribution. However, too short a time scale window can lead to false alarms of shifts, like the example in Figure 8-3.

Some platforms, especially those dealing with real-time data analytics such as monitoring, provide a merge operation that allows merging statistics from shorter time scale windows to create statistics for larger time scale windows. For example, you can compute the data statistics you care about hourly, then merge these hourly statistics chunks into daily views.

More advanced monitoring platforms even attempt a root cause analysis (RCA) feature that automatically analyzes statistics across various time window sizes to detect exactly the time window where a change in data happened.[40]

39 I'm working on a solution that can handle the minute granularity level.

40 Thanks Goku Mohandas for sharing this tip on the MLOps Discord server (*https://oreil.ly/UOJ8h*).

Addressing Data Distribution Shifts

How companies address data shifts depends on how sophisticated their ML infrastructure setups are. At one end of the spectrum, we have companies that have just started with ML and are still working on getting ML models into production, so they might not have gotten to the point where data shifts are catastrophic to them. However, at some point in the future—maybe three months, maybe six months—they might realize that their initial deployed models have degraded to the point that they do more harm than good. They will then need to adapt their models to the shifted distributions or to replace them with other solutions.

At the same time, many companies assume that data shifts are inevitable, so they periodically retrain their models—once a month, once a week, or once a day—regardless of the extent of the shift. How to determine the optimal frequency to retrain your models is an important decision that many companies still determine based on gut feelings instead of experimental data.[41] We'll discuss more about the retraining frequency in Chapter 9.

To make a model work with a new distribution in production, there are three main approaches. The first is the approach that currently dominates research: train models using massive datasets. The hope here is that if the training dataset is large enough, the model will be able to learn such a comprehensive distribution that whatever data points the model will encounter in production will likely come from this distribution.

The second approach, less popular in research, is to adapt a trained model to a target distribution *without requiring new labels*. Zhang et al. (2013) used causal interpretations together with kernel embedding of conditional and marginal distributions to correct models' predictions for both covariate shifts and label shifts without using labels from the target distribution.[42] Similarly, Zhao et al. (2020) proposed domain-invariant representation learning: an unsupervised domain adaptation technique that can learn data representations invariant to changing distributions.[43] However, this area of research is heavily underexplored and hasn't found wide adoption in industry.[44]

---

41 As Han-chung Lee, one early reviewer, pointed out, this is also because smaller companies don't have enough data on their models. When you don't have a lot of data, it's better to have a time-based regimen than to overfit your regime to insufficient data.

42 Kun Zhang, Bernhard Schölkopf, Krikamol Muandet, and Zhikun Wang, "Domain Adaptation under Target and Conditional Shift," *Proceedings of the 30th International Conference on Machine Learning* (2013), *https://oreil.ly/C123l*.

43 Han Zhao, Remi Tachet Des Combes, Kun Zhang, and Geoffrey Gordon, "On Learning Invariant Representations for Domain Adaptation," *Proceedings of Machine Learning Research* 97 (2019): 7523–32, *https://oreil.ly/W78hH*.

44 Zachary C. Lipton, Yu-Xiang Wang, and Alex Smola, "Detecting and Correcting for Label Shift with Black Box Predictors," *arXiv*, February 12, 2018, *https://oreil.ly/zKSlj*.

The third approach is what is usually done in the industry today: retrain your model using the labeled data from the target distribution. However, retraining your model is not so straightforward. Retraining can mean retraining your model from scratch on both the old and new data or continuing training the existing model on new data. The latter approach is also called fine-tuning.

If you want to retrain your model, there are two questions. First, whether to train your model from scratch (stateless retraining) or continue training it from the last checkpoint (stateful training). Second, what data to use: data from the last 24 hours, last week, last 6 months, or from the point when data has started to drift. You might need to run experiments to figure out which retraining strategy works best for you.[45]

In this book, we use "retraining" to refer to both training from scratch and fine-tuning. We'll discuss more about retraining strategy in the next chapter.

Readers familiar with data shift literature might often see data shifts mentioned along with domain adaptation and transfer learning. If you consider a distribution to be a domain, then the question of how to adapt your model to new distributions is similar to the question of how to adapt your model to different domains.

Similarly, if you consider learning a joint distribution $P(X, Y)$ as a task, then adapting a model trained on one joint distribution for another joint distribution can be framed as a form of transfer learning. As discussed in Chapter 4, transfer learning refers to the family of methods where a model developed for a task is reused as the starting point for a model on a second task. The difference is that with transfer learning, you don't retrain the base model from scratch for the second task. However, to adapt your model to a new distribution, you might need to retrain your model from scratch.

Addressing data distribution shifts doesn't have to start after the shifts have happened. It's possible to design your system to make it more robust to shifts. A system uses multiple features, and different features shift at different rates. Consider that you're building a model to predict whether a user will download an app. You might be tempted to use that app's ranking in the app store as a feature since higher-ranking apps tend to be downloaded more. However, app ranking changes very quickly. You might want to instead bucket each app's ranking into general categories such as top 10, between 11 and 100, between 101 and 1,000, between 1,001 and 10,000, and so on. At the same time, an app's categories might change a lot less frequently, but they might have less power to predict whether a user will download that app. When choosing features for your models, you might want to consider the trade-off between the performance and the stability of a feature: a feature might be really good for accuracy but deteriorate quickly, forcing you to train your model more often.

45 Some monitoring vendors claim that their solutions are able to detect not only when your model should be retrained, but also what data to retrain on. I haven't been able to verify the validity of these claims.

You might also want to design your system to make it easier for it to adapt to shifts. For example, housing prices might change a lot faster in major cities like San Francisco than in rural Arizona, so a housing price prediction model serving rural Arizona might need to be updated less frequently than a model serving San Francisco. If you use the same model to serve both markets, you'll have to use data from both markets to update your model at the rate demanded by San Francisco. However, if you use a separate model for each market, you can update each of them only when necessary.

Before we move on to the next section, I want to reiterate that not all performance degradation of models in production requires ML solutions. Many ML failures today are still caused by human errors. If your model failure is caused by human errors, you'd first need to find those errors to fix them. Detecting a data shift is hard, but determining what causes a shift can be even harder.

Monitoring and Observability

As the industry realized that many things can go wrong with an ML system, many companies started investing in monitoring and observability for their ML systems in production.

Monitoring and observability are sometimes used exchangeably, but they are different. Monitoring refers to the act of tracking, measuring, and logging different metrics that can help us determine when something goes wrong. Observability means setting up our system in a way that gives us visibility into our system to help us investigate what went wrong. The process of setting up our system in this way is also called "instrumentation." Examples of instrumentation are adding timers to your functions, counting NaNs in your features, tracking how inputs are transformed through your systems, logging unusual events such as unusually long inputs, etc. Observability is part of monitoring. Without some level of observability, monitoring is impossible.

Monitoring is all about metrics. Because ML systems are software systems, the first class of metrics you'd need to monitor are the operational metrics. These metrics are designed to convey the health of your systems. They are generally divided into three levels: the network the system is run on, the machine the system is run on, and the application that the system runs. Examples of these metrics are latency; throughput; the number of prediction requests your model receives in the last minute, hour, day; the percentage of requests that return with a 2xx code; CPU/GPU utilization; memory utilization; etc. No matter how good your ML model is, if the system is down, you're not going to benefit from it.

Let's look at an example. One of the most important characteristics of a software system in production is availability—how often the system is available to offer reasonable performance to users. This characteristic is measured by *uptime*, the percentage of time a system is up. The conditions to determine whether a system is up are defined in the service level objectives (SLOs) or service level agreements (SLAs). For example, an SLA may specify that the service is considered to be up if it has a median latency of less than 200 ms and a 99th percentile under 2 s.

A service provider might offer an SLA that specifies their uptime guarantee, such as 99.99% of the time, and if this guarantee is not met, they'll give their customers back money. For example, as of October 2021, AWS EC2 service offers a monthly uptime percentage of at least 99.99% (four nines), and if the monthly uptime percentage is lower than that, they'll give you back a service credit toward future EC2 payments.[46] A 99.99% monthly uptime means the service is only allowed to be down a little over 4 minutes a month, and 99.999% means only 26 seconds a month!

However, for ML systems, the system health extends beyond the system uptime. If your ML system is up but its predictions are garbage, your users aren't going to be happy. Another class of metrics you'd want to monitor are ML-specific metrics that tell you the health of your ML models.

ML-Specific Metrics

Within ML-specific metrics, there are generally four artifacts to monitor: a model's accuracy-related metrics, predictions, features, and raw inputs. These are artifacts generated at four different stages of an ML system pipeline, as shown in Figure 8-5. The deeper into the pipeline an artifact is, the more transformations it has gone through, which makes a change in that artifact more likely to be caused by errors in one of those transformations. However, the more transformations an artifact has gone through, the more structured it's become and the closer it is to the metrics you actually care about, which makes it easier to monitor. We'll look at each of these artifacts in detail in the following sections.

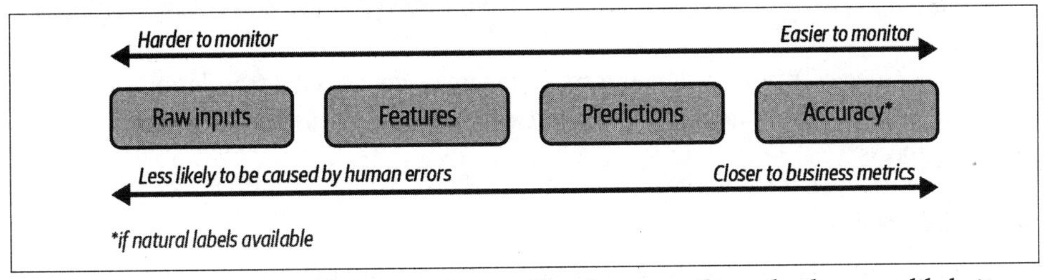

Figure 8-5. The more transformations an artifact has gone through, the more likely its changes are to be caused by errors in one of those transformations

46 "Amazon Compute Service Level Agreement," Amazon Web Services, last updated August 24, 2021, *https://oreil.ly/5bjx9*.

Monitoring accuracy-related metrics

If your system receives any type of user feedback for the predictions it makes—click, hide, purchase, upvote, downvote, favorite, bookmark, share, etc.—you should definitely log and track it. Some feedback can be used to infer natural labels, which can then be used to calculate your model's accuracy-related metrics. Accuracy-related metrics are the most direct metrics to help you decide whether a model's performance has degraded.

Even if the feedback can't be used to infer natural labels directly, it can be used to detect changes in your ML model's performance. For example, when you're building a system to recommend to users what videos to watch next on YouTube, you want to track not only whether the users click on a recommended video (click-through rate), but also the duration of time users spend on that video and whether they complete watching it (completion rate). If, over time, the click-through rate remains the same but the completion rate drops, it might mean that your recommender system is getting worse.[47]

It's also possible to engineer your system so that you can collect users' feedback. For example, Google Translate has the option for users to upvote or downvote a translation, as shown in Figure 8-6. If the number of downvotes the system receives suddenly goes up, there might be issues. These downvotes can also be used to guide the labeling process, such as getting human experts to generate new translations for the samples with downvotes, to train the next iteration of their models.

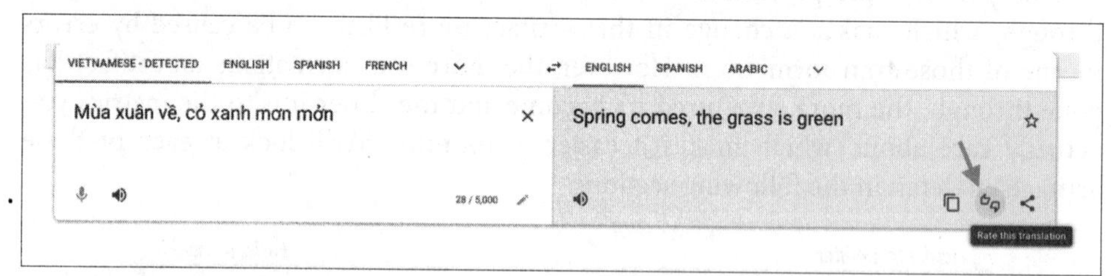

Figure 8-6. Google Translate allows users to upvote or downvote a translation. These votes will be used to evaluate their translation model's quality as well as to guide the labeling process.

Monitoring predictions

Prediction is the most common artifact to monitor. If it's a regression task, each prediction is a continuous value (e.g., the predicted price of a house), and if it's a classification task, each prediction is a discrete value corresponding to the predicted category. Because each prediction is usually just a number (low dimension),

47 Be careful when using the completion rate as a metric to optimize for, as it might bias your recommender system toward short videos.

predictions are easy to visualize, and their summary statistics are straightforward to compute and interpret.

You can monitor predictions for distribution shifts. Because predictions are low dimensional, it's also easier to compute two-sample tests to detect whether the prediction distribution has shifted. Prediction distribution shifts are also a proxy for input distribution shifts. Assuming that the function that maps from input to output doesn't change—the weights and biases of your model haven't changed—then a change in the prediction distribution generally indicates a change in the underlying input distribution.

You can also monitor predictions for anything odd happening, such as predicting an unusual number of False in a row. There could be a long delay between predictions and ground truth labels, as discussed in the section "Natural Labels" on page 91. Changes in accuracy-related metrics might not become obvious for days or weeks, whereas a model predicting all False for 10 minutes can be detected immediately.

Monitoring features

ML monitoring solutions in the industry focus on tracking changes in features, both the features that a model uses as inputs and the intermediate transformations from raw inputs into final features. Feature monitoring is appealing because compared to raw input data, features are well structured following a predefined schema. The first step of feature monitoring is *feature validation*: ensuring that your features follow an expected schema. The expected schemas are usually generated from training data or from common sense. If these expectations are violated in production, there might be a shift in the underlying distribution. For example, here are some of the things you can check for a given feature:

- If the min, max, or median values of a feature are within an acceptable range
- If the values of a feature satisfy a regular expression format
- If all the values of a feature belong to a predefined set
- If the values of a feature are always greater than the values of another feature

Because features are often organized into tables—each column representing a feature and each row representing a data sample—feature validation is also known as table testing or table validation. Some call them unit tests for data. There are many open source libraries that help you do basic feature validation, and the two most common are Great Expectations (*https://oreil.ly/vBa35*) and Deequ (*https://oreil.ly/OWoIB*), which is by AWS. Figure 8-7 shows some of the built-in feature validation functions by Great Expectations and an example of how to use them.

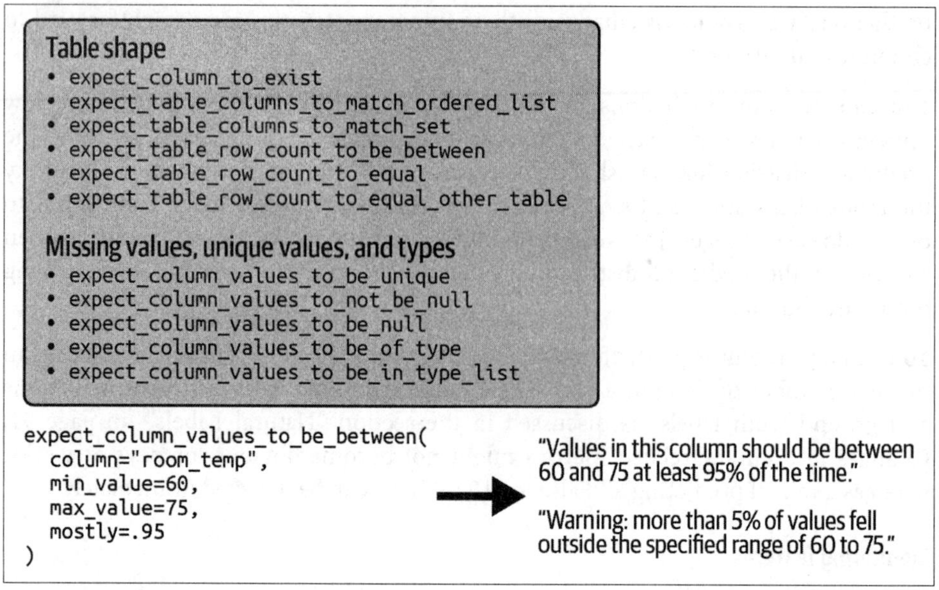

Table shape
- expect_column_to_exist
- expect_table_columns_to_match_ordered_list
- expect_table_columns_to_match_set
- expect_table_row_count_to_be_between
- expect_table_row_count_to_equal
- expect_table_row_count_to_equal_other_table

Missing values, unique values, and types
- expect_column_values_to_be_unique
- expect_column_values_to_not_be_null
- expect_column_values_to_be_null
- expect_column_values_to_be_of_type
- expect_column_values_to_be_in_type_list

```
expect_column_values_to_be_between(
  column="room_temp",
  min_value=60,
  max_value=75,
  mostly=.95
)
```

"Values in this column should be between 60 and 75 at least 95% of the time."

"Warning: more than 5% of values fell outside the specified range of 60 to 75."

Figure 8-7. Some of the built-in feature validation functions by Great Expectations and an example of how to use them. Source: Adapted from content in the Great Expectations GitHub repository

Beyond basic feature validation, you can also use two-sample tests to detect whether the underlying distribution of a feature or a set of features has shifted. Since a feature or a set of features can be high-dimensional, you might need to reduce their dimension before performing the test on them, which can make the test less effective.

There are four major concerns when doing feature monitoring:

A company might have hundreds of models in production, and each model uses hundreds, if not thousands, of features.
: Even something as simple as computing summary statistics for all these features every hour can be expensive, not only in terms of compute required but also memory used. Tracking, i.e., constantly computing, too many metrics can also slow down your system and increase both the latency that your users experience and the time it takes for you to detect anomalies in your system.

While tracking features is useful for debugging purposes, it's not very useful for detecting model performance degradation.
: In theory, a small distribution shift can cause catastrophic failure, but in practice, an individual feature's minor changes might not harm the model's performance at all. Feature distributions shift all the time, and most of these changes are

benign.[48] If you want to be alerted whenever a feature seems to have drifted, you might soon be overwhelmed by alerts and realize that most of these alerts are false positives. This can cause a phenomenon called "alert fatigue" where the monitoring team stops paying attention to the alerts because they are so frequent. The problem of feature monitoring becomes the problem of trying to decide which feature shifts are critical and which are not.

Feature extraction is often done in multiple steps (such as filling missing values and standardization), using multiple libraries (such as pandas, Spark), on multiple services (such as BigQuery or Snowflake).
You might have a relational database as an input to the feature extraction process and a NumPy array as the output. Even if you detect a harmful change in a feature, it might be impossible to detect whether this change is caused by a change in the underlying input distribution or whether it's caused by an error in one of the multiple processing steps.

The schema that your features follow can change over time.
If you don't have a way to version your schemas and map each of your features to its expected schema, the cause of the reported alert might be due to the mismatched schema rather than a change in the data.

These concerns are not to dismiss the importance of feature monitoring; changes in the feature space are a useful source of signals to understand the health of your ML systems. Hopefully, thinking about these concerns can help you choose a feature monitoring solution that works for you.

Monitoring raw inputs

As discussed in the previous section, a change in the features might be caused by problems in processing steps and not by changes in data. What if we monitor the raw inputs before they are processed? The raw input data might not be easier to monitor, as it can come from multiple sources in different formats, following multiple structures. The way many ML workflows are set up today also makes it impossible for ML engineers to get direct access to raw input data, as the raw input data is often managed by a data platform team who processes and moves the data to a location like a data warehouse, and the ML engineers can only query for data from that data warehouse where the data is already partially processed. Therefore, monitoring raw inputs is often a responsibility of the data platform team, not the data science or ML team. Therefore, it's out of scope for this book.

48 Rabanser, Günnemann, and Lipton, "Failing Loudly."

So far, we've discussed different types of metrics to monitor, from operational metrics generally used for software systems to ML-specific metrics that help you keep track of the health of your ML models. In the next section, we'll discuss the toolbox you can use to help with metrics monitoring.

Monitoring Toolbox

Measuring, tracking, and interpreting metrics for complex systems is a nontrivial task, and engineers rely on a set of tools to help them do so. It's common for the industry to herald metrics, logs, and traces as the three pillars of monitoring. However, I find their differentiations murky. They seem to be generated from the perspective of people who develop monitoring systems: traces are a form of logs and metrics can be computed from logs. In this section, I'd like to focus on the set of tools from the perspective of users of the monitoring systems: logs, dashboards, and alerts.

Logs

Traditional software systems rely on logs to record events produced at runtime. An event is anything that can be of interest to the system developers, either at the time the event happens or later for debugging and analysis purposes. Examples of events are when a container starts, the amount of memory it takes, when a function is called, when that function finishes running, the other functions that this function calls, the input and output of that function, etc. Also, don't forget to log crashes, stack traces, error codes, and more. In the words of Ian Malpass at Etsy, "If it moves, we track it."[49] They also track things that haven't changed yet, in case they'll move later.

The number of logs can grow very large very quickly. For example, back in 2019, the dating app Badoo was handling 20 billion events a day.[50] When something goes wrong, you'll need to query your logs for the sequence of events that caused it, a process that can feel like searching for a needle in a haystack.

In the early days of software deployment, an application might be one single service. When something happened, you knew where that happened. But today, a system might consist of many different components: containers, schedulers, microservices, polyglot persistence, mesh routing, ephemeral auto-scaling instances, serverless Lambda functions. A request may do 20–30 hops from when it's sent until when a response is received. The hard part might not be in detecting when something happened, but where the problem was.[51]

---

49 Ian Malpass, "Measure Anything, Measure Everything," *Code as Craft*, February 15, 2011, *https://oreil.ly/3KF1K*.

50 Andrew Morgan, "Data Engineering in Badoo: Handling 20 Billion Events Per Day," InfoQ, August 9, 2019, *https://oreil.ly/qnnuV*.

51 Charity Majors, "Observability—A 3-Year Retrospective," *The New Stack*, August 6, 2019, *https://oreil.ly/Logby*.

When we log an event, we want to make it as easy as possible for us to find it later. This practice with microservice architecture is called *distributed tracing*. We want to give each process a unique ID so that, when something goes wrong, the error message will (hopefully) contain that ID. This allows us to search for the log messages associated with it. We also want to record with each event all the metadata necessary: the time when it happens, the service where it happens, the function that is called, the user associated with the process, if any, etc.

Because logs have grown so large and so difficult to manage, there have been many tools developed to help companies manage and analyze logs. The log management market is estimated to be worth USD 2.3 billion in 2021, and it's expected to grow to USD 4.1 billion by 2026.[52]

Analyzing billions of logged events manually is futile, so many companies use ML to analyze logs. An example use case of ML in log analysis is anomaly detection: to detect abnormal events in your system. A more sophisticated model might even classify each event in terms of its priorities such as usual, abnormal, exception, error, and fatal.

Another use case of ML in log analysis is that when a service fails, it might be helpful to know the probability of related services being affected. This could be especially useful when the system is under cyberattack.

Many companies process logs in batch processes. In this scenario, you collect a large number of logs, then periodically query over them looking for specific events using SQL or process them using a batch process like in a Spark or Hadoop or Hive cluster. This makes the processing of logs efficient because you can leverage distributed and MapReduce processes to increase your processing throughput. However, because you process your logs periodically, you can only discover problems periodically.

To discover anomalies in your logs as soon as they happen, you want to process your events as soon as they are logged. This makes log processing a stream processing problem.[53] You can use real-time transport such as Kafka or Amazon Kinesis to transport events as they are logged. To search for events with specific characteristics in real time, you can leverage a streaming SQL engine like KSQL or Flink SQL.

52 "Log Management Market Size, Share and Global Market Forecast to 2026," MarketsandMarkets, 2021, *https://oreil.ly/q0xgh*.

53 For readers unfamiliar with stream processing, please refer to the section "Batch Processing Versus Stream Processing" on page 78.

Dashboards

A picture is worth a thousand words. A series of numbers might mean nothing to you, but visualizing them on a graph might reveal the relationships among these numbers. Dashboards to visualize metrics are critical for monitoring.

Another use of dashboards is to make monitoring accessible to nonengineers. Monitoring isn't just for the developers of a system, but also for nonengineering stakeholders including product managers and business developers.

Even though graphs can help a lot with understanding metrics, they aren't sufficient on their own. You still need experience and statistical knowledge. Consider the two graphs in Figure 8-8. The only thing that is obvious from these graphs is that the loss fluctuates a lot. If there's a distribution shift in any of these two graphs, I can't tell. It's easier to plot a graph to draw a wiggling line than to understand what this wiggly line means.

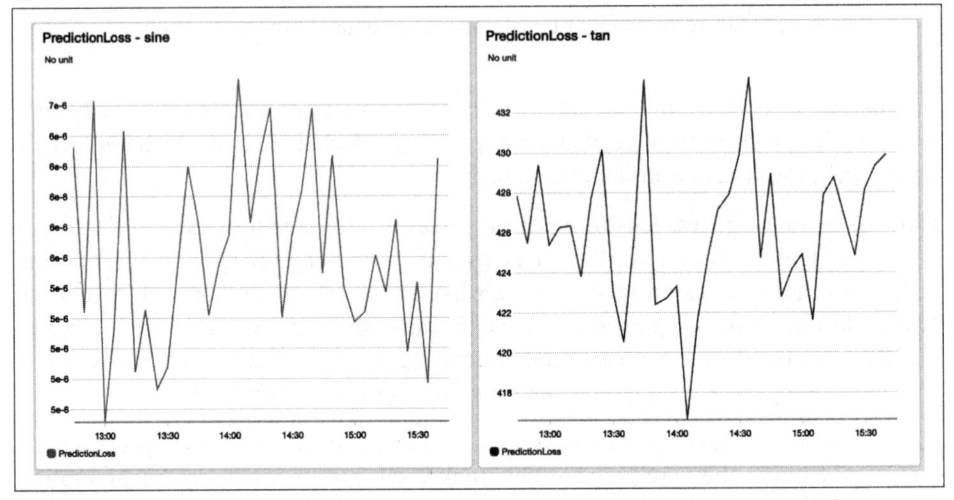

Figure 8-8. Graphs are useful for making sense of numbers, but they aren't sufficient

Excessive metrics on a dashboard can also be counterproductive, a phenomenon known as *dashboard rot*. It's important to pick the right metrics or abstract out lower-level metrics to compute higher-level signals that make better sense for your specific tasks.

Alerts

When our monitoring system detects something suspicious, it's necessary to alert the right people about it. An alert consists of the following three components:

An alert policy
: This describes the condition for an alert. You might want to create an alert when a metric breaches a threshold, optionally over a certain duration. For example, you might want to be notified when a model's accuracy is under 90%, or that the HTTP response latency is higher than a second for at least 10 minutes.

Notification channels
: These describe who is to be notified when the condition is met. The alerts will be shown in the monitoring service you employ, such as Amazon CloudWatch or GCP Cloud Monitoring, but you also want to reach responsible people when they're not on these monitoring services. For example, you might configure your alerts to be sent to an email address such as mlops-monitoring@*[your company email domain]*, or to post to a Slack channel such as #mlops-monitoring or to PagerDuty.

A description of the alert
: This helps the alerted person understand what's going on. The description should be as detailed as possible, such as:

```
## Recommender model accuracy below 90%

${timestamp}: This alert originated from the service ${service-name}
```

Depending on the audience of the alert, it's often necessary to make the alert actionable by providing mitigation instructions or a runbook (*https://oreil.ly/vgLR8*), a compilation of routine procedures and operations that might help with handling the alert.

Alert fatigue is a real phenomenon, as discussed previously in this chapter. Alert fatigue can be demoralizing—nobody likes to be awakened in the middle of the night for something outside of their responsibilities. It's also dangerous—being exposed to trivial alerts can desensitize people to critical alerts. It's important to set meaningful conditions so that only critical alerts are sent out.

Observability

Since the mid-2010s, the industry has started embracing the term "observability" instead of "monitoring." Monitoring makes no assumption about the relationship between the internal state of a system and its outputs. You monitor the external outputs of the system to figure out *when* something goes wrong inside the system—there's no guarantee that the external outputs will help you figure out *what* goes wrong.

In the early days of software deployment, software systems were simple enough that monitoring external outputs was sufficient for software maintenance. A system used to consist of only a few components, and a team used to have control over the entire codebase. If something went wrong, it was possible to make changes to the system to test and figure out what went wrong.

However, software systems have grown significantly more complex over the last decade. Today, a software system consists of many components. Many of these components are services run by other companies—cue all cloud native services—which means that a team doesn't even have control of the inside of all the components of their system. When something goes wrong, a team can no longer just break apart their system to find out. The team has to rely on external outputs of their system to figure out what's going on internally.

Observability is a term used to address this challenge. It's a concept drawn from control theory, and it refers to bringing "better visibility into understanding the complex behavior of software using [outputs] collected from the system at run time."[54]

Telemetry

A system's outputs collected at runtime are also called *telemetry*. Telemetry is another term that has emerged in the software monitoring industry over the last decade. The word "telemetry" comes from the Greek roots *tele*, meaning "remote," and *metron*, meaning "measure." So telemetry basically means "remote measures." In the monitoring context, it refers to logs and metrics collected from remote components such as cloud services or applications run on customer devices.

In other words, observability makes an assumption stronger than traditional monitoring: that the internal states of a system can be inferred from knowledge of its external outputs. Internal states can be current states, such as "the GPU utilization right now," and historical states, such as "the average GPU utilization over the last day."

When something goes wrong with an observable system, we should be able to figure out what went wrong by looking at the system's logs and metrics without having to ship new code to the system. Observability is about instrumenting your system in a way to ensure that sufficient information about a system's runtime is collected and analyzed.

54 Suman Karumuri, Franco Solleza, Stan Zdonik, and Nesime Tatbul, "Towards Observability Data Management at Scale," *ACM SIGMOD Record* 49, no. 4 (December 2020): 18–23, *https://oreil.ly/oS5hn*.

Monitoring centers around metrics, and metrics are usually aggregated. Observability allows more fine-grain metrics, so that you can know not only when a model's performance degrades but also for what types of inputs or what subgroups of users or over what period of time the model degrades. For example, you should be able to query your logs for the answers to questions like: "show me all the users for which model A returned wrong predictions over the last hour, grouped by their zip codes" or "show me the outliers requests in the last 10 minutes" or "show me all the intermediate outputs of this input through the system." To achieve this, you need to have logged your system's outputs using tags and other identifying keywords to allow these outputs to later be sliced and diced along different dimensions of your data.

In ML, observability encompasses interpretability. Interpretability helps us understand how an ML model works, and observability helps us understand how the entire ML system, which includes the ML model, works. For example, when a model's performance degrades over the last hour, being able to interpret which feature contributes the most to all the wrong predictions made over the last hour will help with figuring out what went wrong with the system and how to fix it.[55]

In this section, we've discussed multiple aspects of monitoring, from what data to monitor and what metrics to keep track of to different tools for monitoring and observability. Even though monitoring is a powerful concept, it's inherently *passive*. You wait for a shift to happen to detect it. Monitoring helps unearth the problem without correcting it. In the next section, we'll introduce continual learning, a paradigm that can *actively* help you update your models to address shifts.

Summary

This might have been the most challenging chapter for me to write in this book. The reason is that despite the importance of understanding how and why ML systems fail in production, the literature surrounding it is limited. We usually think of research preceding production, but this is an area of ML where research is still trying to catch up with production.

To understand failures of ML systems, we differentiated between two types of failures: software systems failures (failures that also happen to non-ML systems) and ML-specific failures. Even though the majority of ML failures today are non-ML-specific, as tooling and infrastructure around MLOps matures, this might change.

We discussed three major causes of ML-specific failures: production data differing from training data, edge cases, and degenerate feedback loops. The first two causes are related to data, whereas the last cause is related to system design because it happens when the system's outputs influence the same system's input.

55 See the section "Feature Importance" on page 142.

We zeroed into one failure that has gathered much attention in recent years: data distribution shifts. We looked into three types of shifts: covariate shift, label shift, and concept drift. Even though studying distribution shifts is a growing subfield of ML research, the research community hasn't yet found a standard narrative. Different papers call the same phenomena by different names. Many studies are still based on the assumption that we know in advance how the distribution will shift or have the labels for the data from both the source distribution and the target distribution. However, in reality, we don't know what the future data will be like, and obtaining labels for new data might be costly, slow, or just infeasible.

To be able to detect shifts, we need to monitor our deployed systems. Monitoring is an important set of practices for any software engineering system in production, not just ML, and it's an area of ML where we should learn as much as we can from the DevOps world.

Monitoring is all about metrics. We discussed different metrics we need to monitor: operational metrics—the metrics that should be monitored with any software systems such as latency, throughput, and CPU utilization—and ML-specific metrics. Monitoring can be applied to accuracy-related metrics, predictions, features, and/or raw inputs.

Monitoring is hard because even if it's cheap to compute metrics, understanding metrics isn't straightforward. It's easy to build dashboards to show graphs, but it's much more difficult to understand what a graph means, whether it shows signs of drift, and, if there's drift, whether it's caused by an underlying data distribution change or by errors in the pipeline. An understanding of statistics might be required to make sense of the numbers and graphs.

Detecting model performance's degradation in production is the first step. The next step is how to adapt our systems to changing environments, which we'll discuss in the next chapter.

CHAPTER 9

Continual Learning and Test in Production

In Chapter 8, we discussed various ways an ML system can fail in production. We focused on one especially thorny problem that has generated much discussion among both researchers and practitioners: data distribution shifts. We also discussed multiple monitoring techniques and tools to detect data distribution shifts.

This chapter is a continuation of this discussion: how do we adapt our models to data distribution shifts? The answer is by continually updating our ML models. We'll start with a discussion on what continual learning is and its challenges—spoiler: continual learning is largely an infrastructural problem. Then we'll lay out a four-stage plan to make continual learning a reality.

After you've set up your infrastructure to allow you to update your models as frequently as you want, you might want to consider the question that I've been asked by almost every single ML engineer I've met: "How often should I retrain my models?" This question is the focus of the next section of the book.

If the model is retrained to adapt to the changing environment, evaluating it on a stationary test set isn't enough. We'll cover a seemingly terrifying but necessary concept: test in production. This process is a way to test your systems with live data in production to ensure that your updated model indeed works without catastrophic consequences.

Topics in this chapter and the previous chapter are tightly coupled. Test in production is complementary to monitoring. If monitoring means passively keeping track of the outputs of whatever model is being used, test in production means proactively choosing which model to produce outputs so that we can evaluate it. The goal of both monitoring and test in production is to understand a model's performance and figure out when to update it. The goal of continual learning is to safely and efficiently automate the update. All of these concepts allow us to design an ML system that is maintainable and adaptable to changing environments.

This is the chapter I'm most excited to write about, and I hope that I can get you excited about it too!

Continual Learning

When hearing "continual learning," many people think of the training paradigm where a model updates itself with every incoming sample in production. Very few companies actually do that. First, if your model is a neural network, learning with every incoming sample makes it susceptible to catastrophic forgetting. Catastrophic forgetting refers to the tendency of a neural network to completely and abruptly forget previously learned information upon learning new information.[1]

Second, it can make training more expensive—most hardware backends today were designed for batch processing, so processing only one sample at a time causes a huge waste of compute power and is unable to exploit data parallelism.

Companies that employ continual learning in production update their models in micro-batches. For example, they might update the existing model after every 512 or 1,024 examples—the optimal number of examples in each micro-batch is task dependent.

The updated model shouldn't be deployed until it's been evaluated. This means that you shouldn't make changes to the existing model directly. Instead, you create a replica of the existing model and update this replica on new data, and only replace the existing model with the updated replica if the updated replica proves to be better. The existing model is called the champion model, and the updated replica, the challenger. This process is shown in Figure 9-1. This is an oversimplification of the process for the sake of understanding. In reality, a company might have multiple challengers at the same time, and handling the failed challenger is a lot more sophisticated than simply discarding it.

1 Joan Serrà, Dídac Surís, Marius Miron, and Alexandros Karatzoglou, "Overcoming Catastrophic Forgetting with Hard Attention to the Task," *arXiv*, January 4, 2018, *https://oreil.ly/P95EZ*.

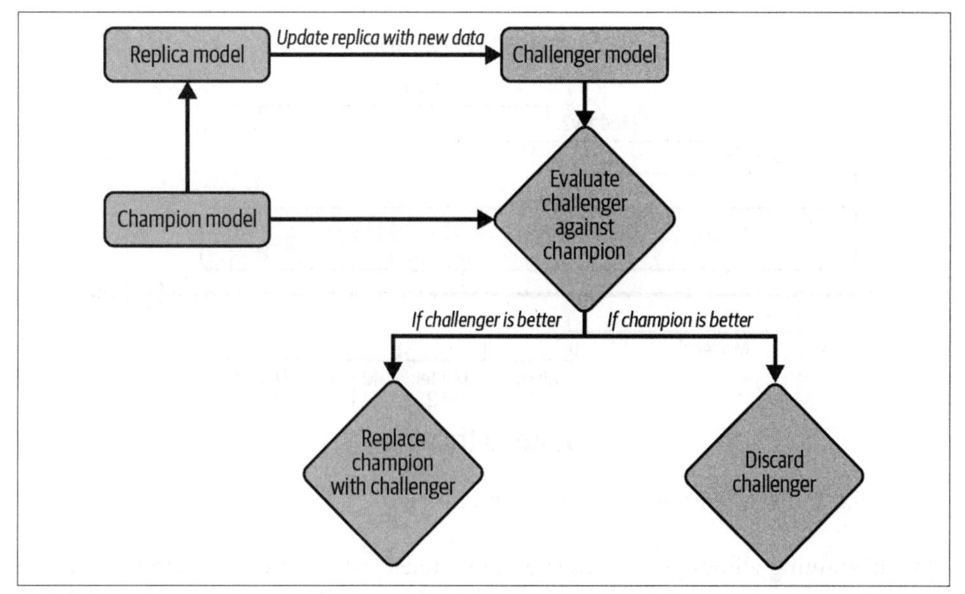

Figure 9-1. A simplification of how continual learning might work in production. In reality, the process of handling the failed challenger is a lot more sophisticated than simply discarding it.

Still, the term "continual learning" makes people imagine updating models very frequently, such as every 5 or 10 minutes. Many people argue that most companies don't need to update their models that frequently because of two reasons. First, they don't have enough traffic (i.e., enough new data) for that retraining schedule to make sense. Second, their models don't decay that fast. I agree with them. If changing the retraining schedule from a week to a day gives no return and causes more overhead, there's no need to do it.

Stateless Retraining Versus Stateful Training

However, continual learning isn't about the retraining frequency, but the manner in which the model is retrained. Most companies do *stateless retraining*—the model is trained from scratch each time. Continual learning means also allowing *stateful training*—the model continues training on new data.[2] Stateful training is also known as fine-tuning or incremental learning. The difference between stateless retraining and stateful training is visualized in Figure 9-2.

2 It's "stateful training" instead of "stateful retraining" because there's no *re*-training here. The model continues training from the last state.

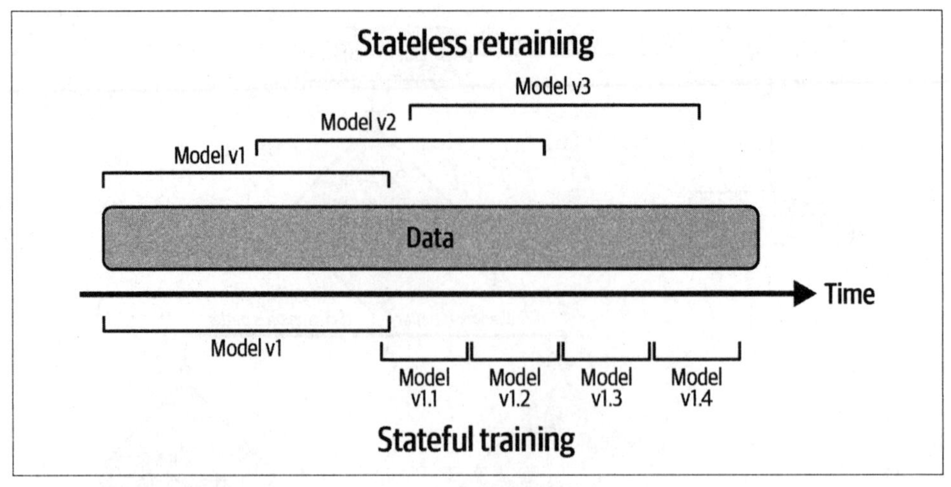

Figure 9-2. Stateless retraining versus stateful training

Stateful training allows you to update your model with less data. Training a model from scratch tends to require a lot more data than fine-tuning the same model. For example, if you retrain your model from scratch, you might need to use all data from the last three months. However, if you fine-tune your model from yesterday's checkpoint, you only need to use data from the last day.

Grubhub found out that stateful training allows their models to converge faster and require much less compute power. Going from daily stateless retraining to daily stateful training reduced their training compute cost 45 times and increased their purchase-through rate by 20%.[3]

One beautiful property that is often overlooked is that with stateful training, it might be possible to avoid storing data altogether. In the traditional stateless retraining, a data sample might be reused during multiple training iterations of a model, which means that data needs to be stored. This isn't always possible, especially for data with strict privacy requirements. In the stateful training paradigm, each model update is trained using only the fresh data, so a data sample is used only once for training, as shown in Figure 9-2. This means that it's possible to train your model without having to store data in permanent storage, which helps eliminate many concerns about data privacy. However, this is overlooked because today's let's-keep-track-of-everything practice still makes many companies reluctant to throw away data.

3 Alex Egg, "Online Learning for Recommendations at Grubhub," *arXiv*, July 15, 2021, *https://oreil.ly/FBBUw*.

Stateful training doesn't mean no training from scratch. The companies that have most successfully used stateful training also occasionally train their model from scratch on a large amount of data to calibrate it. Alternatively, they might also train their model from scratch in parallel with stateful training and then combine both updated models using techniques such as parameter server.[4]

Once your infrastructure is set up to allow both stateless retraining and stateful training, the training frequency is just a knob to twist. You can update your models once an hour, once a day, or whenever a distribution shift is detected. How to find the optimal retraining schedule will be discussed in the section "How Often to Update Your Models" on page 279.

Continual learning is about setting up infrastructure in a way that allows you, a data scientist or ML engineer, to update your models whenever it is needed, whether from scratch or fine-tuning, and to deploy this update quickly.

You might wonder: stateful training sounds cool, but how does this work if I want to add a new feature or another layer to my model? To answer this, we must differentiate two types of model updates:

Model iteration
: A new feature is added to an existing model architecture or the model architecture is changed.

Data iteration
: The model architecture and features remain the same, but you refresh this model with new data.

As of today, stateful training is mostly applied for data iteration, as changing your model architecture or adding a new feature still requires training the resulting model from scratch. There has been research showing that it might be possible to bypass training from scratch for model iteration by using techniques such as knowledge transfer (*https://oreil.ly/lp0GB*) (Google, 2015) and model surgery (*https://oreil.ly/SU0F1*) (OpenAI, 2019). According to OpenAI, "Surgery transfers trained weights from one network to another after a selection process to determine which sections of the model are unchanged and which must be re-initialized."[5] Several large research labs have experimented with this; however, I'm not aware of any clear results in the industry.

---

4 Mu Li, Li Zhou, Zichao Yang, Aaron Li, Fei Xia, David G. Andersen, and Alexander Smola, "Parameter Server for Distributed Machine Learning" (NIPS Workshop on Big Learning, Lake Tahoe, CA, 2013), *https://oreil.ly/xMmru*.

5 Jonathan Raiman, Susan Zhang, and Christy Dennison, "Neural Network Surgery with Sets," *arXiv*, December 13, 2019, *https://oreil.ly/SU0F1*.

Terminology Ambiguity

I use the term "continual learning" instead of "online learning" because when I say "online learning," people usually think of online education. If you enter "online learning" on Google, the top results will likely be about online courses.

Some people use "online learning" to refer to the specific setting where a model learns from each incoming new sample. In this setting, continual learning is a generalization of online learning.

I also use the term "continual learning" instead of "continuous learning." Continuous learning refers to the regime in which your model continuously learns with each incoming sample, whereas with continual learning, the learning is done in a series of batches or micro-batches.

Continuous learning is sometimes used to refer to continuous delivery of ML, which is closely related to continual learning as both help companies to speed up the iteration cycle of their ML models. However, the difference is that "continuous learning," when used in this sense, is from the DevOps perspective about setting up the pipeline for continuous delivery, whereas "continual learning" is from the ML perspective.

Due to the ambiguity of the term "continuous learning," I hope that the community can stay away from this term altogether.

Why Continual Learning?

We discussed that continual learning is about setting up infrastructure so that you can update your models and deploy these changes as fast as you want. But why would you need the ability to update your models as fast as you want?

The first use case of continual learning is to combat data distribution shifts, especially when the shifts happen suddenly. Imagine you're building a model to determine the prices for a ride-sharing service like Lyft.[6] Historically, the ride demand on a Thursday evening in this particular neighborhood is slow, so the model predicts low ride prices, which makes it less appealing for drivers to get on the road. However, on this Thursday evening, there's a big event in the neighborhood, and suddenly the ride demand surges. If your model can't respond to this change quickly enough by increasing its price prediction and mobilizing more drivers to that neighborhood, riders will have to wait a long time for a ride, which causes negative user experience. They might even switch to a competitor, which causes you to lose revenue.

Another use case of continual learning is to adapt to rare events. Imagine you work for an ecommerce website like Amazon. Black Friday is an important shopping event

6 This type of problem is also called "dynamic pricing."

that happens only once a year. There's no way you will be able to gather enough historical data for your model to be able to make accurate predictions on how your customers will behave throughout Black Friday this year. To improve performance, your model should learn throughout the day with fresh data. In 2019, Alibaba acquired Data Artisans, the team leading the development of the stream processing framework Apache Flink, for $103 million so that the team could help them adapt Flink for ML use cases.[7] Their flagship use case was making better recommendations on Singles Day, a shopping occasion in China similar to Black Friday in the US.

A huge challenge for ML production today that continual learning can help overcome is the *continuous cold start* problem. The cold start problem arises when your model has to make predictions for a new user without any historical data. For example, to recommend to a user what movies they might want to watch next, a recommender system often needs to know what that user has watched before. But if that user is new, you won't have their watch history and will have to generate them something generic, e.g., the most popular movies on your site right now.[8]

Continuous cold start is a generalization of the cold start problem,[9] as it can happen not just with new users but also with existing users. For example, it can happen because an existing user switches from a laptop to a mobile phone, and their behavior on a phone is different from their behavior on a laptop. It can happen because users are not logged in—most news sites don't require readers to log in to read.

It can also happen when a user visits a service so infrequently that whatever historical data the service has about this user is outdated. For example, most people only book hotels and flights a few times a year. Coveo, a company that provides search engine and recommender systems to ecommerce websites, found that it is common for an ecommerce site to have more than 70% of their shoppers visit their site less than three times a year.[10]

If your model doesn't adapt quickly enough, it won't be able to make recommendations relevant to these users until the next time the model is updated. By that time, these users might have already left the service because they don't find anything relevant to them.

---

7 Jon Russell, "Alibaba Acquires German Big Data Startup Data Artisans for $103M," *TechCrunch*, January 8, 2019, *https://oreil.ly/4tf5c*. An early reviewer mentioned that it's also possible that the main goal of this acquisition was to increase Alibaba's open source footprint, which is tiny compared to other tech giants.

8 The problem is also equally challenging if you want your model to figure out when to recommend a new movie that no one has watched and given feedback on yet.

9 Lucas Bernardi, Jaap Kamps, Julia Kiseleva, and Melanie J. I. Müller, "The Continuous Cold Start Problem in e-Commerce Recommender Systems," *arXiv*, August 5, 2015, *https://oreil.ly/GWUyD*.

10 Jacopo Tagliabue, Ciro Greco, Jean-Francis Roy, Bingqing Yu, Patrick John Chia, Federico Bianchi, and Giovanni Cassani, "SIGIR 2021 E-Commerce Workshop Data Challenge," *arXiv*, April 19, 2021, *https://oreil.ly/8QxmS*.

If we could make our models adapt to each user within their visiting session, the models would be able to make accurate, relevant predictions to users even on their first visit. TikTok, for example, has successfully applied continual learning to adapt their recommender system to each user within minutes. You download the app and, after a few videos, TikTok's algorithms are able to predict with high accuracy what you want to watch next.[11] I don't think everyone should try to build something as addictive as TikTok, but it's proof that continual learning can unlock powerful predictive potential.

"Why continual learning?" should be rephrased as "why not continual learning?" Continual learning is a superset of batch learning, as it allows you to do everything the traditional batch learning can do. But continual learning also allows you to unlock use cases that batch learning can't.

If continual learning takes the same effort to set up and costs the same to do as batch learning, there's no reason not to do continual learning. As of writing this book, there are still a lot of challenges in setting up continual learning, as we'll go deeper into in the following section. However, MLOps tooling for continual learning is maturing, which means, one day not too far in the future, it might be as easy to set up continual learning as batch learning.

Continual Learning Challenges

Even though continual learning has many use cases and many companies have applied it with great success, continual learning still has many challenges. In this section, we'll discuss three major challenges: fresh data access, evaluation, and algorithms.

Fresh data access challenge

The first challenge is the challenge to get fresh data. If you want to update your model every hour, you need new data every hour. Currently, many companies pull new training data from their data warehouses. The speed at which you can pull data from your data warehouses depends on the speed at which this data is deposited into your data warehouses. The speed can be slow, especially if data comes from multiple sources. An alternative is to allow pull data before it's deposited into data warehouses, e.g., directly from real-time transports such as Kafka and Kinesis that transport data from applications to data warehouses,[12] as shown in Figure 9-3.

11 Catherine Wang, "Why TikTok Made Its User So Obsessive? The AI Algorithm That Got You Hooked," *Towards Data Science*, June 7, 2020, *https://oreil.ly/BDWf8*.

12 See the section "Data Passing Through Real-Time Transport" on page 74.

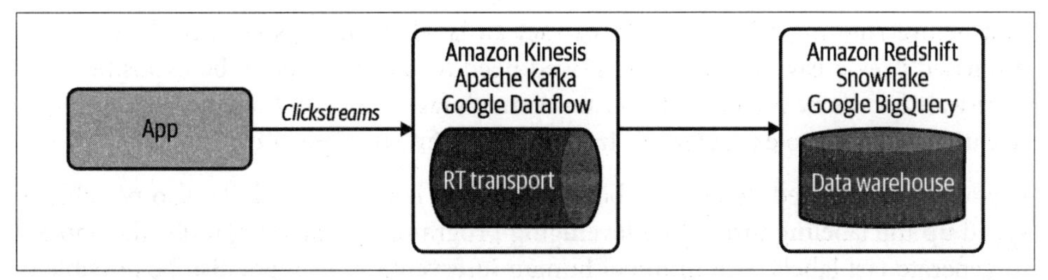

Figure 9-3. Pulling data directly from real-time transports, before it's deposited into data warehouses, can allow you to access fresher data

Being able to pull fresh data isn't enough. If your model needs labeled data to update, as most models today do, this data will need to be labeled as well. In many applications, the speed at which a model can be updated is bottlenecked by the speed at which data is labeled.

The best candidates for continual learning are tasks where you can get natural labels with short feedback loops. Examples of these tasks are dynamic pricing (based on estimated demand and availability), estimating time of arrival, stock price prediction, ads click-through prediction, and recommender systems for online content like tweets, songs, short videos, articles, etc.

However, these natural labels are usually not generated as labels, but rather as behavioral activities that need to be extracted into labels. Let's walk through an example to make this clear. If you run an ecommerce website, your application might register that at 10:33 p.m., user A clicks on the product with the ID of 32345. Your system needs to look back into the logs to see if this product ID was ever recommended to this user, and if yes, then what query prompted this recommendation, so that your system can match this query to this recommendation and label this recommendation as a good recommendation, as shown in Figure 9-4.

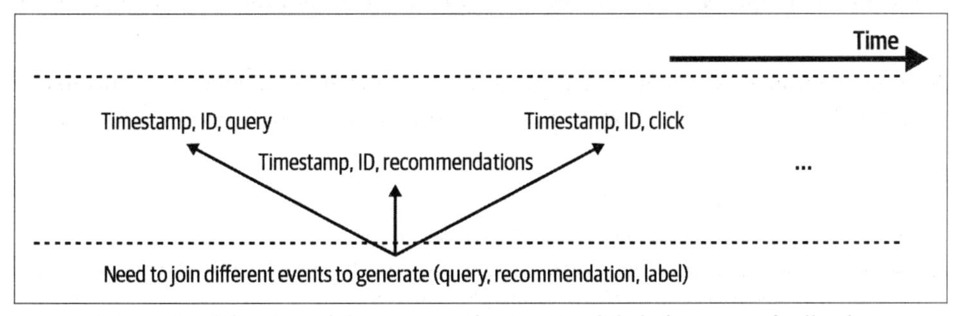

Figure 9-4. A simplification of the process of extracting labels from user feedback

The process of looking back into the logs to extract labels is called label computation. It can be quite costly if the number of logs is large. Label computation can be done with batch processing: e.g., waiting for logs to be deposited into data warehouses first before running a batch job to extract all labels from logs at once. However, as discussed previously, this means that we'd need to wait for data to be deposited first, then wait for the next batch job to run. A much faster approach would be to leverage stream processing to extract labels from the real-time transports directly.[13]

If your model's speed iteration is bottlenecked by labeling speed, it's also possible to speed up the labeling process by leveraging programmatic labeling tools like Snorkel to generate fast labels with minimal human intervention. It might also be possible to leverage crowdsourced labels to quickly annotate fresh data.

Given that tooling around streaming is still nascent, architecting an efficient streaming-first infrastructure for accessing fresh data and extracting fast labels from real-time transports can be engineering-intensive and costly. The good news is that tooling around streaming is growing fast. Confluent, the platform built on top of Kafka, is a $16 billion company as of October 2021. In late 2020, Snowflake started a team focusing on streaming.[14] As of September 2021, Materialize has raised $100 million to develop a streaming SQL database.[15] As tooling around streaming matures, it'll be much easier and cheaper for companies to develop a streaming-first infrastructure for ML.

Evaluation challenge

The biggest challenge of continual learning isn't in writing a function to continually update your model—you can do that by writing a script! The biggest challenge is in making sure that this update is good enough to be deployed. In this book, we've discussed how ML systems make catastrophic failures in production, from millions of minorities being unjustly denied loans, to drivers who trust autopilot too much being involved in fatal crashes.[16]

The risks for catastrophic failures amplify with continual learning. First, the more frequently you update your models, the more opportunities there are for updates to fail.

Second, continual learning makes your models more susceptible to coordinated manipulation and adversarial attack. Because your models learn online from

13 See the section "Batch Processing Versus Stream Processing" on page 78.

14 Tyler Akidau, "Snowflake Streaming: Now Hiring! Help Design and Build the Future of Big Data and Stream Processing," Snowflake blog, October 26, 2020, *https://oreil.ly/Knh2Y*.

15 Arjun Narayan, "Materialize Raises a $60M Series C, Bringing Total Funding to Over $100M," *Materialize*, September 30, 2021, *https://oreil.ly/dqxRb*.

16 Khristopher J. Brooks, "Disparity in Home Lending Costs Minorities Millions, Researchers Find," *CBS News*, November 15, 2019, *https://oreil.ly/SpZ1N*; Lee Brown, "Tesla Driver Killed in Crash Posted Videos Driving Without His Hands on the Wheel," *New York Post*, May 16, 2021, *https://oreil.ly/uku9S*; "A Tesla Driver Is Charged in a Crash Involving Autopilot That Killed 2 People," *NPR*, January 18, 2022, *https://oreil.ly/WWaRA*.

real-world data, it makes it easier for users to input malicious data to trick models into learning wrong things. In 2016, Microsoft released Tay, a chatbot capable of learning through "casual and playful conversation" on Twitter. As soon as Tay launched, trolls started tweeting the bot racist and misogynist remarks. The bot soon began to post inflammatory and offensive tweets, causing Microsoft to shut down the bot 16 hours after its launch.[17]

To avoid similar or worse incidents, it's crucial to thoroughly test each of your model updates to ensure its performance and safety before deploying the updates to a wider audience. We already discussed model offline evaluation in Chapter 6, and will discuss online evaluation (test in production) in this chapter.

When designing the evaluation pipeline for continual learning, keep in mind that evaluation takes time, which can be another bottleneck for model update frequency. For example, a major online payment company I worked with has an ML system to detect fraudulent transactions.[18] The fraud patterns change quickly, so they'd like to update their system quickly to adapt to the changing patterns. They can't deploy the new model before it's been A/B tested against the current model. However, due to the imbalanced nature of the task—most transactions aren't fraud—it takes them approximately two weeks to see enough fraud transactions to be able to accurately assess which model is better.[19] Therefore, they can only update their system every two weeks.

Algorithm challenge

Compared to the fresh data challenge and the evaluation, this is a "softer" challenge as it only affects certain algorithms and certain training frequencies. To be precise, it only affects matrix-based and tree-based models that want to be updated very fast (e.g., hourly).

To illustrate this point, consider two different models: a neural network and a matrix-based model, such as a collaborative filtering model. The collaborative filtering model uses a user-item matrix and a dimension reduction technique.

You can update the neural network model with a data batch of any size. You can even perform the update step with just one data sample. However, if you want to update the collaborative filtering model, you first need to use the entire dataset to build the user-item matrix before performing dimensionality reduction on it. Of course, you can apply dimensionality reduction to your matrix each time you update the matrix

17 James Vincent, "Twitter Taught Microsoft's Friendly AI Chatbot to Be a Racist Asshole in Less Than a Day," *The Verge*, May 24, 2016, *https://oreil.ly/NJEVF*.

18 Their fraud detection system consists of multiple ML models.

19 In the section "Bandits" on page 287, we'll learn about how bandits can be used as a more data-efficient alternative to A/B testing.

with a new data sample, but if your matrix is large, the dimensionality reduction step would be too slow and expensive to perform frequently. Therefore, this model is less suitable for learning with a partial dataset than the preceding neural network model.[20]

It's much easier to adapt models like neural networks than matrix-based and tree-based models to the continual learning paradigm. However, there have been algorithms to create tree-based models that can learn from incremental amounts of data, most notably Hoeffding Tree and its variants Hoeffding Window Tree and Hoeffding Adaptive Tree,[21] but their uses aren't yet widespread.

Not only does the learning algorithm need to work with partial datasets, but the feature extract code has to as well. We discussed in the section "Scaling" on page 126 that it's often necessary to scale your features using statistics such as the min, max, median, and variance. To compute these statistics for a dataset, you often need to do a pass over the entire dataset. When your model can only see a small subset of data at a time, in theory, you can compute these statistics for each subset of data. However, this means that these statistics will fluctuate a lot between different subsets. The statistics computed from one subset might differ wildly from the next subset, making it difficult for the model trained on one subset to generalize to the next subset.

To keep these statistics stable across different subsets, you might want to compute these statistics online. Instead of using the mean or variance from all your data at once, you compute or approximate these statistics incrementally as you see new data, such as the algorithms outlined in "Optimal Quantile Approximation in Streams."[22] Popular frameworks today offer some capacity for computing running statistics—for example, sklearn's StandardScaler has a `partial_fit` that allows a feature scaler to be used with running statistics—but the built-in methods are slow and don't support a wide range of running statistics.

Four Stages of Continual Learning

We've discussed what continual learning is, why continual learning matters, and the challenges of continual learning. Next, we'll discuss how to overcome these challenges and make continual learning happen. As of the writing of this book, continual learning isn't something that companies start out with. The move toward continual

---

20 Some people call this setting "learning with partial information," but learning with partial information refers to another setting, as outlined in the paper "Subspace Learning with Partial Information" (*https://oreil.ly/OuJvG*) by Gonen et al. (2016).

21 Pedro Domingos and Geoff Hulten, "Mining High-Speed Data Streams," in *Proceedings of the Sixth International Conference on Knowledge Discovery and Data Mining* (Boston: ACM Press, 2000), 71–80; Albert Bifet and Ricard Gavaldà, "Adaptive Parameter-free Learning from Evolving Data Streams," 2009, *https://oreil.ly/XIMpl*.

22 Zohar Karnin, Kevin Lang, and Edo Liberty, "Optimal Quantile Approximation in Streams," *arXiv*, March 17, 2016, *https://oreil.ly/bUu4H*.

learning happens in four stages, as outlined next. We'll go over what happens in each stage as well as the requirements necessary to move from a previous stage to this stage.

Stage 1: Manual, stateless retraining

In the beginning, the ML team often focuses on developing ML models to solve as many business problems as possible. For example, if your company is an ecommerce website, you might develop four models in the following succession:

1. A model to detect fraudulent transactions
2. A model to recommend relevant products to users
3. A model to predict whether a seller is abusing a system
4. A model to predict how long it will take to ship an order

Because your team is focusing on developing new models, updating existing models takes a backseat. You update an existing model only when the following two conditions are met: the model's performance has degraded to the point that it's doing more harm than good, and your team has time to update it. Some of your models are being updated once every six months. Some are being updated once a quarter. Some have been out in the wild for a year and haven't been updated at all.

The process of updating a model is manual and ad hoc. Someone, usually a data engineer, has to query the data warehouse for new data. Someone else cleans this new data, extracts features from it, retrains that model from scratch on both the old and new data, and then exports the updated model into a binary format. Then someone else takes that binary format and deploys the updated model. Oftentimes, the code encapsulating data, features, and model logic was changed during the retraining process but these changes failed to be replicated to production, causing bugs that are hard to track down.

If this process sounds painfully familiar to you, you're not alone. A vast majority of companies outside the tech industry—e.g., any company that adopted ML less than three years ago and doesn't have an ML platform team—are in this stage.[23]

Stage 2: Automated retraining

After a few years, your team has managed to deploy models to solve most of the obvious problems. You have anywhere between 5 and 10 models in production. Your priority is no longer to develop new models, but to maintain and improve existing models. The ad hoc, manual process of updating models mentioned from the

23 We'll cover ML platforms in the section "ML Platform" on page 319.

previous stage has grown into a pain point too big to be ignored. Your team decides to write a script to automatically execute all the retraining steps. This script is then run periodically using a batch process such as Spark.

Most companies with somewhat mature ML infrastructure are in this stage. Some sophisticated companies run experiments to determine the optimal retraining frequency. However, for most companies in this stage, the retraining frequency is set based on gut feeling—e.g., "once a day seems about right" or "let's kick off the retraining process each night when we have idle compute."

When creating scripts to automate the retraining process for your system, you need to take into account that different models in your system might require different retraining schedules. For example, consider a recommender system that consists of two models: one model to generate embeddings for all products, and another model to rank the relevance of each product given a query. The embedding model might need to be retrained a lot less frequently than the ranking model. Because products' characteristics don't change that often, you might be able to get away with retraining your embeddings once a week,[24] whereas your ranking models might need to be retrained once a day.

The automating script might get even more complicated if there are dependencies among your models. For example, because the ranking model depends on the embeddings, when the embeddings change, the ranking model should be updated too.

Requirements. If your company has ML models in production, it's likely that your company already has most of the infrastructure pieces needed for automated retraining. The feasibility of this stage revolves around the feasibility of writing a script to automate your workflow and configure your infrastructure to automatically:

1. Pull data.
2. Downsample or upsample this data if necessary.
3. Extract features.
4. Process and/or annotate labels to create training data.
5. Kick off the training process.
6. Evaluate the newly trained model.
7. Deploy it.

How long it will take to write this script depends on many factors, including the script writer's competency. However, in general, the three major factors that will affect the feasibility of this script are: scheduler, data, and model store.

24 You might need to train your embedding model more frequently if you have a lot of new items each day.

A scheduler is basically a tool that handles task scheduling, which we'll cover in the section "Cron, Schedulers, and Orchestrators" on page 311. If you don't already have a scheduler, you'll need time to set up one. However, if you already have a scheduler such as Airflow or Argo, wiring the scripts together shouldn't be that hard.

The second factor is the availability and accessibility of your data. Do you need to gather data yourself into your data warehouse? Will you have to join data from multiple organizations? Do you need to extract a lot of features from scratch? Will you also need to label your data? The more questions you answer yes to, the more time it will take to set up this script. Stefan Krawczyk, ML/data platform manager at Stitch Fix, commented that he suspects most people's time might be spent here.

The third factor you'll need is a model store to automatically version and store all the artifacts needed to reproduce a model. The simplest model store is probably just an S3 bucket that stores serialized blobs of models in some structured manner. However, blob storage like S3 is neither very good at versioning artifacts nor human-readable. You might need a more mature model store like Amazon SageMaker (managed service) and Databricks' MLflow (open source). We'll go into detail on what a model store is and evaluate different model stores in the section "Model Store" on page 321.

Feature Reuse (Log and Wait)

When creating training data from new data to update your model, remember that the new data has already gone through the prediction service. This prediction service has already extracted features from this new data to input into models for predictions. Some companies reuse these extracted features for model retraining, which both saves computation and allows for consistency between prediction and training. This approach is known as "log and wait." It's a classic approach to reduce the train-serving skew discussed in Chapter 8 (see the section "Production data differing from training data" on page 229).

Log and wait isn't yet a popular approach, but it's getting more popular. Faire has a great blog post (*https://oreil.ly/AxFnJ*) discussing the pros and cons of their "log and wait" approach.

Stage 3: Automated, stateful training

In stage 2, each time you retrain your model, you train it from scratch (stateless retraining). It makes your retraining costly, especially for retraining with a higher frequency. You read the section "Stateless Retraining Versus Stateful Training" on page 265 and decide that you want to do stateful training—why train on data from the last three months every day when you can continue training using only data from the last day?

So in this stage, you reconfigure your automatic updating script so that, when the model update is kicked off, it first locates the previous checkpoint and loads it into memory before continuing training on this checkpoint.

Requirements. The main thing you need in this stage is a change in the mindset: retraining from scratch is such a norm—many companies are so used to data scientists handing off a model to engineers to deploy from scratch each time—that many companies don't think about setting up their infrastructure to enable stateful training.

Once you're committed to stateful training, reconfiguring the updating script is straightforward. The main thing you need at this stage is a way to track your data and model lineage. Imagine you first upload model version 1.0. This model is updated with new data to create model version 1.1, and so on to create model 1.2. Then another model is uploaded and called model version 2.0. This model is updated with new data to create model version 2.1. After a while, you might have model version 3.32, model version 2.11, model version 1.64. You might want to know how these models evolve over time, which model was used as its base model, and which data was used to update it so that you can reproduce and debug it. As far as I know, no existing model store has this model lineage capacity, so you'll likely have to build the solution in-house.

If you want to pull fresh data from the real-time transports instead of from data warehouses, as discussed in the section "Fresh data access challenge" on page 270, and your streaming infrastructure isn't mature enough, you might need to revamp your streaming pipeline.

Stage 4: Continual learning

At stage 3, your models are still updated based on a fixed schedule set out by developers. Finding the optimal schedule isn't straightforward and can be situation-dependent. For example, last week, nothing much happened in the market, so your models didn't decay that fast. However, this week, a lot of events happen, so your models decay much faster and require a much faster retraining schedule.

Instead of relying on a fixed schedule, you might want your models to be automatically updated whenever data distributions shift and the model's performance plummets.

The holy grail is when you combine continual learning with edge deployment. Imagine you can ship a base model with a new device—a phone, a watch, a drone, etc.—and the model on that device will continually update and adapt to its environment as needed without having to sync with a centralized server. There will be no need for a centralized server, which means no centralized server cost. There will also be no need to transfer data back and forth between device and cloud, which means better data security and privacy!

Requirements. The move from stage 3 to stage 4 is steep. You'll first need a mechanism to trigger model updates. This trigger can be:

Time-based
: For example, every five minutes

Performance-based
: For example, whenever model performance plummets

Volume-based
: For example, whenever the total amount of labeled data increases by 5%

Drift-based
: For example, whenever a major data distribution shift is detected

For this trigger mechanism to work, you'll need a solid monitoring solution. We discussed in the section "Monitoring and Observability" on page 250 that the hard part is not to detect the changes, but to determine which of these changes matter. If your monitoring solution gives a lot of false alerts, your model will end up being updated much more frequently than it needs to be.

You'll also need a solid pipeline to continually evaluate your model updates. Writing a function to update your models isn't much different from what you'd do in stage 3. The hard part is to ensure that the updated model is working properly. We'll go over various testing techniques you can use in the section "Test in Production" on page 281.

How Often to Update Your Models

Now that your infrastructure has been set up to update a model quickly, you started asking the question that has been haunting ML engineers at companies of all shapes and sizes: "How often should I update my models?" Before attempting to answer that question, we first need to figure out how much gain your model will get from being updated with fresh data. The more gain your model can get from fresher data, the more frequently it should be retrained.

Value of data freshness

The question of how often to update a model becomes a lot easier if we know how much the model performance will improve with updating. For example, if we switch from retraining our model every month to every week, how much performance gain can we get? What if we switch to daily retraining? People keep saying that data distributions shift, so fresher data is better, but how much better is fresher data?

One way to figure out the gain is by training your model on the data from different time windows in the past and evaluating it on the data from today to see how the performance changes. For example, consider that you have data from the year 2020. To measure the value of data freshness, you can experiment with training model version A on the data from January to June 2020, model version B on the data from April to September, and model version C on the data from June to November, then test each of these model versions on the data from December, as shown in Figure 9-5. The difference in the performance of these versions will give you a sense of the performance gain your model can get from fresher data. If the model trained on data from a quarter ago is much worse than the model trained on data from a month ago, you know that you shouldn't wait a quarter to retrain your model.

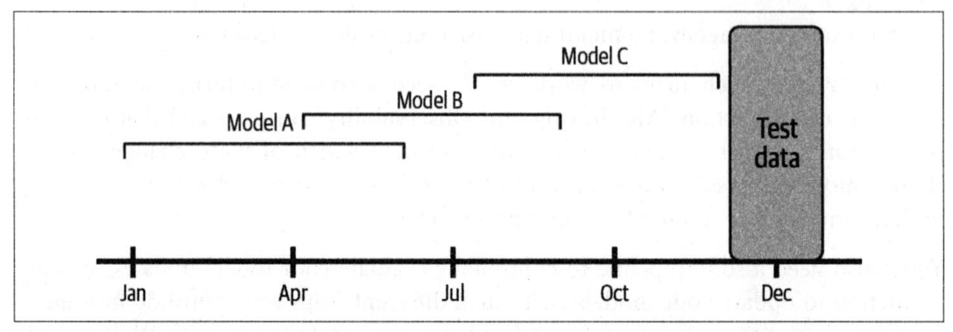

Figure 9-5. To get a sense of the performance gain you can get from fresher data, train your model on data from different time windows in the past and test on data from today to see how the performance changes

This is a simple example to illustrate how the data freshness experiment works. In practice, you might want your experiments to be much more fine-grained, operating not in months but in weeks, days, even hours or minutes. In 2014, Facebook did a similar experiment for ad click-through-rate prediction and found out that they could reduce the model's loss by 1% by going from retraining weekly to retraining daily, and this performance gain was significant enough for them to switch their retraining pipeline from weekly to daily.[25] Given that online contents today are so much more diverse and users' attention online changes much faster, we can imagine that the value of data freshness for ad click-through rate is even higher. Some of the companies with sophisticated ML infrastructure have found enough performance gain to switch their retraining pipeline to every few minutes.[26]

25 Xinran He, Junfeng Pan, Ou Jin, Tianbing Xu, Bo Liu, Tao Xu, Tanxin Shi, et al., "Practical Lessons from Predicting Clicks on Ads at Facebook," in *ADKDD '14: Proceedings of the Eighth International Workshop on Data Mining for Online Advertising* (August 2014): 1–9, *https://oreil.ly/oS16J*.

26 Qian Yu, "Machine Learning with Flink in Weibo," QCon 2019, video, 17:57, *https://oreil.ly/Yia6v*.

Model iteration versus data iteration

We discussed earlier in this chapter that not all model updates are the same. We differentiated between model iteration (adding a new feature to an existing model architecture or changing the model architecture) and data iteration (same model architecture and features but you refresh this model with new data). You might wonder not only how often to update your model, but also what kind of model updates to perform.

In theory, you can do both types of updates, and in practice, you should do both from time to time. However, the more resources you spend in one approach, the fewer resources you can spend in another.

On the one hand, if you find that iterating on your data doesn't give you much performance gain, then you should spend your resources on finding a better model. On the other hand, if finding a better model architecture requires 100X compute for training and gives you 1% performance whereas updating the same model on data from the last three hours requires only 1X compute and also gives 1% performance gain, you'll be better off iterating on data.

Maybe in the near future, we'll get more theoretical understanding to know in what situation an approach will work better (cue "call for research"), but as of today, no book can give you the answer on which approach will work better for your specific model on your specific task. You'll have to do experiments to find out.

The question on how often to update your model is a difficult one to answer, and I hope that this section has sufficiently explained its nuances. In the beginning, when your infrastructure is nascent and the process of updating a model is manual and slow, the answer is: as often as you *can*.

However, as your infrastructure matures and the process of updating a model is partially automated and can be done in a matter of hours, if not minutes, the answer to this question is contingent on the answer to the following question: "How much performance gain would I get from fresher data?" It's important to run experiments to quantify the value of data freshness to your models.

Test in Production

Throughout this book, including this chapter, we've talked about the danger of deploying models that haven't been sufficiently evaluated. To sufficiently evaluate your models, you first need a mixture of offline evaluation discussed in Chapter 6 and online evaluation discussed in this section. To understand why offline evaluation isn't enough, let's go over two major test types for offline evaluation: test splits and backtests.

The first type of model evaluation you might think about is the good old test splits that you can use to evaluate your models offline, as discussed in Chapter 6. These test splits are usually static and have to be static so that you have a trusted benchmark to compare multiple models. It'll be hard to compare the test results of two models if they are tested on different test sets.

However, if you update the model to adapt to a new data distribution, it's not sufficient to evaluate this new model on test splits from the old distribution. Assuming that the fresher the data, the more likely it is to come from the current distribution, one idea is to test your model on the most recent data that you have access to. So, after you've updated your model on the data from the last day, you might want to test this model on the data from the last hour (assuming that data from the last hour wasn't included in the data used to update your model). The method of testing a predictive model on data from a specific period of time in the past is known as a *backtest*.

The question is whether backtests are sufficient to replace static test splits. Not quite. If something went wrong with your data pipeline and some data from the last hour is corrupted, evaluating your model solely on this recent data isn't sufficient.

With backtests, you should still evaluate your model on a static test set that you have extensively studied and (mostly) trust as a form of sanity check.

Because data distributions shift, the fact that a model does well on the data from the last hour doesn't mean that it will continue doing well on the data in the future. The only way to know whether a model will do well in production is to deploy it. This insight led to one seemingly terrifying but necessary concept: test in production. However, test in production doesn't have to be scary. There are techniques to help you evaluate your models in production (mostly) safely. In this section, we'll cover the following techniques: shadow deployment, A/B testing, canary analysis, interleaving experiments, and bandits.

Shadow Deployment

Shadow deployment might be the safest way to deploy your model or any software update. Shadow deployment works as follows:

1. Deploy the candidate model in parallel with the existing model.
2. For each incoming request, route it to both models to make predictions, but only serve the existing model's prediction to the user.
3. Log the predictions from the new model for analysis purposes.

Only when you've found that the new model's predictions are satisfactory do you replace the existing model with the new model.

Because you don't serve the new model's predictions to users until you've made sure that the model's predictions are satisfactory, the risk of this new model doing something funky is low, at least not higher than the existing model. However, this technique isn't always favorable because it's expensive. It doubles the number of predictions your system has to generate, which generally means doubling your inference compute cost.

A/B Testing

A/B testing is a way to compare two variants of an object, typically by testing responses to these two variants, and determining which of the two variants is more effective. In our case, we have the existing model as one variant, and the candidate model (the recently updated model) as another variant. We'll use A/B testing to determine which model is better according to some predefined metrics.

A/B testing has become so prevalent that, as of 2017, companies like Microsoft and Google each conduct over 10,000 A/B tests annually.[27] It is many ML engineers' first response to how to evaluate ML models in production. A/B testing works as follows:

1. Deploy the candidate model alongside the existing model.
2. A percentage of traffic is routed to the new model for predictions; the rest is routed to the existing model for predictions. It's common for both variants to serve prediction traffic at the same time. However, there are cases where one model's predictions might affect another model's predictions—e.g., in ride-sharing's dynamic pricing, a model's predicted prices might influence the number of available drivers and riders, which, in turn, influence the other model's predictions. In those cases, you might have to run your variants alternatively, e.g., serve model A one day and then serve model B the next day.
3. Monitor and analyze the predictions and user feedback, if any, from both models to determine whether the difference in the two models' performance is statistically significant.

To do A/B testing the right way requires doing many things right. In this book, we'll discuss two important things. First, A/B testing consists of a randomized experiment: the traffic routed to each model has to be truly random. If not, the test result will be invalid. For example, if there's a selection bias in the way traffic is routed to the two models, such as users who are exposed to model A are usually on their phones whereas users exposed to model B are usually on their desktops, then if model A has

27 Ron Kohavi and Stefan Thomke, "The Surprising Power of Online Experiments," *Harvard Business Review*, September–October 2017, *https://oreil.ly/OHfj0*.

better accuracy than model B, we can't tell whether it's because A is better than B or whether "being on a phone" influences the prediction quality.

Second, your A/B test should be run on a sufficient number of samples to gain enough confidence about the outcome. How to calculate the number of samples needed for an A/B test is a simple question with a very complicated answer, and I'd recommend readers reference a book on A/B testing to learn more.

The gist here is that if your A/B test result shows that a model is better than another with statistical significance, you can determine which model is indeed better. To measure statistical significance, A/B testing uses statistical hypothesis testing such as two-sample tests. We saw two-sample tests in Chapter 8 when we used them to detect distribution shifts. As a reminder, a two-sample test is a test to determine whether the difference between these two populations is statistically significant. In the distribution shift use case, if a statistical difference suggests that the two populations come from different distributions, this means that the original distribution has shifted. In the A/B testing use case, statistical differences mean that we've gathered sufficient evidence to show that one variant is better than the other variant.

Statistical significance, while useful, isn't foolproof. Say we run a two-sample test and get the result that model A is better than model B with the *p*-value of $p = 0.05$ or 5%, and we define statistical significance as $p \leq 0.5$. This means that if we run the same A/B testing experiment multiple times, (100 – 5 =) 95% of the time, we'll get the result that A is better than B, and the other 5% of the time, B is better than A. So even if the result is statistically significant, it's possible that if we run the experiment again, we'll pick another model.

Even if your A/B test result isn't statistically significant, it doesn't mean that this A/B test fails. If you've run your A/B test with a lot of samples and the difference between the two tested models is statistically insignificant, maybe there isn't much difference between these two models, and it's probably OK for you to use either.

For readers interested in learning more about A/B testing and other statistical concepts important in ML, I recommend Ron Kohav's book *Trustworthy Online Controlled Experiments (A Practical Guide to A/B Testing)* (Cambridge University Press) and Michael Barber's great introduction to statistics for data science (*https://oreil.ly/JdVA0*) (much shorter).

Often, in production, you don't have just one candidate but multiple candidate models. It's possible to do A/B testing with more than two variants, which means we can have A/B/C testing or even A/B/C/D testing.

Canary Release

Canary release is a technique to reduce the risk of introducing a new software version in production by slowly rolling out the change to a small subset of users before rolling it out to the entire infrastructure and making it available to everybody.[28] In the context of ML deployment, canary release works as follows:

1. Deploy the candidate model alongside the existing model. The candidate model is called the canary.
2. A portion of the traffic is routed to the candidate model.
3. If its performance is satisfactory, increase the traffic to the candidate model. If not, abort the canary and route all the traffic back to the existing model.
4. Stop when either the canary serves all the traffic (the candidate model has replaced the existing model) or when the canary is aborted.

The candidate model's performance is measured against the existing model's performance according to the metrics you care about. If the candidate model's key metrics degrade significantly, the canary is aborted and all the traffic will be routed to the existing model.

Canary releases can be used to implement A/B testing due to the similarities in their setups. However, you can do canary analysis without A/B testing. For example, you don't have to randomize the traffic to route to each model. A plausible scenario is that you first roll out the candidate model to a less critical market before rolling out to everybody.

For readers interested in how canary release works in the industry, Netflix and Google have a great shared blog post (*https://oreil.ly/QfBrn*) on how automated canary analysis is used at their companies.

Interleaving Experiments

Imagine you have two recommender systems, A and B, and you want to evaluate which one is better. Each time, a model recommends 10 items users might like. With A/B testing, you'd divide your users into two groups: one group is exposed to A and the other group is exposed to B. Each user will be exposed to the recommendations made by one model.

What if instead of exposing a user to recommendations from a model, we expose that user to recommendations from both models and see which model's recommendations they will click on? That's the idea behind interleaving experiments, originally

28 Danilo Sato, "CanaryRelease," June 25, 2014, MartinFowler.com, *https://oreil.ly/YtKJE*.

proposed by Thorsten Joachims in 2002 for the problems of search rankings.[29] In experiments, Netflix found that interleaving "reliably identifies the best algorithms with considerably smaller sample size compared to traditional A/B testing."[30]

Figure 9-6 shows how interleaving differs from A/B testing. In A/B testing, core metrics like retention and streaming are measured and compared between the two groups. In interleaving, the two algorithms can be compared by measuring user preferences. Because interleaving can be decided by user preferences, there's no guarantee that user preference will lead to better core metrics.

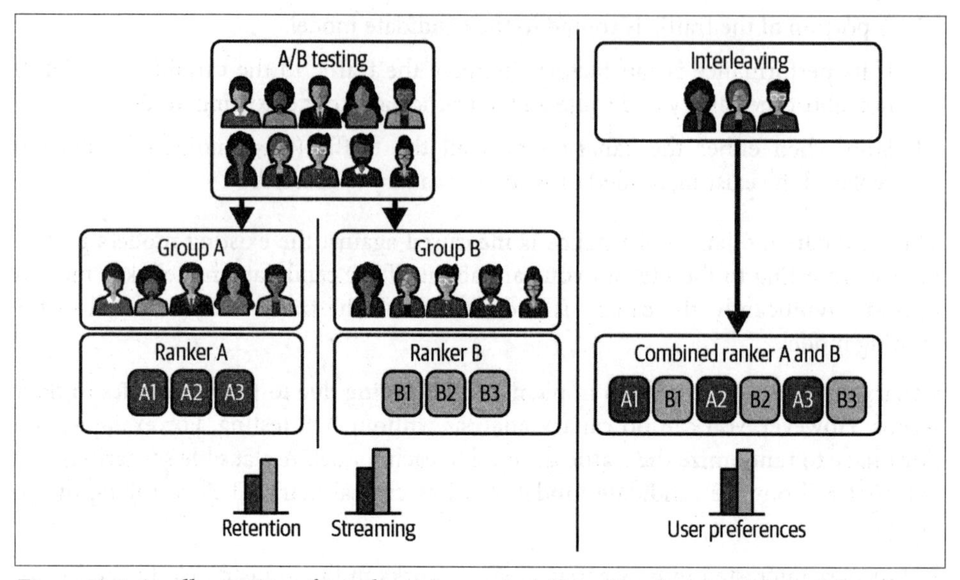

Figure 9-6. An illustration of interleaving versus A/B testing. Source: Adapted from an image by Parks et al.

When we show recommendations from multiple models to users, it's important to note that the position of a recommendation influences how likely a user will click on it. For example, users are much more likely to click on the top recommendation than the bottom recommendation. For interleaving to yield valid results, we must ensure that at any given position, a recommendation is equally likely to be generated by A or B. To ensure this, one method we can use is team-draft interleaving, which mimics the drafting process in sports. For each recommendation position, we randomly select A or B with equal probability, and the chosen model picks the

29 Thorsten Joachims, "Optimizing Search Engines using Clickthrough Data," KDD 2002, *https://oreil.ly/XnH5G*.

30 Joshua Parks, Juliette Aurisset, and Michael Ramm, "Innovating Faster on Personalization Algorithms at Netflix Using Interleaving," *Netflix Technology Blog*, November 29, 2017, *https://oreil.ly/lnvDY*.

top recommendation that hasn't already been picked.[31] A visualization of how this team-drafting method works is shown in Figure 9-7.

Figure 9-7. Interleaving video recommendations from two ranking algorithms using team draft. Source: Parks et al.[32]

Bandits

For those unfamiliar, bandit algorithms originated in gambling. A casino has multiple slot machines with different payouts. A slot machine is also known as a one-armed bandit, hence the name. You don't know which slot machine gives the highest payout. You can experiment over time to find out which slot machine is the best while maximizing your payout. Multi-armed bandits are algorithms that allow you to balance between exploitation (choosing the slot machine that has paid the most in the past) and exploration (choosing other slot machines that may pay off even more).

31 Olivier Chapelle, Thorsten Joachims, Filip Radlinski, and Yisong Yue, "Large-Scale Validation and Analysis of Interleaved Search Evaluation," *ACM Transactions on Information Systems* 30, no. 1 (February 2012): 6, *https://oreil.ly/lccvK*.

32 Parks et al., "Innovating Faster on Personalization Algorithms."

As of today, the standard method for testing models in production is A/B testing. With A/B testing, you randomly route traffic to each model for predictions and measure at the end of your trial which model works better. A/B testing is stateless: you can route traffic to each model without having to know about their current performance. You can do A/B testing even with batch prediction.

When you have multiple models to evaluate, each model can be considered a slot machine whose payout (i.e., prediction accuracy) you don't know. Bandits allow you to determine how to route traffic to each model for prediction to determine the best model while maximizing prediction accuracy for your users. Bandit is stateful: before routing a request to a model, you need to calculate all models' current performance. This requires three things:

- Your model must be able to make online predictions.
- Preferably short feedback loops: you need to get feedback on whether a prediction is good or not. This is usually true for tasks where labels can be determined from users' feedback, like in recommendations—if users click on a recommendation, it's inferred to be good. If the feedback loops are short, you can update the payoff of each model quickly.
- A mechanism to collect feedback, calculate and keep track of each model's performance, and route prediction requests to different models based on their current performance.

Bandits are well-studied in academia and have been shown to be a lot more data-efficient than A/B testing (in many cases, bandits are even optimal). Bandits require less data to determine which model is the best and, at the same time, reduce opportunity cost as they route traffic to the better model more quickly. See discussions on bandits at LinkedIn, Netflix, Facebook, and Dropbox (*https://oreil.ly/vsKsg*), Zillow (*https://oreil.ly/A7KkD*), and Stitch Fix (*https://oreil.ly/2LKZd*). For a more theoretical view, see Chapter 2 of *Reinforcement Learning* (*https://oreil.ly/fpR2H*) (Sutton and Barto 2020).

In an experiment by Google's Greg Rafferty, A/B testing required over 630,000 samples to get a confidence interval of 95%, whereas a simple bandit algorithm (Thompson Sampling) determined that a model was 5% better than the other with less than 12,000 samples.[33]

However, bandits are a lot more difficult to implement than A/B testing because it requires computing and keeping track of models' payoffs. Therefore, bandit algorithms are not widely used in the industry other than at a few big tech companies.

33 Greg Rafferty, "A/B Testing—Is There a Better Way? An Exploration of Multi-Armed Bandits," *Towards Data Science*, January 22, 2020, *https://oreil.ly/MsaAK*.

Bandit Algorithms

Many of the solutions for the multi-armed bandit problem can be used here. The simplest algorithm for exploration is ε-greedy. For a percentage of time, say 90% of the time or $\varepsilon = 0.9$, you route traffic to the model that is currently the best-performing one, and for the other 10% of the time, you route traffic to a random model. This means that for each of the predictions your system generates, 90% of them come from the best-at-that-point-in-time model.

Two of the most popular exploration algorithms are Thompson Sampling and Upper Confidence Bound (UCB). Thompson Sampling selects a model with a probability that this model is optimal given the current knowledge.[34] In our case, it means that the algorithm selects the model based on its probability of having a higher value (better performance) than all other models. On the other hand, UCB selects the item with the highest upper confidence bound.[35] We say that UCB implements *optimism in the face of uncertainty*, it gives an "uncertainty bonus," also called "exploration bonus," to the items it's uncertain about.

Contextual bandits as an exploration strategy

If bandits for model evaluation are to determine the payout (i.e., prediction accuracy) of each model, contextual bandits are to determine the payout of each action. In the case of recommendations/ads, an action is an item/ad to show to users, and the payout is how likely it is a user will click on it. Contextual bandits, like other bandits, are an amazing technique to improve the data efficiency of your model.

Some people also call bandits for model evaluation "contextual bandits." This makes conversations confusing, so in this book, "contextual bandits" refer to exploration strategies to determine the payout of predictions.

Imagine that you're building a recommender system with 1,000 items to recommend, which makes it a 1,000-arm bandit problem. Each time, you can only recommend the top 10 most relevant items to a user. In bandit terms, you'll have to choose the best 10 arms. The shown items get user feedback, inferred via whether the user clicks on them. But you won't get feedback on the other 990 items. This is known as the *partial*

34 William R. Thompson, "On the Likelihood that One Unknown Probability Exceeds Another in View of the Evidence of Two Samples," *Biometrika* 25, no. 3/4 (December 1933): 285–94, *https://oreil.ly/TH1HC*.

35 Peter Auer, "Using Confidence Bounds for Exploitation–Exploration Trade-offs," *Journal of Machine Learning Research* 3 (November 2002): 397–422, *https://oreil.ly/vp9mI*.

feedback problem, also known as *bandit feedback*. You can also think of contextual bandits as a classification problem with bandit feedback.

Let's say that each time a user clicks on an item, this item gets 1 value point. When an item has 0 value points, it could either be because the item has never been shown to a user, or because it's been shown but not clicked on. You want to show users the items with the highest value to them, but if you keep showing users only the items with the most value points, you'll keep on recommending the same popular items, and the never-before-shown items will keep having 0 value points.

Contextual bandits are algorithms that help you balance between showing users the items they will like and showing the items that you want feedback on.[36] It's the same exploration–exploitation trade-off that many readers might have encountered in reinforcement learning. Contextual bandits are also called "one-shot" reinforcement learning problems.[37] In reinforcement learning, you might need to take a series of actions before seeing the rewards. In contextual bandits, you can get bandit feedback right away after an action—e.g., after recommending an ad, you get feedback on whether a user has clicked on that recommendation.

Contextual bandits are well researched and have been shown to improve models' performance significantly (see reports by Twitter (*https://oreil.ly/EqjmB*) and Google (*https://oreil.ly/ipMxd*)). However, contextual bandits are even harder to implement than model bandits, since the exploration strategy depends on the ML model's architecture (e.g., whether it's a decision tree or a neural network), which makes it less generalizable across use cases. Readers interested in combining contextual bandits with deep learning should check out a great paper written by a team at Twitter: "Deep Bayesian Bandits: Exploring in Online Personalized Recommendations" (*https://oreil.ly/Uv03p*) (Guo et al. 2020).

Before we wrap up this section, there's one point I want to emphasize. We've gone through multiple types of tests for ML models. However, it's important to note that a good evaluation pipeline is not only about what tests to run, but also about who should run those tests. In ML, the evaluation process is often owned by data scientists—the same people who developed the model are responsible for evaluating it. Data scientists tend to evaluate their new model ad hoc using the sets of tests that they like. First, this process is imbued with biases—data scientists have contexts about their models that most users don't, which means they probably won't use this model

36 Lihong Li, Wei Chu, John Langford, and Robert E. Schapire, "A Contextual-Bandit Approach to Personalized News Article Recommendation," *arXiv*, February 28, 2010, *https://oreil.ly/uaWHm*.

37 According to Wikipedia, *multi-armed bandit* is a classic reinforcement learning problem that exemplifies the exploration–exploitation trade-off dilemma (s.v., "Multi-armed bandit," *https://oreil.ly/ySjwo*). The name comes from imagining a gambler at a row of slot machines (sometimes known as "one-armed bandits") who has to decide which machines to play, how many times to play each machine and in which order to play them, and whether to continue with the current machine or try a different machine.

in a way most of their users will. Second, the ad hoc nature of the process means that the results might be variable. One data scientist might perform a set of tests and find that model A is better than model B, while another data scientist might report differently.

The lack of a way to ensure models' quality in production has led to many models failing after being deployed, which, in turn, fuels data scientists' anxiety when deploying models. To mitigate this issue, it's important for each team to outline clear pipelines on how models should be evaluated: e.g., the tests to run, the order in which they should run, the thresholds they must pass in order to be promoted to the next stage. Better, these pipelines should be automated and kicked off whenever there's a new model update. The results should be reported and reviewed, similar to the continuous integration/continuous deployment (CI/CD) process for traditional software engineering. It's crucial to understand that a good evaluation process involves not only what tests to run but also who should run those tests.

Summary

This chapter touches on a topic that I believe is among the most exciting yet underexplored topics: how to continually update your models in production to adapt them to changing data distributions. We discussed the four stages a company might go through in the process of modernizing their infrastructure for continual learning: from the manual, training from scratch stage to automated, stateless continual learning.

We then examined the question that haunts ML engineers at companies of all shapes and sizes, "How often *should* I update my models?" by urging them to consider the value of data freshness to their models and the trade-offs between model iteration and data iteration.

Similar to online prediction discussed in Chapter 7, continual learning requires a mature streaming infrastructure. The training part of continual learning can be done in batch, but the online evaluation part requires streaming. Many engineers worry that streaming is hard and costly. It was true three years ago, but streaming technologies have matured significantly since then. More and more companies are providing solutions to make it easier for companies to move to streaming, including Spark Streaming, Snowflake Streaming, Materialize, Decodable, Vectorize, etc.

Continual learning is a problem specific to ML, but it largely requires an infrastructural solution. To be able to speed up the iteration cycle and detect failures in new model updates quickly, we need to set up our infrastructure in the right way. This requires the data science/ML team and the platform team to work together. We'll discuss infrastructure for ML in the next chapter.

CHAPTER 10

Infrastructure and Tooling for MLOps

In Chapters 4 to 6, we discussed the logic for developing ML systems. In Chapters 7 to 9, we discussed the considerations for deploying, monitoring, and continually updating an ML system. Up until now, we've assumed that ML practitioners have access to all the tools and infrastructure they need to implement that logic and carry out these considerations. However, that assumption is far from being true. Many data scientists have told me that they know the right things to do for their ML systems, but they can't do them because their infrastructure isn't set up in a way that enables them to do so.

ML systems are complex. The more complex a system, the more it can benefit from good infrastructure. Infrastructure, when set up right, can help automate processes, reducing the need for specialized knowledge and engineering time. This, in turn, can speed up the development and delivery of ML applications, reduce the surface area for bugs, and enable new use cases. When set up wrong, however, infrastructure is painful to use and expensive to replace. In this chapter, we'll discuss how to set up infrastructure right for ML systems.

Before we dive in, it's important to note that every company's infrastructure needs are different. The infrastructure required for you depends on the number of applications you develop and how specialized the applications are. At one end of the spectrum, you have companies that use ML for ad hoc business analytics such as to project the number of new users they'll have next year to present at their quarterly planning meeting. These companies probably won't need to invest in any infrastructure—Jupyter Notebooks, Python, and Pandas would be their best friends. If you have only one simple ML use case, such as an Android app for object detection to show your friends, you probably won't need any infrastructure either—you just need an Android-compatible ML framework like TensorFlow Lite.

At the other end of the spectrum, there are companies that work on applications with unique requirements. For example, self-driving cars have unique accuracy and latency requirements—the algorithm must be able to respond within milliseconds and its accuracy must be near-perfect since a wrong prediction can lead to serious accidents. Similarly, Google Search has a unique scale requirement since most companies don't process 63,000 search queries a second, which translates to 234 million search queries an hour, like Google does.[1] These companies will likely need to develop their own highly specialized infrastructure. Google developed a large part of their internal infrastructure for search; so did self-driving car companies like Tesla and Waymo.[2] It's common that part of specialized infrastructure is later made public and adopted by other companies. For example, Google extended their internal cloud infrastructure to the public, resulting in Google Cloud Platform (*https://oreil.ly/0g02L*).

In the middle of the spectrum are the majority of companies, those who use ML for multiple common applications—a fraud detection model, a price optimization model, a churn prediction model, a recommender system, etc.—at reasonable scale. "Reasonable scale" refers to companies that work with data in the order of gigabytes and terabytes, instead of petabytes, a day. Their data science team might range from 10 to hundreds of engineers.[3] This category might include any company from a 20-person startup to a company at Zillow's scale, but not at FAAAM scale.[4] For example, back in 2018, Uber was adding tens of terabytes of data a day to their data lake, and Zillow's biggest dataset was bringing in 2 terabytes of uncompressed data a day.[5] In contrast, even back in 2014, Facebook was generating *4 petabytes* of data a day.[6]

1 Kunal Shah, "This Is What Makes SEO Important for Every Business," *Entrepreneur India*, May 11, 2020, *https://oreil.ly/teQlX*.

2 For a sneak peek into Tesla's compute infrastructure for ML, I highly recommend watching the recording of Tesla AI Day 2021 on YouTube (*https://oreil.ly/etH9C*).

3 The definition for "reasonable scale" was inspired by Jacopo Tagliabue in his paper "You Do Not Need a Bigger Boat: Recommendations at Reasonable Scale in a (Mostly) Serverless and Open Stack," *arXiv*, July 15, 2021, *https://oreil.ly/YNRZQ*. For more discussion on reasonable scale, see "ML and MLOps at a Reasonable Scale" (*https://oreil.ly/goPrb*) by Ciro Greco (October 2021).

4 FAAAM is short for Facebook, Apple, Amazon, Alphabet, Microsoft.

5 Reza Shiftehfar, "Uber's Big Data Platform: 100+ Petabytes with Minute Latency," *Uber Engineering*, October 17, 2018, *https://oreil.ly/6Ykd3*; Kaushik Krishnamurthi, "Building a Big Data Pipeline to Process Clickstream Data," Zillow, April 6, 2018, *https://oreil.ly/SGmNe*.

6 Nathan Bronson and Janet Wiener, "Facebook's Top Open Data Problems," Meta, October 21, 2014, *https://oreil.ly/p6QjX*.

Companies in the middle of the spectrum will likely benefit from generalized ML infrastructure that is being increasingly standardized (see Figure 10-1). In this book, we'll focus on the infrastructure for the vast majority of ML applications at a reasonable scale.

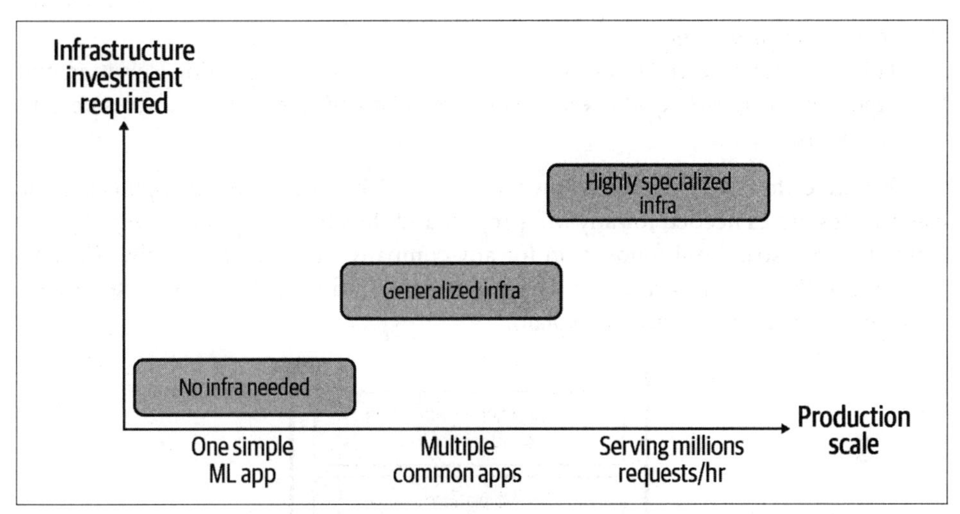

Figure 10-1. Infrastructure requirements for companies at different production scales

In order to set up the right infrastructure for your needs, it's important to understand exactly what infrastructure means and what it consists of. According to Wikipedia, in the physical world, "infrastructure is the set of fundamental facilities and systems that support the sustainable functionality of households and firms."[7] In the ML world, infrastructure is the set of fundamental facilities that support the development and maintenance of ML systems. What should be considered the "fundamental facilities" varies greatly from company to company, as discussed earlier in this chapter. In this section, we will examine the following four layers:

Storage and compute
: The storage layer is where data is collected and stored. The compute layer provides the compute needed to run your ML workloads such as training a model, computing features, generating features, etc.

Resource management
: Resource management comprises tools to schedule and orchestrate your workloads to make the most out of your available compute resources. Examples of tools in this category include Airflow, Kubeflow, and Metaflow.

7 Wikipedia, s.v. "Infrastructure," *https://oreil.ly/YaIk8*.

ML platform

This provides tools to aid the development of ML applications such as model stores, feature stores, and monitoring tools. Examples of tools in this category include SageMaker and MLflow.

Development environment

This is usually referred to as the dev environment; it is where code is written and experiments are run. Code needs to be versioned and tested. Experiments need to be tracked.

These four different layers are shown in Figure 10-2. Data and compute are the essential resources needed for any ML project, and thus the *storage and compute layer* forms the infrastructural foundation for any company that wants to apply ML. This layer is also the most abstract to a data scientist. We'll discuss this layer first because these resources are the easiest to explain.

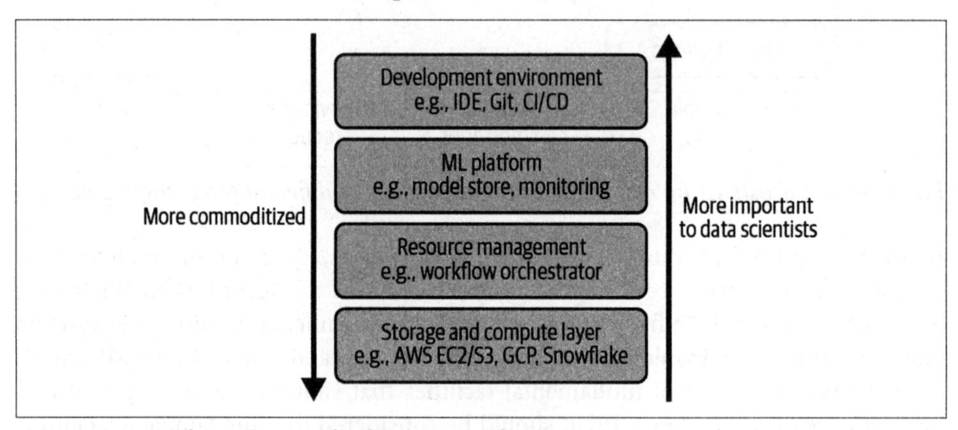

Figure 10-2. Different layers of infrastructure for ML

The dev environment is what data scientists have to interact with daily, and therefore, it is the least abstract to them. We'll discuss this category next, then we'll discuss resource management, a contentious topic among data scientists—people are still debating whether a data scientist needs to know about this layer or not. Because "ML platform" is a relatively new concept with its different components still maturing, we'll discuss this category last, after we've familiarized ourselves with all other categories. An ML platform requires up-front investment from a company, but if it's done right, it can make the life of data scientists across business use cases at that company so much easier.

Even if two companies have the exact same infrastructure needs, their resulting infrastructure might look different depending on their approaches to build versus buy decisions—i.e., what they want to build in-house versus what they want to outsource to other companies. We'll discuss the build versus buy decisions in the last part of this chapter, where we'll also discuss the hope for standardized and unified abstractions for ML infrastructure.

Let's dive in!

Storage and Compute

ML systems work with a lot of data, and this data needs to be stored somewhere. The *storage layer* is where data is collected and stored. At its simplest form, the storage layer can be a hard drive disk (HDD) or a solid state disk (SSD). The storage layer can be in one place, e.g., you might have all your data in Amazon S3 or in Snowflake, or spread out over multiple locations.[8] Your storage layer can be on-prem in a private data center or on the cloud. In the past, companies might have tried to manage their own storage layer. However, in the last decade, the storage layer has been mostly commoditized and moved to the cloud. Data storage has become so cheap that most companies just store all the data they have without the cost.[9] We've covered the data layer intensively in Chapter 3, so in this chapter, we'll focus on the compute layer.

The *compute layer* refers to all the compute resources a company has access to and the mechanism to determine how these resources can be used. The amount of compute resources available determines the scalability of your workloads. You can think of the compute layer as the engine to execute your jobs. At its simplest form, the compute layer can just be a single CPU core or a GPU core that does all your computation. Its most common form is cloud compute managed by a cloud provider such as AWS Elastic Compute Cloud (EC2) or GCP.

The compute layer can usually be sliced into smaller compute units to be used concurrently. For example, a CPU core might support two concurrent threads; each thread is used as a compute unit to execute its own job. Or multiple CPU cores might be joined together to form a larger compute unit to execute a larger job. A compute unit can be created for a specific short-lived job such as an AWS Step Function or a GCP Cloud Run—the unit will be eliminated after the job finishes. A compute unit can also be created to be more "permanent," aka without being tied to a job, like a virtual machine. A more permanent compute unit is sometimes called an "instance."

8 I've seen a company whose data is spread over Amazon Redshift and GCP BigQuery, and their engineers are not very happy about it.

9 We only discuss data storage here since we've discussed data systems in Chapter 2.

However, the compute layer doesn't always use threads or cores as compute units. There are compute layers that abstract away the notions of cores and use other units of computation. For example, computation engines like Spark and Ray use "job" as their unit, and Kubernetes uses "pod," a wrapper around containers, as its smallest deployable unit. While you can have multiple containers in a pod, you can't independently start or stop different containers in the same pod.

To execute a job, you first need to load the required data into your compute unit's memory, then execute the required operations—addition, multiplication, division, convolution, etc.—on that data. For example, to add two arrays, you will first need to load these two arrays into memory, and then perform addition on the two arrays. If the compute unit doesn't have enough memory to load these two arrays, the operation will be impossible without an algorithm to handle out-of-memory computation. Therefore, a compute unit is mainly characterized by two metrics: how much memory it has and how fast it runs an operation.

The memory metric can be specified using units like GB, and it's generally straightforward to evaluate: a compute unit with 8 GB of memory can handle more data in memory than a compute unit with only 2 GB, and it is generally more expensive.[10] Some companies care not only how much memory a compute unit has but also how fast it is to load data in and out of memory, so some cloud providers advertise their instances as having "high bandwidth memory" or specify their instances' I/O bandwidth.

The operation speed is more contentious. The most common metric is FLOPS—floating point operations per second. As the name suggests, this metric denotes the number of float point operations a compute unit can run per second. You might see a hardware vendor advertising that their GPUs or TPUs or IPUs (intelligence processing units) have teraFLOPS (one trillion FLOPS) or another massive number of FLOPS.

However, this metric is contentious because, first, companies that measure this metric might have different ideas on what is counted as an operation. For example, if a machine fuses two operations into one and executes this fused operation,[11] does this count as one operation or two? Second, just because a compute unit is capable of doing a trillion FLOPS doesn't mean you'll be able to execute your job at the speed of a trillion FLOPS. The ratio of the number of FLOPS a job can run to the number of FLOPs a compute unit is capable of handling is called utilization.[12] If an instance

10 As of writing this book, an ML workload typically requires between 4 GB and 8 GB of memory; 16 GB of memory is enough to handle most ML workloads.

11 See operation fusion in the section "Model optimization" on page 216.

12 "What Is FLOP/s and Is It a Good Measure of Performance?," Stack Overflow, last updated October 7, 2020, *https://oreil.ly/M8jPP*.

is capable of doing a million FLOPs and your job runs with 0.3 million FLOPS, that's a 30% utilization rate. Of course, you'd want to have your utilization rate as high as possible. However, it's near impossible to achieve 100% utilization rate. Depending on the hardware backend and the application, the utilization rate of 50% might be considered good or bad. Utilization also depends on how fast you can load data into memory to perform the next operations—hence the importance of I/O bandwidth.[13]

When evaluating a new compute unit, it's important to evaluate how long it will take this compute unit to do common workloads. For example, MLPerf (*https://oreil.ly/XuVka*) is a popular benchmark for hardware vendors to measure their hardware performance by showing how long it will take their hardware to train a ResNet-50 model on the ImageNet dataset or use a BERT-large model to generate predictions for the SQuAD dataset.

Because thinking about FLOPS is not very useful, to make things easier, when evaluating compute performance, many people just look into the number of cores a compute unit has. So you might use an instance with 4 CPU cores and 8 GB of memory. Keep in mind that AWS uses the concept of vCPU, which stands for virtual CPU and which, for practical purposes, can be thought of as half a physical core.[14] You can see the number of cores and memory offered by some AWS EC2 and GCP instances in Figure 10-3.

Some GPU instances on AWS

Instance	GPUs	vCPU	Mem (GiB)	GPU Mem (GiB)
p3.2xlarge	1	8	61	16
p3.8xlarge	4	32	244	64
p3.16xlarge	8	64	488	128
p3dn.24xlarge	8	96	768	256

Some TPU instances on GCP

TPU type (v2)	v2 cores	Total memory
v2-8	8	64 GiB

TPU type (v3)	v3 cores	Total memory
v3-8	8	128 GiB

Figure 10-3. Examples of GPU and TPU instances available on AWS and GCP as of February 2022. Source: Screenshots of AWS and GCP websites

13 For readers interested in FLOPS and bandwidth and how to optimize them for deep learning models, I recommend the post "Making Deep Learning Go Brrrr From First Principles" (*https://oreil.ly/zvVFB*) (He 2022).

14 According to Amazon, "EC2 instances support multithreading, which enables multiple threads to run concurrently on a single CPU core. Each thread is represented as a virtual CPU (vCPU) on the instance. An instance has a default number of CPU cores, which varies according to instance type. For example, an m5.xlarge instance type has two CPU cores and two threads per core by default—four vCPUs in total" ("Optimize CPU Options," Amazon Web Services, last accessed April 2020, *https://oreil.ly/eeOtd*).

Public Cloud Versus Private Data Centers

Like data storage, the compute layer is largely commoditized. This means that instead of setting up their own data centers for storage and compute, companies can pay cloud providers like AWS and Azure for the exact amount of compute they use. Cloud compute makes it extremely easy for companies to start building without having to worry about the compute layer. It's especially appealing to companies that have variable-sized workloads. Imagine if your workloads need 1,000 CPU cores one day of the year and only 10 CPU cores the rest of the year. If you build your own data centers, you'll need to pay for 1,000 CPU cores up front. With cloud compute, you only need to pay for 1,000 CPU cores one day of the year and 10 CPU cores the rest of the year. It's convenient to be able to just add more compute or shut down instances as needed—most cloud providers even do that automatically for you—reducing engineering operational overhead. This is especially useful in ML as data science workloads are bursty. Data scientists tend to run experiments a lot for a few weeks during development, which requires a surge of compute power. Later on, during production, the workload is more consistent.

Keep in mind that cloud compute is elastic but not magical. It doesn't actually have infinite compute. Most cloud providers offer limits (*https://oreil.ly/TzUOv*) on the compute resources you can use at a time. Some, but not all, of these limits can be raised through petitions. For example, as of writing this book, AWS EC2's largest instance is X1e (*https://oreil.ly/29lsT*) with 128 vCPUs and almost 4 TB of memory.[15] Having a lot of compute resources doesn't mean that it's always easy to use them, especially if you have to work with spot instances to save cost.[16]

Due to the cloud's elasticity and ease of use, more and more companies are choosing to pay for the cloud over building and maintaining their own storage and compute layer. Synergy Research Group's research shows that in 2020, "enterprise spending on cloud infrastructure services [grew] by 35% to reach almost $130 billion" while "enterprise spending on data [centers] dropped by 6% to under $90 billion,"[17] as shown in Figure 10-4.

15 Which costs $26.688/hour.

16 On-demand instances are instances that are available when you request them. Spot instances are instances that are available when nobody else is using them. Cloud providers tend to offer spot instances at a discount compared to on-demand instances.

17 Synergy Research Group, "2020—The Year That Cloud Service Revenues Finally Dwarfed Enterprise Spending on Data Centers," March 18, 2021, *https://oreil.ly/uPx94*.

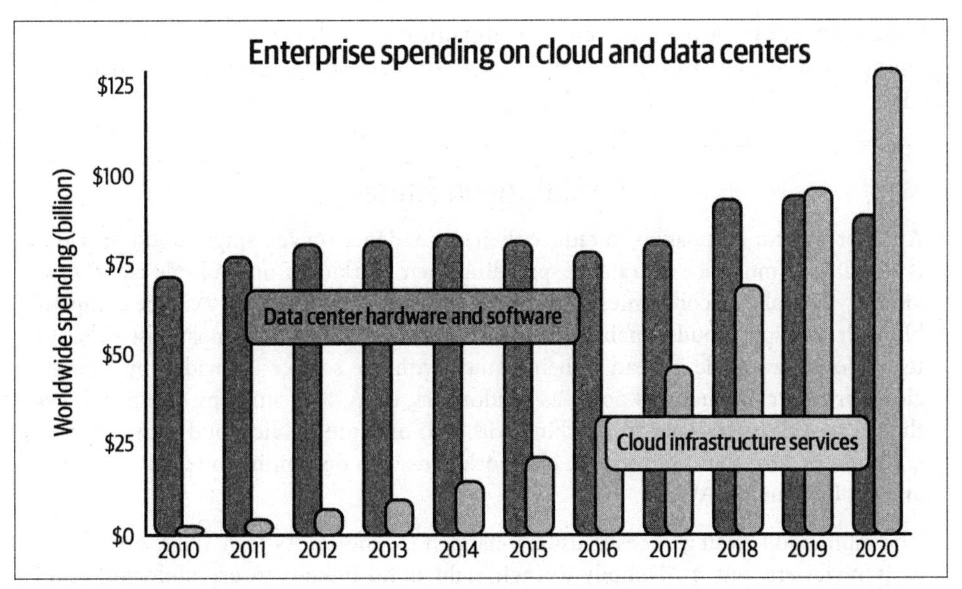

Figure 10-4. In 2020, enterprise spending on cloud infrastructure services grew by 35% while spending on data centers dropped by 6%. Source: Adapted from an image by Synergy Research Group

While leveraging the cloud tends to give companies higher returns than building their own storage and compute layers early on, this becomes less defensible as a company grows. Based on disclosed cloud infrastructure spending by public software companies, the venture capital firm a16z shows that cloud spending accounts for approximately 50% cost of revenue of these companies.[18]

The high cost of the cloud has prompted companies to start moving their workloads back to their own data centers, a process called "cloud repatriation." Dropbox's S-1 filing in 2018 (*https://oreil.ly/zRm9j*) shows that the company was able to save $75M over the two years prior to IPO due to their infrastructure optimization overhaul—a large chunk of it consisted of moving their workloads from public cloud to their own data centers. Is the high cost of cloud unique to Dropbox because Dropbox is in the data storage business? Not quite. In the aforementioned analysis, a16z estimated that "across 50 of the top public software companies currently utilizing cloud infrastructure, an estimated $100B of market value is being lost among them due to cloud impact on margins—relative to running the infrastructure themselves."[19]

While getting started with the cloud is easy, moving away from the cloud is hard. Cloud repatriation requires nontrivial up-front investment in both commodities and

18 Sarah Wang and Martin Casado, "The Cost of Cloud, a Trillion Dollar Paradox," a16z, *https://oreil.ly/3nWU3*.

19 Wang and Casado, "The Cost of Cloud."

engineering effort. More and more companies are following a hybrid approach: keeping most of their workloads on the cloud but slowly increasing their investment in data centers.

On Multicloud Strategy

Another way for companies to reduce their dependence on any single cloud provider is to follow a multicloud strategy: spreading their workloads on multiple cloud providers.[20] This allows companies to architect their systems so that they can be compatible with multiple clouds, enabling them to leverage the best and most cost-effective technologies available instead of being stuck with the services provided by a single cloud provider, a situation known as vendor lock-in. A 2019 study by Gartner shows that 81% of organizations are working with two or more public cloud providers.[21] A common pattern that I've seen for ML workloads is to do training on GCP or Azure, and deployment on AWS.

The multicloud strategy doesn't usually happen by choice. As Josh Wills, one of our early reviewers, put it: "Nobody in their right mind intends to use multicloud." It's incredibly hard to move data and orchestrate workloads across clouds.

Often, multicloud just happens because different parts of the organization operate independently, and each part makes their own cloud decision. It can also happen following an acquisition—the acquired team is already on a cloud different from the host organization, and migrating hasn't happened yet.

In my work, I've seen multicloud happen due to strategic investments. Microsoft and Google are big investors in the startup ecosystem, and several companies that I work with that were previously on AWS have moved to Azure/GCP after Microsoft/Google invested in them.

Development Environment

The dev environment is where ML engineers write code, run experiments, and interact with the production environment where champion models are deployed and challenger models evaluated. The dev environment consists of the following components: IDE (integrated development environment), versioning, and CI/CD.

If you're a data scientist or ML engineer who writes code daily, you're probably very familiar with all these tools and might wonder what there is to say about them. In my experience, outside of a handful of tech companies, the dev environment is severely

20 Laurence Goasduff, "Why Organizations Choose a Multicloud Strategy," Gartner, May 7, 2019, *https://oreil.ly/ZiqzQ*.

21 Goasduff, "Why Organizations Choose a Multicloud Strategy."

underrated and underinvested in at most companies. According to Ville Tuulos in his book *Effective Data Science Infrastructure*, "you would be surprised to know how many companies have well-tuned, scalable production infrastructure but the question of how the code is developed, debugged, and tested in the first place is solved in an ad-hoc manner."[22]

He suggested that "if you have time to set up only one piece of infrastructure well, make it the development environment for data scientists." Because the dev environment is where engineers work, improvements in the dev environment translate directly into improvements in engineering productivity.

In this section, we'll first cover different components of the dev environment, then we'll discuss the standardization of the dev environment before we discuss how to bring your changes from the dev environment to the production environment with containers.

Dev Environment Setup

The dev environment should be set up to contain all the tools that can make it easier for engineers to do their job. It should also consist of tools for *versioning*. As of this writing, companies use an ad hoc set of tools to version their ML workflows, such as Git to version control code, DVC to version data, Weights & Biases or Comet.ml to track experiments during development, and MLflow to track artifacts of models when deploying them. Claypot AI is working on a platform that can help you version and track all your ML workflows in one place. Versioning is important for any software engineering projects, but even more so for ML projects because of both the sheer number of things you can change (code, parameters, the data itself, etc.) and the need to keep track of prior runs to reproduce later on. We've covered this in the section "Experiment Tracking and Versioning" on page 162.

The dev environment should also be set up with a *CI/CD* test suite to test your code before pushing it to the staging or production environment. Examples of tools to orchestrate your CI/CD test suite are GitHub Actions and CircleCI. Because CI/CD is a software engineering concern, it's beyond the scope of this book.

In this section, we'll focus on the place where engineers write code: the IDE.

IDE

The *IDE* is the editor where you write your code. IDEs tend to support multiple programming languages. IDEs can be native apps like VS Code or Vim. IDEs can be browser-based, which means they run in browsers, such as AWS Cloud9.

22 Ville Tuulos, *Effective Data Science Infrastructure* (Manning, 2022).

Many data scientists write code not just in IDEs but also in notebooks like Jupyter Notebooks and Google Colab.[23] Notebooks are more than just places to write code. You can include arbitrary artifacts such as images, plots, data in nice tabular formats, etc., which makes notebooks very useful for exploratory data analysis and analyzing model training results.

Notebooks have a nice property: they are stateful—they can retain states after runs. If your program fails halfway through, you can rerun from the failed step instead of having to run the program from the beginning. This is especially helpful when you have to deal with large datasets that might take a long time to load. With notebooks, you only need to load your data once—notebooks can retain this data in memory—instead of having to load it each time you want to run your code. As shown in Figure 10-5, if your code fails at step 4 in a notebook, you'll only need to rerun step 4 instead of from the beginning of your program.

```
In [1]: import pandas as pd

In [2]: fname = "large-dataset.csv"

In [3]: df = pd.read_csv(fname)

In [4]: features = df["Timestamp", "Cost"]

        ---------------------------------------------------------------------------
        KeyError                                  Traceback (most recent call last)
        ~/miniconda3/envs/stove39/lib/python3.9/site-packages/pandas/core/indexes/base.py
        ance)
           3360             try:
        -> 3361                 return self._engine.get_loc(casted_key)
           3362             except KeyError as err:
```

Figure 10-5. In Jupyter Notebooks, if step 4 fails, you only need to run step 4 again, instead of having to run steps 1 to 4 again

Note that this statefulness can be a double-edged sword, as it allows you to execute your cells out of order. For example, in a normal script, cell 4 must run after cell 3 and cell 3 must run after cell 2. However, in notebooks, you can run cell 2, 3, then 4 or cell 4, 3, then 2. This makes notebook reproducibility harder unless your notebook comes with an instruction on the order in which to run your cells. This difficulty is captured in a joke by Chris Albon (see Figure 10-6).

23 As of writing this book, Google Colab even offers free GPUs (*https://oreil.ly/9ij7E*) for their users.

Chris Albon @chrisalbon · 1d

Me explaining the execution order of my Jupyter notebook cells.

Figure 10-6. Notebooks' statefulness allows you to execute cells out of order, making it hard to reproduce a notebook

Because notebooks are so useful for data exploration and experiments, notebooks have become an indispensable tool for data scientists and ML. Some companies have made notebooks the center of their data science infrastructure. In their seminal article, "Beyond Interactive: Notebook Innovation at Netflix," Netflix included a list of infrastructure tools that can be used to make notebooks even more powerful.[24] The list includes:

Papermill (https://oreil.ly/569ot)
: For spawning multiple notebooks with different parameter sets—such as when you want to run different experiments with different sets of parameters and execute them concurrently. It can also help summarize metrics from a collection of notebooks.

Commuter (https://oreil.ly/dFlYV)
: A notebook hub for viewing, finding, and sharing notebooks within an organization.

---

24 Michelle Ufford, M. Pacer, Matthew Seal, and Kyle Kelley, "Beyond Interactive: Notebook Innovation at Netflix," *Netflix Technology Blog*, August 16, 2018, *https://oreil.ly/EHvAe*.

Another interesting project aimed at improving the notebook experience is nbdev (*https://nbdev.fast.ai*), a library on top of Jupyter Notebooks that encourages you to write documentation and tests in the same place.

Standardizing Dev Environments

The first thing about the dev environment is that it should be standardized, if not company-wide, then at least team-wide. We'll go over a story to understand what it means to have the dev environment standardized and why that is needed.

In the early days of our startup, we each worked from our own computer. We had a bash file that a new team member could run to create a new virtual environment—in our case, we use conda for virtual environments—and install the required packages needed to run our code. The list of the required packages was the good old *requirements.txt* that we kept adding to as we started using a new package. Sometimes, one of us got lazy and we just added a package name (e.g., `torch`) without specifying which version of the package it was (e.g., `torch==1.10.0+cpu`). Occasionally, a new pull request would run well on my computer but not another coworker's computer,[25] and we usually quickly figured out that it was because we used different versions of the same package. We resolved to always specify the package name together with the package version when adding a new package to the *requirements.txt*, and that removed a lot of unnecessary headaches.

One day, we ran into this weird bug that only happened during some runs and not others. I asked my coworker to look into it, but he wasn't able to reproduce the bug. I told him that the bug only happened some of the time, so he might have to run the code around 20 times just to be sure. He ran the code 20 times and still found nothing. We compared our packages and everything matched. After a few hours of hair-pulling frustration, we discovered that it was a concurrency issue that is only an issue for Python version 3.8 or earlier. I had Python 3.8 and my coworker had Python 3.9, so he didn't see the bug. We resolved to have everyone on the same Python version, and that removed some more headaches.

Then one day, my coworker got a new laptop. It was a MacBook with the then new M1 chip. He tried to follow our setup steps on this new laptop but ran into difficulty. It was because the M1 chip was new, and some of the tools we used, including Docker, weren't working well with M1 chips yet. After seeing him struggling with setting the environment up for a day, we decided to move to a cloud dev environment. This means that we still standardize the virtual environment and tools and packages, but now everyone uses the virtual environment and tools and packages on the same type of machine too, provided by a cloud provider.

25 For the uninitiated, a new pull request can be understood as a new piece of code being added to the codebase.

When using a cloud dev environment, you can use a cloud dev environment that also comes with a cloud IDE like AWS Cloud9 (*https://oreil.ly/xFEZx*) (which has no built-in notebooks) and Amazon SageMaker Studio (*https://oreil.ly/m1yFZ*) (which comes with hosted JupyterLab). As of writing this book, Amazon SageMaker Studio seems more widely used than Cloud9. However, most engineers I know who use cloud IDEs do so by installing IDEs of their choice, like Vim, on their cloud instances.

A much more popular option is to use a cloud dev environment with a local IDE. For example, you can use VS Code installed on your computer and connect the local IDE to the cloud environment using a secure protocol like Secure Shell (SSH).

While it's generally agreed upon that tools and packages should be standardized, some companies are hesitant to standardize IDEs. Engineers can get emotionally attached to IDEs, and some have gone to great length to defend their IDE of choice,[26] so it'll be hard forcing everyone to use the same IDE. However, over the years, some IDEs have emerged to be the most popular. Among them, VS Code is a good choice since it allows easy integration with cloud dev instances.

At our startup, we chose GitHub Codespaces (*https://oreil.ly/bQdUW*) as our cloud dev environment, but an AWS EC2 or a GCP instance that you can SSH into is also a good option. Before moving to cloud environments, like many other companies, we were worried about the cost—what if we forgot to shut down our instances when not in use and they kept charging us money? However, this worry has gone away for two reasons. First, tools like GitHub Codespaces automatically shut down your instance after 30 minutes of inactivity. Second, some instances are pretty cheap. For example, an AWS instance with 4 vCPUs and 8 GB of memory costs around $0.1/hour, which comes to approximately $73/month if you never shut it down. Because engineering time is expensive, if a cloud dev environment can help you save a few hours of engineering time a month, it's worth it for many companies.

Moving from local dev environments to cloud dev environments has many other benefits. First, it makes IT support so much easier—imagine having to support 1,000 different local machines instead of having to support only one type of cloud instance. Second, it's convenient for remote work—you can just SSH into your dev environment wherever you go from any computer. Third, cloud dev environments can help with security. For example, if an employee's laptop is stolen, you can just revoke access to cloud instances from that laptop to prevent the thief from accessing your codebase and proprietary information. Of course, some companies might not be able to move to cloud dev environments also because of security concerns. For example, they aren't allowed to have their code or data on the cloud.

26 See editor war (*https://oreil.ly/OOkqJ*), the decade-long, heated debate on Vim versus Emacs.

The fourth benefit, which I would argue is the biggest benefit for companies that do production on the cloud, is that having your dev environment on the cloud reduces the gap between the dev environment and the production environment. If your production environment is in the cloud, bringing your dev environment to the cloud is only natural.

Occasionally, a company has to move their dev environments to the cloud not only because of the benefits, but also out of necessity. For the use cases where data can't be downloaded or stored on a local machine, the only way to access it is via a notebook in the cloud (SageMaker Studio) that can read the data from S3, provided it has the right permissions.

Of course, cloud dev environments might not work for every company due to cost, security, or other concerns. Setting up cloud dev environments also requires some initial investments, and you might need to educate your data scientists on cloud hygiene, including establishing secure connections to the cloud, security compliance, or avoiding wasteful cloud usage. However, standardization of dev environments might make your data scientists' lives easier and save you money in the long run.

From Dev to Prod: Containers

During development, you might usually work with a fixed number of machines or instances (usually one) because your workloads don't fluctuate a lot—your model doesn't suddenly change from serving only 1,000 requests an hour to 1 million requests an hour.

A production service, on the other hand, might be spread out on multiple instances. The number of instances changes from time to time depending on the incoming workloads, which can be unpredictable at times. For example, a celebrity tweets about your fledgling app and suddenly your traffic spikes 10x. You will have to turn on new instances as needed, and these instances will need to be set up with required tools and packages to execute your workloads.

Previously, you'd have to spin up and shut down instances yourself, but most public cloud providers have taken care of the autoscaling part. However, you still have to worry about setting up new instances.

When you consistently work with the same instance, you can install dependencies once and use them whenever you use this instance. In production, if you dynamically allocate instances as needed, your environment is inherently stateless. When a new instance is allocated for your workload, you'll need to install dependencies using a list of predefined instructions.

A question arises: how do you re-create an environment on any new instance? Container technology—of which Docker is the most popular—is designed to answer this question. With Docker, you create a Dockerfile with step-by-step instructions on how to re-create an environment in which your model can run: install this package, download this pretrained model, set environment variables, navigate into a folder, etc. These instructions allow hardware anywhere to run your code.

Two key concepts in Docker are image and container. Running all the instructions in a Dockerfile gives you a Docker image. If you run this Docker image, you get back a Docker container. You can think of a Dockerfile as the recipe to construct a mold, which is a Docker image. From this mold, you can create multiple running instances; each is a Docker container.

You can build a Docker image either from scratch or from another Docker image. For example, NVIDIA might provide a Docker image that contains TensorFlow and all necessary libraries to optimize TensorFlow for GPUs. If you want to build an application that runs TensorFlow on GPUs, it's not a bad idea to use this Docker image as your base and install dependencies specific to your application on top of this base image.

A container registry is where you can share a Docker image or find an image created by other people to be shared publicly or only with people inside their organizations. Common container registries include Docker Hub and AWS ECR (Elastic Container Registry).

Here's an example of a simple Dockerfile that runs the following instructions. The example is to show how Dockerfiles work in general, and might not be executable.

1. Download the latest PyTorch base image.
2. Clone NVIDIA's apex repository on GitHub, navigate to the newly created *apex* folder, and install apex.
3. Set *fancy-nlp-project* to be the working directory.
4. Clone Hugging Face's transformers repository on GitHub, navigate to the newly created *transformers* folder, and install transformers.

```
FROM pytorch/pytorch:latest
RUN git clone https://github.com/NVIDIA/apex
RUN cd apex && \
    python3 setup.py install && \
    pip install -v --no-cache-dir --global-option="--cpp_ext" \
    --global-option="--cuda_ext" ./

WORKDIR /fancy-nlp-project
RUN git clone https://github.com/huggingface/transformers.git && \
    cd transformers && \
    python3 -m pip install --no-cache-dir.
```

If your application does anything interesting, you will probably need more than one container. Consider the case where your project consists of the featurizing code that is fast to run but requires a lot of memory, and the model training code that is slow to run but requires less memory. If you run both parts of the code on the same GPU instances, you'll need GPU instances with high memory, which can be very expensive. Instead, you can run your featurizing code on CPU instances and the model training code on GPU instances. This means you'll need one container for featurizing and another container for training.

Different containers might also be necessary when different steps in your pipeline have conflicting dependencies, such as your featurizer code requires NumPy 0.8 but your model requires NumPy 1.0.

If you have 100 microservices and each microservice requires its own container, you might have 100 containers running at the same time. Manually building, running, allocating resources for, and stopping 100 containers might be a painful chore. A tool to help you manage multiple containers is called container orchestration. Docker Compose is a lightweight container orchestrator that can manage containers on a single host.

However, each of your containers might run on its own host, and this is where Docker Compose is at its limits. Kubernetes (K8s) is a tool for exactly that. K8s creates a network for containers to communicate and share resources. It can help you spin up containers on more instances when you need more compute/memory as well as shutting down containers when you no longer need them, and it helps maintain high availability for your system.

K8s was one of the fastest-growing technologies in the 2010s. Since its inception in 2014, it's become ubiquitous in production systems today. Jeremy Jordan has a great introduction to K8s (*https://oreil.ly/QLAC3*) for readers interested in learning more. However, K8s is not the most data-scientist-friendly tool, and there have been many discussions on how to move data science workloads away from it.[27] We'll go more into K8s in the next section.

27 Chip Huyen, "Why Data Scientists Shouldn't Need to Know Kubernetes," September 13, 2021, *https://huyenchip.com/2021/09/13/data-science-infrastructure.html*; Neil Conway and David Hershey, "Data Scientists Don't Care About Kubernetes," Determined AI, November 30, 2020, *https://oreil.ly/FFDQW*; I Am Developer on Twitter (@iamdevloper): "I barely understand my own feelings how am I supposed to understand kubernetes," June 26, 2021, *https://oreil.ly/T2eQE*.

Resource Management

In the pre-cloud world (and even today in companies that maintain their own data centers), storage and compute were finite. Resource management back then centered around how to make the most out of limited resources. Increasing resources for one application could mean decreasing resources for other applications, and complex logic was required to maximize resource utilization, even if that meant requiring more engineering time.

However, in the cloud world where storage and compute resources are much more elastic, the concern has shifted from how to maximize resource utilization to how to use resources cost-effectively. Adding more resources to an application doesn't mean decreasing resources for other applications, which significantly simplifies the allocation challenge. Many companies are OK with adding more resources to an application as long as the added cost is justified by the return, e.g., extra revenue or saved engineering time.

In the vast majority of the world, where engineers' time is more valuable than compute time, companies are OK using more resources if this means it can help their engineers become more productive. This means that it might make sense for companies to invest in automating their workloads, which might make using resources less efficient than manually planning their workloads, but free their engineers to focus on work with higher returns. Often, if a problem can be solved by either using more non-human resources (e.g., throwing more compute at it) or using more human resources (e.g., requiring more engineering time to redesign), the first solution might be preferred.

In this section, we'll discuss how to manage resources for ML workflows. We'll focus on cloud-based resources; however, the discussed ideas can also be applicable for private data centers.

Cron, Schedulers, and Orchestrators

There are two key characteristics of ML workflows that influence their resource management: repetitiveness and dependencies.

In this book, we've discussed at length how developing ML systems is an iterative process. Similarly, ML workloads are rarely one-time operations but something repetitive. For example, you might train a model every week or generate a new batch of predictions every four hours. These repetitive processes can be scheduled and orchestrated to run smoothly and cost-effectively using available resources.

Scheduling repetitive jobs to run at fixed times is exactly what *cron* does. This is also all that cron does: run a script at a predetermined time and tell you whether the job succeeds or fails. It doesn't care about the dependencies between the jobs it runs—you can run job A after job B with cron but you can't schedule anything complicated like run B if A succeeds and run C if A fails.

This leads us to the second characteristic: dependencies. Steps in an ML workflow might have complex *dependency* relationships with each other. For example, an ML workflow might consist of the following steps:

1. Pull last week's data from data warehouses.
2. Extract features from this pulled data.
3. Train two models, A and B, on the extracted features.
4. Compare A and B on the test set.
5. Deploy A if A is better; otherwise deploy B.

Each step depends on the success of the previous step. Step 5 is what we call conditional dependency: the action for this step depends on the outcome of the previous step. The order of execution and dependencies among these steps can be represented using a graph, as shown in Figure 10-7.

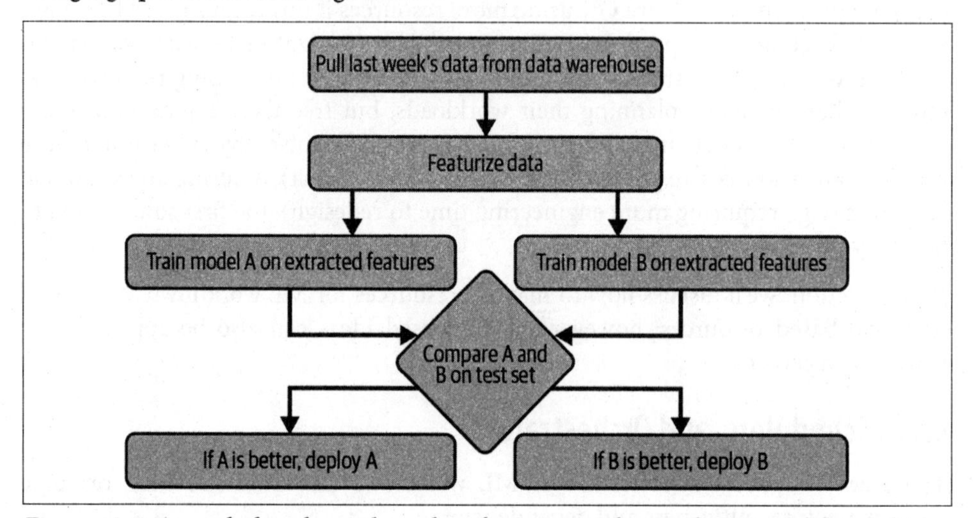

Figure 10-7. A graph that shows the order of execution of a simple ML workflow, which is essentially a DAG (directed acyclic graph)

Many readers might recognize that Figure 10-7 is a DAG: directed acyclic graph. It has to be directed to express the dependencies among steps. It can't contain cycles because, if it does, the job will just keep on running forever. DAG is a common way to represent computing workflows in general, not just ML workflows. Most workflow management tools require you to specify your workflows in a form of DAGs.

Schedulers are cron programs that can handle dependencies. It takes in the DAG of a workflow and schedules each step accordingly. You can even schedule to start a job based on an event-based trigger, e.g., start a job whenever an event X happens. Schedulers also allow you to specify what to do if a job fails or succeeds, e.g., if it fails, how many times to retry before giving up.

Schedulers tend to leverage queues to keep track of jobs. Jobs can be queued, prioritized, and allocated resources needed to execute. This means that schedulers need to be aware of the resources available and the resources needed to run each job—the resources needed are either specified as options when you schedule a job or estimated by the scheduler. For instance, if a job requires 8 GB of memory and two CPUs, the scheduler needs to find among the resources it manages an instance with 8 GB of memory and two CPUs and wait until the instance is not executing other jobs to run this job on the instance.

Here's an example of how to schedule a job with the popular scheduler Slurm, where you specify the job name, the time when the job needs to be executed, and the amount of memory and CPUs to be allocated for the job:

```
#!/bin/bash
#SBATCH -J JobName
#SBATCH --time=11:00:00        # When to start the job
#SBATCH --mem-per-cpu=4096   # Memory, in MB, to be allocated per CPU
#SBATCH --cpus-per-task=4           # Number of cores per task
```

Schedulers should also optimize for resource utilization since they have information on resources available, jobs to run, and resources needed for each job to run. However, the number of resources specified by users is not always correct. For example, I might estimate, and therefore specify, that a job needs 4 GB of memory, but this job only needs 3 GB of memory or needs 4 GB of memory at peak and only 1–2 GB of memory otherwise. Sophisticated schedulers like Google's Borg estimate how many resources a job will actually need and reclaim unused resources for other jobs,[28] further optimizing resource utilization.

Designing a general-purpose scheduler is hard, since this scheduler will need to be able to manage almost any number of concurrent machines and workflows. If your scheduler is down, every single workflow that this scheduler touches will be interrupted.

If schedulers are concerned with *when* to run jobs and what resources are needed to run those jobs, orchestrators are concerned with *where* to get those resources. Schedulers deal with job-type abstractions such as DAGs, priority queues, user-level

28 Abhishek Verma, Luis Pedrosa, Madhukar Korupolu, David Oppenheimer, Eric Tune, and John Wilkes, "Large-Scale Cluster Management at Google with Borg," *EuroSys '15: Proceedings of the Tenth European Conference on Computer Systems* (April 2015): 18, *https://oreil.ly/9TeTM*.

quotas (i.e., the maximum number of instances a user can use at a given time), etc. Orchestrators deal with lower-level abstractions like machines, instances, clusters, service-level grouping, replication, etc. If the orchestrator notices that there are more jobs than the pool of available instances, it can increase the number of instances in the available instance pool. We say that it "provisions" more computers to handle the workload. Schedulers are often used for periodical jobs, whereas orchestrators are often used for services where you have a long-running server that responds to requests.

The most well-known orchestrator today is undoubtedly Kubernetes, the container orchestrator we discussed in the section "From Dev to Prod: Containers" on page 308. K8s can be used on-prem (even on your laptop via minikube). However, I've never met anyone who enjoys setting up their own K8s clusters, so most companies use K8s as a hosted service managed by their cloud providers, such as AWS's Elastic Kubernetes Service (EKS) or Google Kubernetes Engine (GKE).

Many people use schedulers and orchestrators interchangeably because schedulers usually run on top of orchestrators. Schedulers like Slurm and Google's Borg have some orchestrating capacity, and orchestrators like HashiCorp Nomad and K8s come with some scheduling capacity. But you can have separate schedulers and orchestrators, such as running Spark's job scheduler on top of Kubernetes or AWS Batch scheduler on top of EKS. Orchestrators such as HashiCorp Nomad and data science–specific orchestrators including Airflow, Argo, Prefect, and Dagster have their own schedulers.

Data Science Workflow Management

We've discussed the differences between schedulers and orchestrators and how they can be used to execute workflows in general. Readers familiar with workflow management tools aimed especially at data science like Airflow, Argo, Prefect, Kubeflow, Metaflow, etc. might wonder where they fit in this scheduler versus orchestrator discussion. We'll go into this topic here.

In its simplest form, workflow management tools manage workflows. They generally allow you to specify your workflows as DAGs, similar to the one in Figure 10-7. A workflow might consist of a featurizing step, a model training step, and an evaluation step. Workflows can be defined using either code (Python) or configuration files (YAML). Each step in a workflow is called a task.

Almost all workflow management tools come with some schedulers, and therefore, you can think of them as schedulers that, instead of focusing on individual jobs, focus on the workflow as a whole. Once a workflow is defined, the underlying scheduler usually works with an orchestrator to allocate resources to run the workflow, as shown in Figure 10-8.

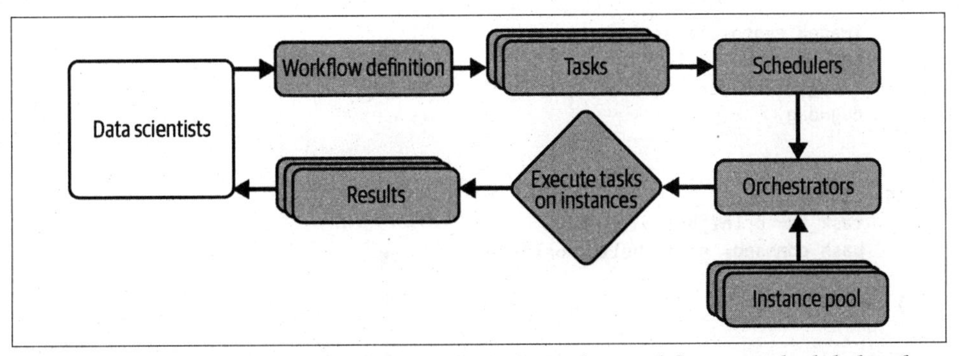

Figure 10-8. After a workflow is defined, the tasks in this workflow are scheduled and orchestrated

There are many articles online comparing different data science workflow management tools. In this section, we'll go over five of the most common tools: Airflow, Argo, Prefect, Kubeflow, and Metaflow. This section isn't meant to be a comprehensive comparison of those tools, but to give you an idea of different features a workflow management tool might need.

Originally developed at Airbnb and released in 2014, Airflow is one of the earliest workflow orchestrators. It's an amazing task scheduler that comes with a huge library of operators that makes it easy to use Airflow with different cloud providers, databases, storage options, and so on. Airflow is a champion of the "configuration as code" (*https://oreil.ly/aNVdq*) principle. Its creators believed that data workflows are complex and should be defined using code (Python) instead of YAML or other declarative language. Here's an example of an Airflow workflow, drawn from the platform's GitHub repository (*https://oreil.ly/Ubgf1*):

```
from datetime import datetime, timedelta

from airflow import DAG
from airflow.operators.bash import BashOperator
from airflow.providers.docker.operators.docker import DockerOperator

dag = DAG(
    'docker_sample',
    default_args={'retries': 1},
    schedule_interval=timedelta(minutes=10),
    start_date=datetime(2021, 1, 1),
    catchup=False,
)

t1 = BashOperator(task_id='print_date', bash_command='date', dag=dag)
t2 = BashOperator(task_id='sleep', bash_command='sleep 5', retries=3, dag=dag)
t3 = DockerOperator(
    docker_url='tcp://localhost:2375',  # Set your docker URL
    command='/bin/sleep 30',
```

```
        image='centos:latest',
        network_mode='bridge',
        task_id='docker_op_tester',
        dag=dag,
    )

    t4 = BashOperator(
        task_id='print_hello',
        bash_command='echo "hello world!!!"',
        dag=dag
    )

    t1 >> t2
    t1 >> t3
    t3 >> t4
```

However, because Airflow was created earlier than most other tools, it had no tool to learn lessons from and suffers from many drawbacks, as discussed in detail in a blog post by Uber Engineering (*https://oreil.ly/U7gkM*). Here, we'll go over only three to give you an idea.

First, Airflow is monolithic, which means it packages the entire workflow into one container. If two different steps in your workflow have different requirements, you can, in theory, create different containers for them using Airflow's `DockerOperator` (*https://oreil.ly/NwVFF*), but it's not that easy to do so.

Second, Airflow's DAGs are not parameterized, which means you can't pass parameters into your workflows. So if you want to run the same model with different learning rates, you'll have to create different workflows.

Third, Airflow's DAGs are static, which means it can't automatically create new steps at runtime as needed. Imagine you're reading from a database and you want to create a step to process each record in the database (e.g., to make a prediction), but you don't know in advance how many records there are in the database. Airflow won't be able to handle that.

The next generation of workflow orchestrators (Argo, Prefect) were created to address different drawbacks of Airflow.

Prefect's CEO, Jeremiah Lowin, was a core contributor of Airflow. Their early marketing campaign drew intense comparison (*https://oreil.ly/E19Pg*) between Prefect and Airflow. Prefect's workflows are parameterized and dynamic, a vast improvement compared to Airflow. It also follows the "configuration as code" principle so workflows are defined in Python.

However, like Airflow, containerized steps aren't the first priority of Prefect. You can run each step in a container, but you'll still have to deal with Dockerfiles and register your docker with your workflows in Prefect.

Argo addresses the container problem. Every step in an Argo workflow is run in its own container. However, Argo's workflows are defined in YAML, which allows you to define each step and its requirements in the same file. The following code sample, drawn from the Argo GitHub repository (*https://oreil.ly/Su1XX*), demonstrates how to create a workflow to show a coin flip:

```
apiVersion: argoproj.io/v1alpha1
kind: Workflow
metadata:
  generateName: coinflip-
  annotations:
    workflows.argoproj.io/description: |
      This is an example of coin flip defined as a sequence of conditional steps.
      You can also run it in Python:
      https://couler-proj.github.io/couler/examples/#coin-flip
spec:
  entrypoint: coinflip
  templates:
  - name: coinflip
    steps:
    - - name: flip-coin
        template: flip-coin
    - - name: heads
        template: heads
        when: "{{steps.flip-coin.outputs.result}} == heads"
      - name: tails
        template: tails
        when: "{{steps.flip-coin.outputs.result}} == tails"

    - name: flip-coin
      script:
        image: python:alpine3.6
        command: [python]
        source: |
          import random
          result = "heads" if random.randint(0,1) == 0 else "tails"
          print(result)
    - name: heads
      container:
        image: alpine:3.6
        command: [sh, -c]
        args: ["echo \"it was heads\""]

    - name: tails
      container:
        image: alpine:3.6
        command: [sh, -c]
        args: ["echo \"it was tails\""]
```

The main drawback of Argo, other than its messy YAML files, is that it can only run on K8s clusters, which are only available in production. If you want to test the same workflow locally, you'll have to use minikube to simulate a K8s on your laptop, which can get messy.

Enter Kubeflow and Metaflow, the two tools that aim to help you run the workflow in both dev and prod environments by abstracting away infrastructure boilerplate code usually needed to run Airflow or Argo. They promise to give data scientists access to the full compute power of the prod environment from local notebooks, which effectively allows data scientists to use the same code in both dev and prod environments.

Even though both tools have some scheduling capacity, they are meant to be used with a bona fide scheduler and orchestrator. One component of Kubeflow is Kubeflow Pipelines, which is built on top of Argo, and it's meant to be used on top of K8s. Metaflow can be used with AWS Batch or K8s.

Both tools are fully parameterized and dynamic. Currently, Kubeflow is the more popular one. However, from a user experience perspective, Metaflow is superior, in my opinion. In Kubeflow, while you can define your workflow in Python, you still have to write a Dockerfile and a YAML file to specify the specs of each component (e.g., process data, train, deploy) before you can stitch them together in a Python workflow. Basically, Kubeflow helps you abstract away other tools' boilerplate by making you write Kubeflow boilerplate.

In Metaflow, you can use a Python decorator `@conda` to specify the requirements for each step—required libraries, memory and compute requirements—and Metaflow will automatically create a container with all these requirements to execute the step. You save on Dockerfiles or YAML files.

Metaflow allows you to work seamlessly with both dev and prod environments from the same notebook/script. You can run experiments with small datasets on local machines, and when you're ready to run with the large dataset on the cloud, simply add `@batch` decorator to execute it on AWS Batch (*https://aws.amazon.com/batch*). You can even run different steps in the same workflow in different environments. For example, if a step requires a small memory footprint, it can run on your local machine. But if the next step requires a large memory footprint, you can just add `@batch` to execute it on the cloud.

```
# Example: sketch of a recommender system that uses an ensemble of two models.
# Model A will be run on your local machine and model B will be run on AWS.

class RecSysFlow(FlowSpec):
    @step
    def start(self):
        self.data = load_data()
        self.next(self.fitA, self.fitB)
```

```
    # fitA requires a different version of NumPy compared to fitB
    @conda(libraries={"scikit-learn":"0.21.1", "numpy":"1.13.0"})
    @step
    def fitA(self):
        self.model = fit(self.data, model="A")
        self.next(self.ensemble)

    @conda(libraries={"numpy":"0.9.8"})
    # Requires 2 GPU of 16GB memory
    @batch(gpu=2, memory=16000)
    @step
    def fitB(self):
        self.model = fit(self.data, model="B")
        self.next(self.ensemble)

    @step
    def ensemble(self, inputs):
        self.outputs = (
                   (inputs.fitA.model.predict(self.data) +
                    inputs.fitB.model.predict(self.data)) / 2
                   for input in inputs
        )
        self.next(self.end)

    def end(self):
        print(self.outputs)
```

ML Platform

The manager of the ML platform team at a major streaming company told me the story of how his team got started. He originally joined the company to work on their recommender systems. To deploy their recommender systems, they needed to build out tools such as feature management, model management, monitoring, etc. Last year, his company realized that these same tools could be used by other ML applications, not just recommender systems. They created a new team, the ML platform team, with the goal of providing shared infrastructure across ML applications. Because the recommender system team had the most mature tool, their tools were adopted by other teams, and some members from the recommender system team were asked to join the new ML platform team.

This story represents a growing trend since early 2020. As each company finds uses for ML in more and more applications, there's more to be gained by leveraging the same set of tools for multiple applications instead of supporting a separate set of tools for each application. This shared set of tools for ML deployment makes up the ML platform.

Because ML platforms are relatively new, what exactly constitutes an ML platform varies from company to company. Even within the same company, it's an ongoing discussion. Here, I'll focus on the components that I most often see in ML platforms, which include model development, model store, and feature store.

Evaluating a tool for each of these categories depends on your use case. However, here are two general aspects you might want to keep in mind:

Whether the tool works with your cloud provider or allows you to use it on your own data center

You'll need to run and serve your models from a compute layer, and usually tools only support integration with a handful of cloud providers. Nobody likes having to adopt a new cloud provider for another tool.

Whether it's open source or a managed service

If it's open source, you can host it yourself and have to worry less about data security and privacy. However, self-hosting means extra engineering time required to maintain it. If it's managed service, your models and likely some of your data will be on its service, which might not work for certain regulations. Some managed services work with virtual private clouds, which allows you to deploy your machines in your own cloud clusters, helping with compliance. We'll discuss this more in the section "Build Versus Buy" on page 327.

Let's start with the first component: model deployment.

Model Deployment

Once a model is trained (and hopefully tested), you want to make its predictive capability accessible to users. In Chapter 7, we talked at length on how a model can serve its predictions: online or batch prediction. We also discussed how the simplest way to deploy a model is to push your model and its dependencies to a location accessible in production then expose your model as an endpoint to your users. If you do online prediction, this endpoint will provoke your model to generate a prediction. If you do batch prediction, this endpoint will fetch a precomputed prediction.

A deployment service can help with both pushing your models and their dependencies to production and exposing your models as endpoints. Since deploying is the name of the game, deployment is the most mature among all ML platform components, and many tools exist for this. All major cloud providers offer tools for deployment: AWS with SageMaker (*https://oreil.ly/S7IR4*), GCP with Vertex AI (*https://oreil.ly/JNnGr*), Azure with Azure ML (*https://oreil.ly/7deF1*), Alibaba with Machine Learning Studio (*https://oreil.ly/jzQfg*), and so on. There are also a myriad of startups that offer model deployment tools such as MLflow Models (*https://oreil.ly/tUJz9*), Seldon (*https://www.seldon.io*), Cortex (*https://oreil.ly/UpnsA*), Ray Serve (*https://oreil.ly/WNEL5*), and so on.

When looking into a deployment tool, it's important to consider how easy it is to do both online prediction and batch prediction with the tool. While it's usually straightforward to do online prediction at a smaller scale with most deployment services, doing batch prediction is usually trickier.[29] Some tools allow you to batch requests together for online prediction, which is different from batch prediction. Many companies have separate deployment pipelines for online prediction and batch prediction. For example, they might use Seldon for online prediction but leverage Databricks for batch prediction.

An open problem with model deployment is how to ensure the quality of a model before it's deployed. In Chapter 9, we talked about different techniques for test in production such as shadow deployment, canary release, A/B testing, and so on. When choosing a deployment service, you might want to check whether this service makes it easy for you to perform the tests that you want.

Model Store

Many companies dismiss model stores because they sound simple. In the section "Model Deployment" on page 320, we talked about how, to deploy a model, you have to package your model and upload it to a location accessible in production. Model store suggests that it stores models—you can do so by uploading your models to storage like S3. However, it's not quite that simple. Imagine now that your model's performance dropped for a group of inputs. The person who was alerted to the problem is a DevOps engineer, who, after looking into the problem, decided that she needed to inform the data scientist who created this model. But there might be 20 data scientists in the company; who should she ping?

Imagine now that the right data scientist is looped in. The data scientist first wants to reproduce the problems locally. She still has the notebook she used to generate this model and the final model, so she starts the notebook and uses the model with the problematic sets of inputs. To her surprise, the outputs the model produces locally are different from the outputs produced in production. Many things could have caused this discrepancy; here are just a few examples:

- The model being used in production right now is not the same model that she has locally. Perhaps she uploaded the wrong model binary to production?
- The model being used in production is correct, but the list of features used is wrong. Perhaps she forgot to rebuild the code locally before pushing it to production?

29 When doing online prediction at a smaller scale, you can just hit an endpoint with payloads and get back predictions. Batch prediction requires setting up batch jobs and storing predictions.

- The model is correct, the feature list is correct, but the featurization code is outdated.
- The model is correct, the feature list is correct, the featurization code is correct, but something is wrong with the data processing pipeline.

Without knowing the cause of the problem, it'll be very difficult to fix it. In this simple example, we assume that the data scientist responsible still has access to the code used to generate the model. What if that data scientist no longer has access to that notebook, or she has already quit or is on vacation?

Many companies have realized that storing the model alone in blob storage isn't enough. To help with debugging and maintenance, it's important to track as much information associated with a model as possible. Here are eight types of artifacts that you might want to store. Note that many artifacts mentioned here are information that should be included in the model card, as discussed in the section "Create model cards" on page 351.

Model definition
: This is the information needed to create the shape of the model, e.g., what loss function it uses. If it's a neural network, this includes how many hidden layers it has and how many parameters are in each layer.

Model parameters
: These are the actual values of the parameters of your model. These values are then combined with the model's shape to re-create a model that can be used to make predictions. Some frameworks allow you to export both the parameters and the model definition together.

Featurize and predict functions
: Given a prediction request, how do you extract features and input these features into the model to get back a prediction? The featurize and predict functions provide the instruction to do so. These functions are usually wrapped in endpoints.

Dependencies
: The dependencies—e.g., Python version, Python packages—needed to run your model are usually packaged together into a container.

Data
: The data used to train this model might be pointers to the location where the data is stored or the name/version of your data. If you use tools like DVC to version your data, this can be the DVC commit that generated the data.

Model generation code

This is the code that specifies how your model was created, such as:

- What frameworks it used
- How it was trained
- The details on how the train/valid/test splits were created
- The number of experiments run
- The range of hyperparameters considered
- The actual set of hyperparameters that final model used

Very often, data scientists generate models by writing code in notebooks. Companies with more mature pipelines make their data scientists commit the model generation code into their Git repos on GitHub or GitLab. However, in many companies, this process is ad hoc, and data scientists don't even check in their notebooks. If the data scientist responsible for the model loses the notebook or quits or goes on vacation, there's no way to map a model in production to the code that generated it for debugging or maintenance.

Experiment artifacts

These are the artifacts generated during the model development process, as discussed in the section "Experiment Tracking and Versioning" on page 162. These artifacts can be graphs like the loss curve. These artifacts can be raw numbers like the model's performance on the test set.

Tags

This includes tags to help with model discovery and filtering, such as owner (the person or the team who is the owner of this model) or task (the business problem this model solves, like fraud detection).

Most companies store a subset, but not all, of these artifacts. The artifacts a company stores might not be in the same place but scattered. For example, model definitions and model parameters might be in S3. Containers that contain dependencies might be in ECS (Elastic Container Service). Data might be in Snowflake. Experiment artifacts might be in Weights & Biases. Featurize and prediction functions might be in AWS Lambda. Some data scientists might manually keep track of these locations in, say, a README, but this file can be easily lost.

A model store that can store sufficient general use cases is far from being a solved problem. As of writing this book, MLflow is undoubtedly the most popular model store that isn't associated with a major cloud provider. Yet three out of the six top MLflow questions on Stack Overflow are about storing and accessing artifacts in MLflow, as shown in Figure 10-9. Model stores are due for a makeover, and I hope that in the near future a startup will step up and solve this.

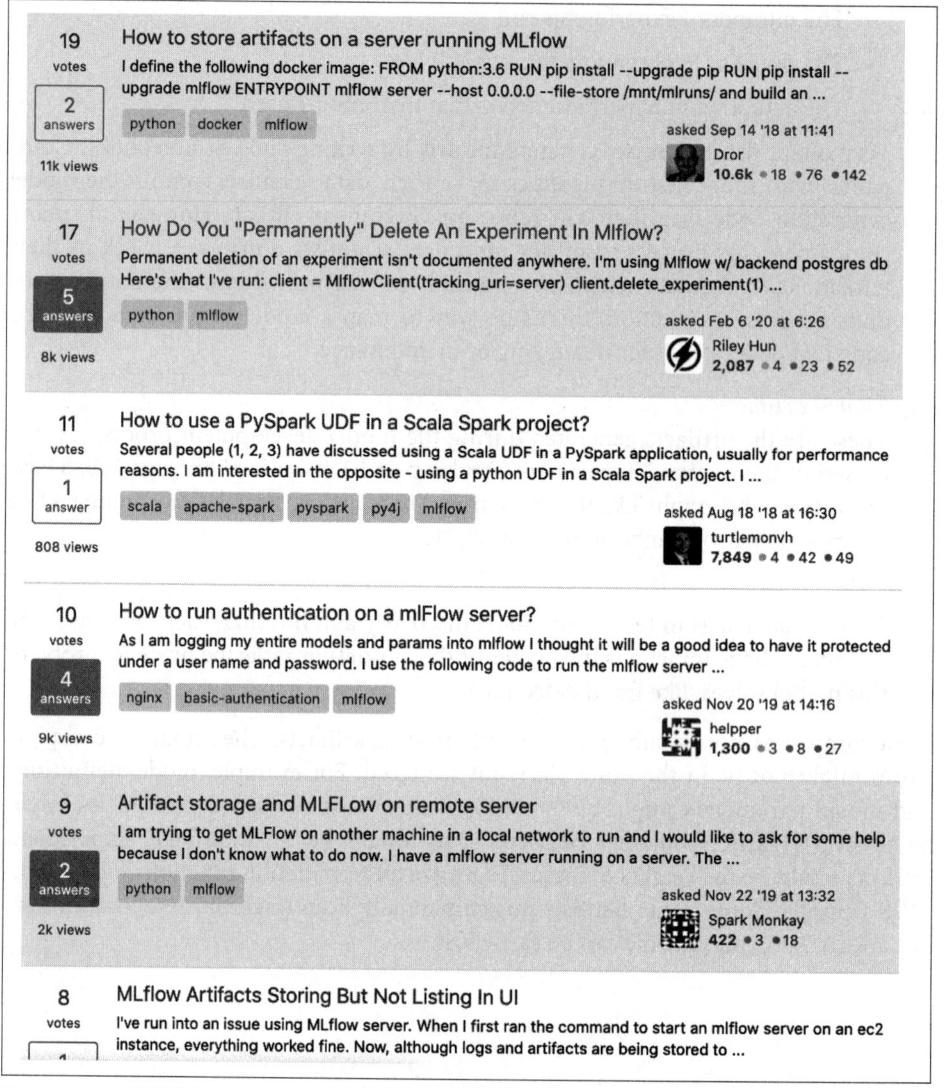

Figure 10-9. MLflow is the most popular model store, yet it's far from solving the artifact problem. Three out of the six top MLflow questions on Stack Overflow are about storing and accessing artifacts in MLflow. Source: Screenshot of Stack Overflow page

Because of the lack of a good model store solution, companies like Stitch Fix resolve to build their own model store. Figure 10-10 shows the artifacts that Stitch Fix's model store tracks. When a model is uploaded to their model store, this model comes with the link to the serialized model, the dependencies needed to run the model (Python environment), the Git commit where the model code generation is created (Git information), tags (to at least specify the team that owns the model), etc.

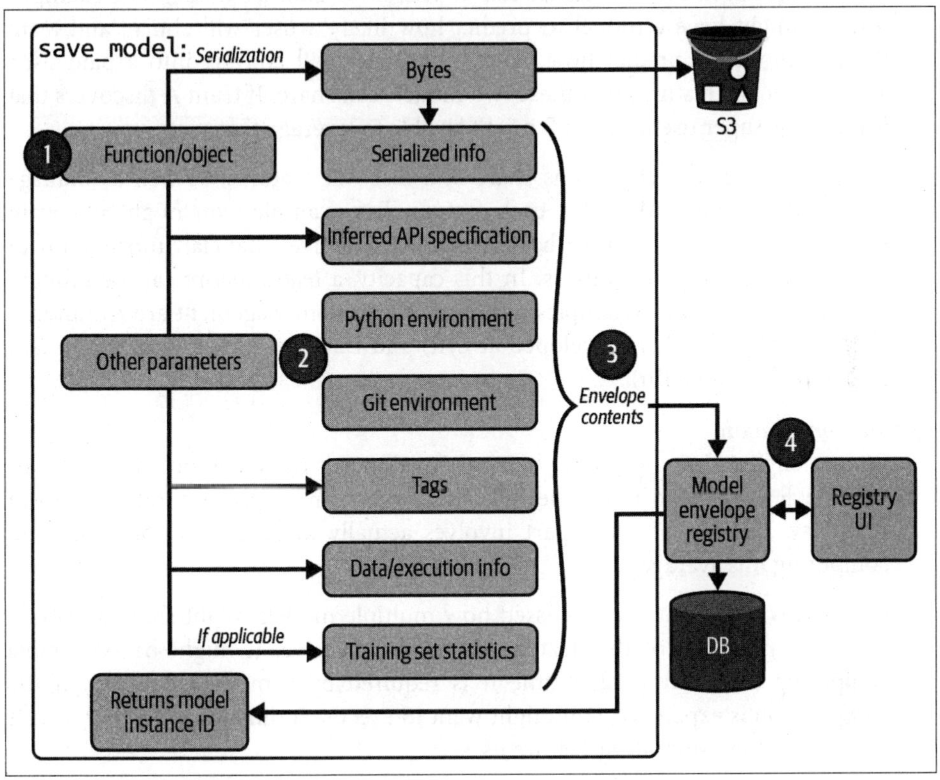

Figure 10-10. Artifacts that Stitch Fix's model store tracks. Source: Adapted from a slide by Stefan Krawczyk for CS 329S (Stanford) (https://oreil.ly/zWQM9).

Feature Store

"Feature store" is an increasingly loaded term that can be used by different people to refer to very different things. There have been many attempts by ML practitioners to define what features a feature store should have.[30] At its core, there are three main problems that a feature store can help address: feature management, feature

30 Neal Lathia, "Building a Feature Store," December 5, 2020, *https://oreil.ly/DgsvA*; Jordan Volz, "Why You Need a Feature Store," *Continual*, September 28, 2021, *https://oreil.ly/kQPMb*; Mike Del Balso, "What Is a Feature Store?" *Tecton*, October 20, 2020, *https://oreil.ly/pzy0I*.

transformation, and feature consistency. A feature store solution might address one or a combination of these problems:

Feature management
: A company might have multiple ML models, each model using a lot of features. Back in 2017, Uber had about 10,000 features across teams![31] It's often the case that features used for one model can be useful for another model. For example, team A might have a model to predict how likely a user will churn, and team B has a model to predict how likely a free user will convert into a paid user. There are many features that these two models can share. If team A discovers that feature X is super useful, team B might be able to leverage that too.

  A feature store can help teams share and discover features, as well as manage roles and sharing settings for each feature. For example, you might not want everyone in the company to have access to sensitive financial information of either the company or its users. In this capacity, a feature store can be thought of as a feature catalog. Examples of tools for feature management are Amundsen (*https://oreil.ly/Cm5Xe*) (developed at Lyft) and DataHub (*https://oreil.ly/ApXeL*) (developed at LinkedIn).

Feature computation[32]
: Feature engineering logic, after being defined, needs to be computed. For example, the feature logic might be: use the average meal preparation time from yesterday. The computation part involves actually looking into your data and computing this average.

  In the previous point, we discussed how multiple models might share a feature. If the computation of this feature isn't too expensive, it might be acceptable computing this feature each time it is required by a model. However, if the computation is expensive, you might want to execute it only once the first time it is required, then store it for feature uses.

  A feature store can help with both performing feature computation and storing the results of this computation. In this capacity, a feature store acts like a data warehouse.

Feature consistency
: In Chapter 7, we talked about the problem of having two separate pipelines for the same model: the training pipeline extracts batch features from historical data and the inference pipeline extracts streaming features. During development, data scientists might define features and create models using Python. Production

31 Jeremy Hermann and Mike Del Balso, "Meet Michelangelo: Uber's Machine Learning Platform," *Uber Engineering*, September 5, 2017, *https://oreil.ly/XteNy*.

32 Some people use the term "feature transformation."

code, however, might be written in another language, such as Java or C, for performance.

This means that feature definitions written in Python during development might need to be converted into the languages used in production. So you have to write the same features twice, once for training and once for inference. First, it's annoying and time-consuming. Second, it creates extra surface for bugs since one or more features in production might differ from their counterparts in training, causing weird model behaviors.

A key selling point of modern feature stores is that they unify the logic for both batch features and streaming features, ensuring the consistency between features during training and features during inference.

Feature store is a newer category that only started taking off around 2020. While it's generally agreed that feature stores should manage feature definitions and ensure feature consistency, their exact capacities vary from vendor to vendor. Some feature stores only manage feature definitions without computing features from data; some feature stores do both. Some feature stores also do feature validation, i.e., detecting when a feature doesn't conform to a predefined schema, and some feature stores leave that aspect to a monitoring tool.

As of writing this book, the most popular open source feature store is Feast. However, Feast's strength is in batch features, not streaming features. Tecton is a fully managed feature store that promises to be able to handle both batch features and online features, but their actual traction is slow because they require deep integration. Platforms like SageMaker and Databricks also offer their own interpretations of feature stores. Out of 95 companies I surveyed in January 2022, only around 40% of them use a feature store. Out of those who use a feature store, half of them build their own feature store.

Build Versus Buy

At the beginning of this chapter, we discussed how difficult it is to set up the right infrastructure for your ML needs. What infrastructure you need depends on the applications you have and the scale at which you run these applications.

How much you need to invest into infrastructure also depends on what you want to build in-house and what you want to buy. For example, if you want to use fully managed Databricks clusters, you probably need only one engineer. However, if you want to host your own Spark Elastic MapReduce clusters, you might need five more people.

At one extreme, you can outsource all your ML use cases to a company that provides ML applications end-to-end, and then perhaps the only piece of infrastructure you need is for data movement: moving your data from your applications to your vendor, and moving predictions from that vendor back to your users. The rest of your infrastructure is managed by your vendor.

At the other extreme, if you're a company that handles sensitive data that prevents you from using services managed by another company, you might need to build and maintain all your infrastructure in-house, even having your own data centers.

Most companies, however, are in neither of these extremes. If you work for one of these companies, you'll likely have some components managed by other companies and some components developed in-house. For example, your compute might be managed by AWS EC2 and your data warehouse managed by Snowflake, but you have your own feature store and your own monitoring dashboards.

Your build versus buy decisions depend on many factors. Here, we'll discuss three common ones that I often encounter when talking with heads of infrastructures on how they evaluate these decisions:

The stage your company is at
: In the beginning, you might want to leverage vendor solutions to get started as quickly as possible so that you can focus your limited resources on the core offerings of your product. As your use cases grow, however, vendor costs might become exorbitant and it might be cheaper for you to invest in your own solution.

What you believe to be the focus or the competitive advantages of your company
: Stefan Krawczyk, manager of the ML platform team at Stitch Fix, explained to me his build versus buy decision: "If it's something we want to be really good at, we'll manage that in-house. If not, we'll use a vendor." For the vast majority of companies outside the technology sector—e.g., companies in retail, banking, manufacturing—ML infrastructure isn't their focus, so they tend to bias toward buying. When I talk to these companies, they prefer managed services, even point solutions (e.g., solutions that solve a business problem for them, like a demand forecasting service). For many tech companies where technology is their competitive advantage, and whose strong engineering teams prefer to have control over their stacks, they tend to bias toward building. If they use a managed service, they might prefer that service to be modular and customizable, so that they can plug and play with any component.

The maturity of the available tools
: For example, your team might decide that you need a model store, and you'd have preferred to use a vendor, but there's no vendor mature enough for your needs, so you have to build your own feature store, perhaps on top of an open source solution.

This is what happens in the early days of ML adoption in the industry. Companies that are early adopters, i.e., big tech companies, build out their own infrastructure because there are no solutions mature enough for their needs. This leads to the situation where every company's infrastructure is different. A few years later, solution offerings mature. However, these offerings find it difficult to sell to big tech companies because it's impossible to create a solution that works with the majority of custom infrastructure.

As we're building out Claypot AI, other founders have actually advised us to avoid selling to big tech companies because, if we do, we'll get sucked into what they call "integration hell"—spending more time integrating our solution with custom infrastructure instead of building out our core features. They advised us to focus on startups with much cleaner slates to build on.

Some people think that building is cheaper than buying, which is not necessarily the case. Building means that you'll have to bring on more engineers to build and maintain your own infrastructure. It can also come with future cost: the cost of innovation. In-house, custom infrastructure makes it hard to adopt new technologies available because of the integration issues.

The build versus buy decisions are complex, highly context-dependent, and likely what heads of infrastructure spend much time mulling over. Erik Bernhardsson, ex-CTO of Better.com, said in a tweet that "one of the most important jobs of a CTO is vendor/product selection and the importance of this keeps going up rapidly every year since the infrastructure space grows so fast."[33] There's no way that a small section can address all its nuances. But I hope that this section provides you with some pointers to start the discussion.

Summary

If you've stayed with me until now, I hope you agree that bringing ML models to production is an infrastructural problem. To enable data scientists to develop and deploy ML models, it's crucial to have the right tools and infrastructure set up.

In this chapter, we covered different layers of infrastructure needed for ML systems. We started from the storage and compute layer, which provides vital resources for any engineering project that requires intensive data and compute resources like ML projects. The storage and compute layer is heavily commoditized, which means that most companies pay cloud services for the exact amount of storage and compute they use instead of setting up their own data centers. However, while cloud providers make it easy for a company to get started, their cost becomes prohibitive as this

33 Erik Bernhardsson on Twitter (@bernhardsson), September 29, 2021, *https://oreil.ly/GnxOH*.

company grows, and more and more large companies are looking into repatriating from the cloud to private data centers.

We then continued on to discuss the development environment where data scientists write code and interact with the production environment. Because the dev environment is where engineers spend most of their time, improvements in the dev environment translate directly into improvements in productivity. One of the first things a company can do to improve the dev environment is to standardize the dev environment for data scientists and ML engineers working on the same team. We discussed in this chapter why standardization is recommended and how to do so.

We then discussed an infrastructural topic whose relevance to data scientists has been debated heavily in the last few years: resource management. Resource management is important to data science workflows, but the question is whether data scientists should be expected to handle it. In this section, we traced the evolution of resource management tools from cron to schedulers to orchestrators. We also discussed why ML workflows are different from other software engineering workflows and why they need their own workflow management tools. We compared various workflow management tools such as Airflow, Argo, and Metaflow.

ML platform is a team that has emerged recently as ML adoption matures. Since it's an emerging concept, there are still disagreements on what an ML platform should consist of. We chose to focus on the three sets of tools that are essential for most ML platforms: deployment, model store, and feature store. We skipped monitoring of the ML platform since it's already covered in Chapter 8.

When working on infrastructure, a question constantly haunts engineering managers and CTOs alike: build or buy? We ended this chapter with a few discussion points that I hope can provide you or your team with sufficient context to make those difficult decisions.

CHAPTER 11

The Human Side of Machine Learning

Throughout this book, we've covered many technical aspects of designing an ML system. However, ML systems aren't just technical. They involve business decision makers, users, and, of course, developers of the systems. We've discussed stakeholders and their objectives in Chapters 1 and 2. In this chapter, we'll discuss how users and developers of ML systems might interact with these systems.

We'll first consider how user experience might be altered and affected due to the probabilistic nature of ML models. We'll continue to discuss organizational structure to allow different developers of the same ML system to work together effectively. We'll end the chapter with how ML systems can affect the society as a whole in the section "Responsible AI" on page 339.

User Experience

We've discussed at length how ML systems behave differently from traditional software systems. First, ML systems are probabilistic instead of deterministic. Usually, if you run the same software on the same input twice at different times, you can expect the same result. However, if you run the same ML system twice at different times on the exact same input, you might get different results.[1] Second, due to this probabilistic nature, ML systems' predictions are mostly correct, and the hard part is we usually don't know for what inputs the system will be correct! Third, ML systems can also be large and might take an unexpectedly long time to produce a prediction.

These differences mean that ML systems can affect user experience differently, especially for users that have so far been used to traditional software. Due to the relatively

1 Sometimes, you can get different results if you run the same model on the same input twice *at the exact same time.*

new usage of ML in the real world, how ML systems affect user experience is still not well studied. In this section, we'll discuss three challenges that ML systems pose to good user experience and how to address them.

Ensuring User Experience Consistency

When using an app or a website, users expect a certain level of consistency. For example, I'm used to Chrome having their "minimize" button on the top left corner on my MacBook. If Chrome moved this button to the right, I'd be confused, even frustrated.

ML predictions are probabilistic and inconsistent, which means that predictions generated for one user today might be different from what will be generated for the same user the next day, depending on the context of the predictions. For tasks that want to leverage ML to improve users' experience, the inconsistency in ML predictions can be a hindrance.

To make this concrete, consider a case study (*https://oreil.ly/qBLV2*) published by Booking.com in 2020. When you book accommodations on Booking.com, there are about 200 filters you can use to specify your preferences, such as "breakfast included," "pet friendly," and "non-smoking rooms." There are so many filters that it takes time for users to find the filters that they want. The applied ML team at Booking.com wanted to use ML to automatically suggest filters that a user might want, based on the filters they've used in a given browsing session.

The challenge they encountered is that if their ML model kept suggesting different filters each time, users could get confused, especially if they couldn't find a filter that they had already applied before. The team resolved this challenge by creating a rule to specify the conditions in which the system must return the same filter recommendations (e.g., when the user has applied a filter) and the conditions in which the system can return new recommendations (e.g., when the user changes their destination). This is known as the consistency–accuracy trade-off, since the recommendations deemed most accurate by the system might not be the recommendations that can provide user consistency.

Combatting "Mostly Correct" Predictions

In the previous section, we talked about the importance of ensuring the consistency of a model's predictions. In this section, we'll talk about how, in some cases, we want less consistency and more diversity in a model's predictions.

Since 2018, the large language model GPT (*https://oreil.ly/sY39d*) and its successors, GPT-2 (*https://oreil.ly/TttNU*) and GPT-3 (*https://oreil.ly/ug9P4*), have been taking the world by storm. An advantage of these large language models is that they're able to generate predictions for a wide range of tasks with little to no task-specific training data required. For example, you can use the requirements for a web page as an input to the model, and it'll output the React code needed to create that web page, as shown in Figure 11-1.

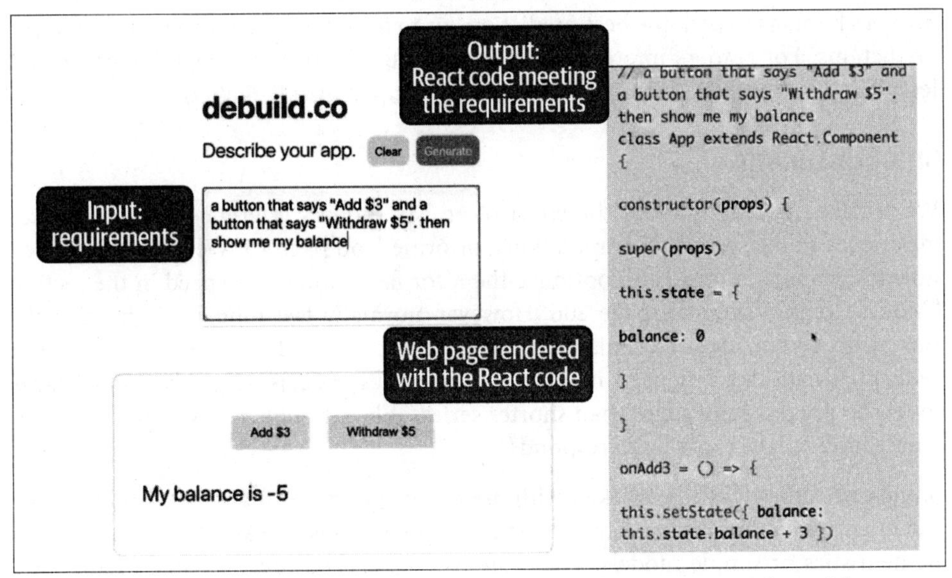

Figure 11-1. GPT-3 can help you write code for your website. Source: Adapted from screenshots of a video by Sharif Shameem (https://oreil.ly/VEuml)

However, a drawback of these models is that these predictions are not always correct, and it's very expensive to fine-tune them on task-specific data to improve their predictions. These mostly correct predictions can be useful for users who can easily correct them. For example, in the case of customer support, for each customer request, ML systems can produce mostly correct responses and the human operators can quickly edit those responses. This can speed up the response compared to having to write the response from scratch.

However, these mostly correct predictions won't be very useful if users don't know how to or can't correct the responses. Consider the same task of using a language model to generate React code for a web page. The generated code might not work, or if it does, it might not render to a web page that meets the specified requirements. A React engineer might be able to fix this code quickly, but many users of this application might not know React. And this application might attract a lot of users who don't know React—that's why they needed this app in the first place!

To overcome this, an approach is to show users multiple resulting predictions for the same input to increase the chance of at least one of them being correct. These predictions should be rendered in a way that even nonexpert users can evaluate them. In this case, given a set of requirements input by users, you can have the model produce multiple snippets of React code. The code snippets are rendered into visual web pages so that nonengineering users can evaluate which one is the best for them.

This approach is very common and is sometimes called "human-in-the-loop" AI, as it involves humans to pick the best predictions or to improve on the machine-generated predictions. For readers interested in human-in-the-loop AI, I'd highly recommend Jessy Lin's "Rethinking Human-AI Interaction" (*https://oreil.ly/6o4pu*).

Smooth Failing

We've talked at length about the effect of an ML model's inference latency on user experience in the section "Computational priorities" on page 15. We've also discussed how to compress models and optimize them for faster inference speed in the section "Model Compression" on page 206. However, normally fast models might still take time with certain queries. This can happen especially with models that deal with sequential data like language models or time-series models—e.g., the model takes longer to process long series than shorter series. What should we do with the queries where models take too long to respond?

Some companies that I've worked with use a backup system that is less optimal than the main system but is guaranteed to generate predictions quickly. These systems can be heuristics or simple models. They can even be cached precomputed predictions. This means that you might have a rule that specifies: if the main model takes longer than *X* milliseconds to generate predictions, use the backup model instead. Some companies, instead of having this simple rule, have another model to predict how long it'll take the main model to generate predictions for a given query, and route that prediction to either the main model or the backup model accordingly. Of course, this added model might also add extra inference latency to your system.

This is related to the speed–accuracy trade-off: a model might have worse performance than another model but can do inference much faster. This less-optimal but fast model might give users worse predictions but might still be preferred in situations where latency is crucial. Many companies have to choose one model over another, but with a backup system, you can do both.

Team Structure

An ML project involves not only data scientists and ML engineers, but also other types of engineers such as DevOps engineers and platform engineers as well as nondeveloper stakeholders like subject matter experts (SMEs). Given a diverse set of stakeholders, the question is what is the optimal structure when organizing ML teams. We'll focus on two aspects: cross-functional teams collaboration and the much debated role of an end-to-end data scientist.

Cross-functional Teams Collaboration

SMEs (doctors, lawyers, bankers, farmers, stylists, etc.) are often overlooked in the design of ML systems, but many ML systems wouldn't work without subject matter expertise. They're not only users but also developers of ML systems.

Most people only think of subject matter expertise during the data labeling phase—e.g., you'd need trained professionals to label whether a CT scan of a lung shows signs of cancer. However, as training ML models becomes an ongoing process in production, labeling and relabeling might also become an ongoing process spanning the entire project lifecycle. An ML system would benefit a lot to have SMEs involved in the rest of the lifecycle, such as problem formulation, feature engineering, error analysis, model evaluation, reranking predictions, and user interface: how to best present results to users and/or to other parts of the system.

There are many challenges that arise from having multiple different profiles working on a project. For example, how do you explain ML algorithms' limitations and capacities to SMEs who might not have engineering or statistical backgrounds? To build an ML system, we want everything to be versioned, but how do you translate domain expertise (e.g., if there's a small dot in this region between X and Y then it might be a sign of cancer) into code and version that?

Good luck trying to get your doctor to use Git.

It's important to involve SMEs early on in the project planning phase and empower them to make contributions without having to burden engineers to give them access. For example, to help SMEs get more involved in the development of ML systems, many companies are building no-code/low-code platforms that allow people to make changes without writing code. Most of the no-code ML solutions for SMEs are currently at the labeling, quality assurance, and feedback stages, but more platforms are being developed to aid in other critical junctions such as dataset creation and views for investigating issues that require SME input.

End-to-End Data Scientists

Through this book, I hope I've convinced you that ML production is not just an ML problem but also an infrastructure problem. To do MLOps, we need not only ML expertise but also Ops (operational) expertise, especially around deployment, containerization, job orchestration, and workflow management.

To be able to bring all these areas of expertise into an ML project, companies tend to follow one of the two following approaches: have a separate team to manage all the Ops aspects or include data scientists on the team and have them own the entire process.

Let's take a closer look at how each of these approaches works in practice.

Approach 1: Have a separate team to manage production

In this approach, the data science/ML team develops models in the dev environment. Then a separate team, usually the Ops/platform/ML engineering team, productionizes the models in prod. This approach makes hiring easier as it's easier to hire people with one set of skills instead of people with multiple sets of skills. It might also make life easier for each person involved, as they only have to focus on one concern (e.g., developing models or deploying models). However, this approach has many drawbacks:

Communication and coordination overhead
: A team can become blockers for other teams. According to Frederick P. Brooks, "What one programmer can do in one month, two programmers can do in two months."

Debugging challenges
: When something fails, you don't know whether your team's code or some other team's code might have caused it. It might not have been because of your company's code at all. You need cooperation from multiple teams to figure out what's wrong.

Finger-pointing
: Even when you've figured out what went wrong, each team might think it's another team's responsibility to fix it.

Narrow context
: No one has visibility into the entire process to optimize/improve it. For example, the platform team has ideas on how to improve the infrastructure but they can only act on requests from data scientists, but data scientists don't have to deal with infrastructure so they have less incentives to proactively make changes to it.

Approach 2: Data scientists own the entire process

In this approach, the data science team also has to worry about productionizing models. Data scientists become grumpy unicorns, expected to know everything about the process, and they might end up writing more boilerplate code than data science.

About a year ago, I tweeted (*https://oreil.ly/DPpt0*) about a set of skills I thought was important to become an ML engineer or data scientist, as shown in Figure 11-2. The list covers almost every part of the workflow: querying data, modeling, distributed training, and setting up endpoints. It even includes tools like Kubernetes and Airflow.

Things I'd prioritize learning if I was to study to become a ML engineer again:

1. Version control
2. SQL + NoSQL
3. Python
4. Pandas/Dask
5. Data structures
6. Prob & stats
7. ML algos
8. Parallel computing
9. REST API
10. Kubernetes + Airflow
11. Unit/integration tests

6:30 AM · Oct 11, 2020 · Twitter Web App

View Tweet analytics

1,246 Retweets **62** Quote Tweets **6,927** Likes

Figure 11-2. I used to think that a data scientist would need to know all these things

The tweet seems to resonate with my audience. Eugene Yan also wrote about how "data scientists should be more end-to-end."[2] Eric Colson, Stitch Fix's chief algorithms officer (who previously was also VP data science and engineering at Netflix), wrote a post on "the power of the full-stack data science generalist and the perils of division of labor through function."[3]

---

2 Eugene Yan, "Unpopular Opinion—Data Scientists Should be More End-to-End," EugeneYan.com, August 9, 2020, *https://oreil.ly/A6oPi*.

3 Eric Colson, "Beware the Data Science Pin Factory: The Power of the Full-Stack Data Science Generalist and the Perils of Division of Labor Through Function," MultiThreaded, March 11, 2019, *https://oreil.ly/m6WWu*.

When I wrote that tweet, I believed that Kubernetes was essential to the ML workflow. This sentiment came from the frustration at my own job—my life as an ML engineer would've been much easier if I was more fluent with K8s.

However, as I learned more about low-level infrastructure, I realized how unreasonable it is to expect data scientists to know about it. Infrastructure requires a very different set of skills from data science. In theory, you can learn both sets of skills. In practice, the more time you spend on one means the less time you spend on the other. I love Erik Bernhardsson's analogy that expecting data scientists to know about infrastructure is like expecting app developers to know about how Linux kernels work.[4] I joined an ML company because I wanted to spend more time with data, not with spinning up AWS instances, writing Dockerfiles, scheduling/scaling clusters, or debugging YAML configuration files.

For data scientists to own the entire process, we need good tools. In other words, we need good infrastructure.

What if we have an abstraction to allow data scientists to own the process end-to-end without having to worry about infrastructure?

What if I can just tell this tool, "Here's where I store my data (S3), here are the steps to run my code (featurizing, modeling), here's where my code should run (EC2 instances, serverless stuff like AWS Batch, Function, etc.), here's what my code needs to run at each step (dependencies)," and then this tool manages all the infrastructure stuff for me?

According to both Stitch Fix and Netflix, the success of a full-stack data scientist relies on the tools they have. They need tools that "abstract the data scientists from the complexities of containerization, distributed processing, automatic failover, and other advanced computer science concepts."[5]

In Netflix's model, the specialists—people who originally owned a part of the project—first create tools that automate their parts, as shown in Figure 11-3. Data scientists can leverage these tools to own their projects end-to-end.

4 Erik Bernhardsson on Twitter (@bernhardsson), July 20, 2021, *https://oreil.ly/7X4J9*.

5 Colson, "Beware the Data Science Pin Factory."

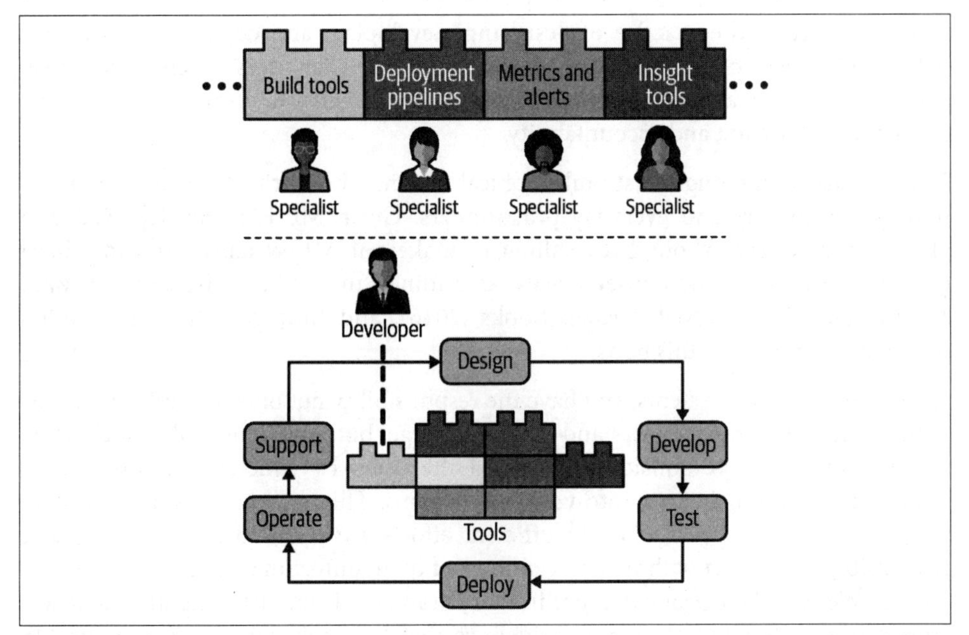

Figure 11-3. Full-cycle developers at Netflix. Source: Adapted from an image by Netflix[6]

We've talked about how ML systems might affect user experience and how organizational structure might influence productivity of ML projects. In the second half of this chapter, we'll focus on an even more crucial consideration: how ML systems might affect society and what ML system developers should do to ensure that the systems they develop do more good than harm.

Responsible AI

This section was written with generous contributions from Abhishek Gupta (*https://oreil.ly/AGJHF*), founder and principal researcher at the Montreal AI Ethics Institute (*https://montrealethics.ai*). His work focuses on applied technical and policy measures to build ethical, safe, and inclusive AI systems.

The question of how to make intelligent systems responsible is relevant not only to ML systems but also general artificial intelligence (AI) systems. AI is a broader term that includes ML. Therefore, in this section, we use AI instead of ML.

6 "Full Cycle Developers at Netflix—Operate What You Build," *Netflix Technology Blog*, May 17, 2018, *https://oreil.ly/iYgQs*.

Responsible AI is the practice of designing, developing, and deploying AI systems with good intention and sufficient awareness to empower users, to engender trust, and to ensure fair and positive impact to society. It consists of areas like fairness, privacy, transparency, and accountability.

These terms are no longer just philosophical musings, but serious considerations for both policy makers and everyday practitioners. Given ML is being deployed into almost every aspect of our lives, failing to make our ML systems fair and ethical can lead to catastrophic consequences, as outlined in the book *Weapons of Math Destruction* (Cathy O'Neil, Crown Books, 2016), and through other case studies mentioned throughout this book.

As developers of ML systems, you have the responsibility not only to think about how your systems will impact users and society at large, but also to help all stakeholders better realize their responsibilities toward the users by concretely implementing ethics, safety, and inclusivity into your ML systems. This section is a brief introduction to what can happen when insufficient efforts are spent to make ML systems responsible. We'll start with two case studies of quite unfortunate and public failures of ML. We will then propose a preliminary framework for data scientists and ML engineers to select the tools and guidelines that best help with making your ML systems responsible.

Disclaimer: Responsible AI is a complex topic with growing literature that deserves its own coverage and can easily span multiple books. This section is far from an exhaustive guide. We only aim to give ML developers an overview to effectively navigate the developments in this field. Those interested in further reading are highly recommended to check out the following resources:

- NIST Special Publication 1270: Towards a Standard for Identifying and Managing Bias in Artificial Intelligence (*https://oreil.ly/Glvnp*)
- ACM Conference on Fairness, Accountability, and Transparency (ACM FAccT) publications (*https://facctconference.org*)
- Trustworthy ML's list of recommended resources and fundamental papers (*https://oreil.ly/NmLxU*) for researchers and practitioners who want to learn more about trustworthy ML
- Sara Hooker's awesome slide deck (*https://oreil.ly/upBxx*) on fairness, security, and governance in machine learning (2022)
- Timnit Gebru and Emily Denton's tutorials (*https://oreil.ly/jdAyF*) on fairness, accountability, transparency, and ethics (2020)

Irresponsible AI: Case Studies

We'll start this section off by looking at two failures of AI systems that led to severe harm for not only the users of these systems but also to the organizations who developed the systems. We'll trace some of the places where the organizations went wrong and what the practitioners could have done to potentially anticipate these points of failure. These highlights will serve as background as we dive into the engineering framework for responsible AI.

There are other interesting examples of "AI incidents" logged at the AI Incident Database (*https://incidentdatabase.ai*). Keep in mind that while the following two examples and the ones logged at AI Incident Database are the ones that caught attention, there are many more instances of irresponsible AI that happen silently.

Case study I: Automated grader's biases

In the summer of 2020, the United Kingdom canceled A levels, the high-stakes exams that determine college placement, due to the COVID-19 pandemic. Ofqual, the regulatory body for education and examinations in the UK, sanctioned the use of an automated system to assign final A-level grades to students—without them taking the test. According to Jones and Safak from Ada Lovelace Institute, "Awarding students' grades based on teacher assessment was originally rejected by Ofqual on the grounds of unfairness between schools, incomparability across generations and devaluing of results because of grade inflation. The fairer option, Ofqual surmised, was to combine previous attainment data and teacher assessment to assign grades, using a particular statistical model—an 'algorithm.'"[7]

The results published by this algorithm, however, turned out to be unjust and untrustworthy. They quickly led to public outcries to get rid of it, with hundreds of students chanting in protest.[8]

What caused the public outcries? The first glance seems to point at the algorithm's poor performance. Ofqual stated that their model, tested on 2019 data, had about 60% average accuracy across A-level subjects.[9] This means that they expected 40% of the grades assigned by this model to be different from the students' actual grades.

7 Elliot Jones and Cansu Safak, "Can Algorithms Ever Make the Grade?" *Ada Lovelace Institute Blog*, 2020, *https://oreil.ly/ztTxR*.

8 Tom Simonite, "Skewed Grading Algorithms Fuel Backlash Beyond the Classroom," *Wired*, August 19, 2020, *https://oreil.ly/GFRet*.

9 Ofqual, "Awarding GCSE, AS & A Levels in Summer 2020: Interim Report," Gov.uk, August 13, 2020, *https://oreil.ly/r22iz*.

While the model's accuracy seems low, Ofqual defended their algorithm as being broadly comparable to the accuracy of human graders. When comparing an examiner's grades with those made by a senior examiner, the agreement is also around 60%.[10] The accuracy by both human examiners and the algorithm exposes the underlying uncertainty in assessing students at a single point in time,[11] further fueling the frustration of the public.

If you've read this book thus far, you know that coarse-grained accuracy alone is nowhere close to being sufficient to evaluate a model's performance, especially for a model whose performance can influence the future of so many students. A closer look into this algorithm reveals at least three major failures along the process of designing and developing this automated grading system:

- Failure to set the right objective
- Failure to perform fine-grained evaluation to discover potential biases
- Failure to make the model transparent

We'll go into detail about each of these failures. Keep in mind that even if these failures are addressed, the public might still be upset with the auto-grading system.

Failure 1: Setting the wrong objective. We discussed in Chapter 2 how the objective of an ML project will affect the resulting ML system's performance. When developing an automated system to grade students, you would've thought that the objective of this system would be "grading accuracy for students."

However, the objective that Ofqual seemingly chose to optimize was "maintaining standards" across schools—fitting the model's predicted grades to historical grade distributions from each school. For example, if school A had historically outperformed school B in the past, Ofqual wanted an algorithm that, on average, also gives students from school A higher grades than students from school B. Ofqual prioritized fairness between schools over fairness between students—they preferred a model that gets school-level results right over another model that gets each individual's grades right.

Due to this objective, the model disproportionately downgraded high-performing cohorts from historically low-performing schools. A students from classes where students had historically received straight Ds were downgraded to Bs and Cs.[12]

10 Ofqual, "Awarding GCSE, AS & A levels."

11 Jones and Safak, "Can Algorithms Ever Make the Grade?"

12 Jones and Safak, "Can Algorithms Ever Make the Grade?"

Ofqual failed to take into account the fact that schools with more resources tend to outperform schools with fewer resources. By prioritizing schools' historical performance over students' current performance, this auto-grader punished students from low resource schools, which tend to have more students from underprivileged backgrounds.

Failure 2: Insufficient fine-grained model evaluation to discover biases. Bias against students from historically low-performing schools is only one of the many biases discovered about this model after the results were brought to the public. The automated grading system took into account teachers' assessments as inputs but failed to address teachers' inconsistency in evaluation across demographic groups. It also "does not take into consideration the impact of multiple disadvantages for some protected groups [under the] 2010 Equalities Act, who will be double/triple disadvantaged by low teacher expectations, [and] racial discrimination that is endemic in some schools."[13]

Because the model took into account each school's historical performance, Ofqual acknowledged that their model didn't have enough data for small schools. For these schools, instead of using this algorithm to assign final grades, they only used teacher-assessed grades. In practice, this led to "better grades for private school students who tend to have smaller classes."[14]

It might have been possible to discover these biases through the public release of the model's predicted grades with fine-grained evaluation to understand their model's performance for different slices of data—e.g., evaluating the model's accuracy for schools of different sizes and for students from different backgrounds.

Failure 3: Lack of transparency. Transparency is the first step in building trust in systems, yet Ofqual failed to make important aspects of their auto-grader public before it was too late. For example, they didn't let the public know that the objective of their system was to maintain fairness between schools until the day the grades were published. The public, therefore, couldn't express their concern over this objective as the model was being developed.

Further, Ofqual didn't let teachers know how their assessments would be used by the auto-grader until after the assessments and student ranking had been submitted. Ofqual's rationale was to avoid teachers attempting to alter their assessments to influence the model's predictions. Ofqual chose not to release the exact model being used until results day to ensure that everyone would find out their results at the same time.

13 Ofqual, "Awarding GCSE, AS & A Levels."

14 Jones and Safak, "Can Algorithms Ever Make the Grade?"

These considerations came from good intention; however, Ofqual's decision to keep their model development in the dark meant that their system didn't get sufficient independent, external scrutiny. Any system that operates on the trust of the public should be reviewable by independent experts trusted by the public. The Royal Statistical Society (RSS), in their inquiry into the development of this auto-grader, expressed concerns over the composition of the "technical advisory group" that Ofqual put together to evaluate the model. RSS indicated that "without a stronger procedural basis to ensure statistical rigor, and greater transparency about the issues that Ofqual is examining,"[15] the legitimacy of Ofqual's statistical model is questionable.

This case study shows the importance of transparency when building a model that can make a direct impact on the lives of so many people, and what the consequences can be for failing to disclose important aspects of your model at the right time. It also shows the importance of choosing the right objective to optimize, as the wrong objective (e.g., prioritizing fairness among schools) can not only lead you to choose a model that underperforms for the right objective, but also perpetuate biases.

It also exemplifies the currently mucky boundary between what should be automated by algorithms and what should not. There must be people in the UK government who think it's OK for A-level grading to be automated by algorithms, but it's also possible to argue that due to the potential for catastrophic consequences of the A-level grading, it should never have been automated in the first place. Until there is a clearer boundary, there will be more cases of misusing AI algorithms. A clearer boundary can only be achieved with more investments in time and resources as well as serious considerations from AI developers, the public, and the authorities.

Case study II: The danger of "anonymized" data

This case study is interesting to me because here, the algorithm is not an explicit culprit. Rather it's how the interface and collection of data is designed that allows the leakage of sensitive data. Since the development of ML systems relies heavily on the quality of data, it's important for user data to be collected. The research community needs access to high-quality datasets to develop new techniques. Practitioners and companies require access to data to discover new use cases and develop new AI-powered products.

15 "Royal Statistical Society Response to the House of Commons Education Select Committee Call for Evidence: The Impact of COVID-19 on Education and Children's Services Inquiry," Royal Statistical Society, June 8, 2020, *https://oreil.ly/ernho*.

However, collecting and sharing datasets might violate the privacy and security of the users whose data is part of these datasets. To protect users, there have been calls for anonymization of personally identifiable information (PII). According to the US Department of Labor, PII is defined as "any representation of information that permits the identity of an individual to whom the information applies to be reasonably inferred by either direct or indirect means" such as name, address, or telephone number.[16]

However, anonymization may not be a sufficient guarantee for preventing data misuse and erosion of privacy expectations. In 2018, online fitness tracker Strava published a heatmap showing the paths it records of its users around the world as they exercise, e.g., running, jogging, or swimming. The heatmap was aggregated from one billion activities recorded between 2015 and September 2017, covering 27 billion kilometers of distance. Strava stated that the data used had been anonymized, and "excludes activities that have been marked as private and user-defined privacy zones."[17]

Since Strava was used by military personnel, their public data, despite anonymization, allowed people to discover patterns that expose activities of US military bases overseas, including the "forward operating bases in Afghanistan, Turkish military patrols in Syria, and a possible guard patrol in the Russian operating area of Syria."[18] An example of these discriminating patterns is shown in Figure 11-4. Some analysts even suggested that the data could reveal the names and heart rates of individual Strava users.[19]

So where did the anonymization go wrong? First, Strava's default privacy setting was "opt-out," meaning that it requires users to manually opt out if they don't want their data to be collected. However, users have pointed out that these privacy settings aren't always clear and can cause surprises to users.[20] Some of the privacy settings can only be changed through the Strava website rather than in its mobile app. This shows the importance of educating users about your privacy settings. Better, data opt-in (data collecting isn't by default), not opt-out, should be the default.

---

16 "Guidance on the Protection of Personal Identifiable Information," US Department of Labor, *https://oreil.ly/FokAV*.

17 Sasha Lekach, "Strava's Fitness Heatmap Has a Major Security Problem for the Military," *Mashable*, January 28, 2018, *https://oreil.ly/9ogYx*.

18 Jeremy Hsu, "The Strava Heat Map and the End of Secrets," *Wired*, January 29, 2018, *https://oreil.ly/mBOGD*.

19 Matt Burgess, "Strava's Heatmap Data Lets Anyone See the Names of People Exercising on Military Bases," *Wired*, January 30, 2018, *https://oreil.ly/eJPdj*.

20 Matt Burgess, "Strava's Heatmap Data Lets Anyone See"; Rosie Spinks, "Using a Fitness App Taught Me the Scary Truth About Why Privacy Settings Are a Feminist Issue," *Quartz*, August 1, 2017, *https://oreil.ly/DO3WR*.

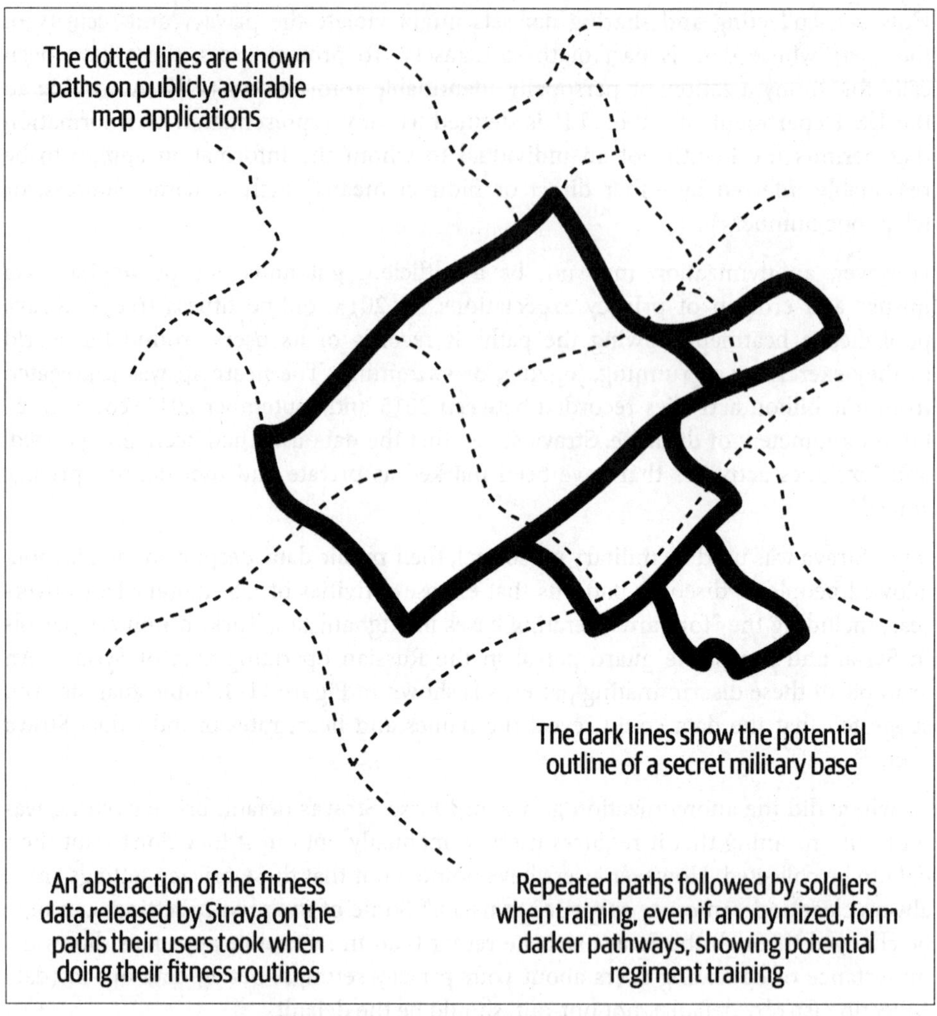
The dotted lines are known paths on publicly available map applications
The dark lines show the potential outline of a secret military base
An abstraction of the fitness data released by Strava on the paths their users took when doing their fitness routines
Repeated paths followed by soldiers when training, even if anonymized, form darker pathways, showing potential regiment training

Figure 11-4. Image created based on analysis done by BBC News[21]

When this issue with the Strava heatmap became public, some of the responsibilities were shifted toward users: e.g., how military personnel shouldn't use non-military-issue devices with GPS tracking and how location services should be turned off.[22]

21 "Fitness App Strava Lights Up Staff at Military Bases," *BBC News*, January 29, 2018, *https://oreil.ly/hXwpN*.

22 Matt Burgess, "Strava's Heatmap Data Lets Anyone See."

However, privacy settings and users' choices only address the problem at a surface level. The underlying problem is that the devices we use today are constantly collecting and reporting data on us. This data has to be moved and stored somewhere, creating opportunities for it to be intercepted and misused. The data that Strava has is small compared to much more widely used applications like Amazon, Facebook, Google, etc. Strava's blunder might have exposed military bases' activities, but other privacy failures might cause even more dangers not only to individuals but also to society at large.

Collecting and sharing data is essential for the development of data-driven technologies like AI. However, this case study shows the hidden danger of collecting and sharing data, even when data is supposedly anonymized and was released with good intention. Developers of applications that gather user data must understand that their users might not have the technical know-how and privacy awareness to choose the right privacy settings for themselves, and so developers must proactively work to make the right settings the default, even at the cost of gathering less data.

A Framework for Responsible AI

In this section, we will lay down the foundations for you, as an ML practitioner, to audit model behavior and set out guidelines that best help you meet the needs of your projects. This framework is not sufficient for every use case. There are certain applications where the use of AI might altogether be inappropriate or unethical (e.g., criminal sentencing decisions, predictive policing), regardless of which framework you follow.

Discover sources for model biases

As someone who has been following the discussions around ML systems design, you know that biases can creep in your system through the entire workflow. Your first step is to discover how these biases can creep in. The following are some examples of the sources of data, but keep in mind that this list is far from being exhaustive. One of the reasons why biases are so hard to combat is that biases can come from any step during a project lifecycle.

Training data

Is the data used for developing your model representative of the data your model will handle in the real world? If not, your model might be biased against the groups of users with less data represented in the training data.

Labeling

If you use human annotators to label your data, how do you measure the quality of these labels? How do you ensure that annotators follow standard guidelines instead of relying on subjective experience to label your data? The more annotators have to rely on their subjective experience, the more room for human biases.

Feature engineering

Does your model use any feature that contains sensitive information? Does your model cause a disparate impact on a subgroup of people? Disparate impact occurs "when a selection process has widely different outcomes for different groups, even as it appears to be neutral."[23] This can happen when a model's decision relies on information correlated with legally protected classes (e.g., ethnicity, gender, religious practice) even when this information isn't used in training the model directly. For example, a hiring process can cause disparate impact by race if it leverages variables correlated with race such as zip code and high school diplomas. To mitigate this potential disparate impact, you might want to use disparate impact remover techniques proposed by Feldman et al. in "Certifying and Removing Disparate Impact" (*https://oreil.ly/a9vxm*) or to use the function `DisparateImpactRemover` (*https://oreil.ly/6LyA8*) implemented by AI Fairness 360 (*https://oreil.ly/TjavU*) (AIF360). You can also identify hidden bias in variables (which can then be removed from the training set) using the Infogram method (*https://oreil.ly/JFZCL*), implemented in H2O.

Model's objective

Are you optimizing your model using an objective that enables fairness to all users? For example, are you prioritizing your model's performance on all users, which skews your model toward the majority group of users?

Evaluation

Are you performing adequate, fine-grained evaluation to understand your model's performance on different groups of users? This is covered in the section "Slice-based evaluation" on page 185. Fair, adequate evaluation depends on the existence of fair, adequate evaluation data.

---

23 Michael Feldman, Sorelle Friedler, John Moeller, Carlos Scheidegger, and Suresh Venkatasubramanian, "Certifying and Removing Disparate Impact," *arXiv*, July 16, 2015, *https://oreil.ly/FjSve*.

Understand the limitations of the data-driven approach

ML is a data-driven approach to solving problems. However, it's important to understand that data isn't enough. Data concerns people in the real world, with socioeconomic and cultural aspects to consider. We need to gain a better understanding of the blind spots caused by too much reliance on data. This often means crossing over disciplinary and functional boundaries, both within and outside the organization, so that we can account for the lived experiences of those who will be impacted by the systems that we build.

As an example, to build an equitable automated grading system, it's essential to work with domain experts to understand the demographic distribution of the student population and how socioeconomic factors get reflected in the historical performance data.

Understand the trade-offs between different desiderata

When building an ML system, there are different properties you might want this system to have. For example, you might want your system to have low inference latency, which could be obtained by model compression techniques like pruning. You might also want your model to have high predictive accuracy, which could be achieved by adding more data. You might also want your model to be fair and transparent, which could require the model and the data used to develop this model to be made accessible for public scrutiny.

Often, ML literature makes the unrealistic assumption that optimizing for one property, like model accuracy, holds all others static. People might discuss techniques to improve a model's fairness with the assumption that this model's accuracy or latency will remain the same. However, in reality, improving one property can cause other properties to degrade. Here are two examples of these trade-offs:

Privacy versus accuracy trade-off

According to Wikipedia, differential privacy is "a system for publicly sharing information about a dataset by describing the patterns of groups within the dataset while withholding information about individuals in the dataset. The idea behind differential privacy is that if the effect of making an arbitrary single substitution in the database is small enough, the query result cannot be used to infer much about any single individual, and therefore provides privacy."[24]

Differential privacy is a popular technique used on training data for ML models. The trade-off here is that the higher the level of privacy that differential privacy can provide, the lower the model's accuracy. However, this accuracy reduction isn't equal for all samples. As pointed out by Bagdasaryan and Shmatikov (2019),

24 Wikipedia, s.v. "Differential privacy," *https://oreil.ly/UcxzZ*.

"the accuracy of differential privacy models drops much more for the underrepresented classes and subgroups."[25]

Compactness versus fairness trade-off

In Chapter 7, we talked at length about various techniques for model compression such as pruning and quantization. We learned that it's possible to reduce a model's size significantly with minimal cost of accuracy, e.g., reducing a model's parameter count by 90% with minimal accuracy cost.

The minimal accuracy cost is indeed minimal if it's spread uniformly across all classes, but what if the cost is concentrated in only a few classes? In their 2019 paper, "What Do Compressed Deep Neural Networks Forget?," Hooker et al. found that "models with radically different numbers of weights have comparable top-line performance metrics but diverge considerably in behavior on a narrow subset of the dataset."[26] For example, they found that compression techniques amplify algorithmic harm when the protected feature (e.g., sex, race, disability) is in the long tail of the distribution. This means that compression disproportionately impacts underrepresented features.[27]

Another important finding from their work is that while all compression techniques they evaluated have a nonuniform impact, not all techniques have the same level of disparate impact. Pruning incurs a far higher disparate impact than is observed for the quantization techniques that they evaluated.[28]

Similar trade-offs continue to be discovered. It's important to be aware of these trade-offs so that we can make informed design decisions for our ML systems. If you are working with a system that is compressed or differentially private, allocating more resources to auditing model behavior is recommended to avoid unintended harm.

Act early

Consider a new building being constructed downtown. A contractor has been called upon to build something that will stand for the next 75 years. To save costs, the contractor uses poor-quality cement. The owner doesn't invest in supervision since they want to avoid overhead to be able to move fast. The contractor continues building on top of that poor foundation and finishes the building on time.

---

25 Eugene Bagdasaryan and Vitaly Shmatikov, "Differential Privacy Has Disparate Impact on Model Accuracy," *arXiv*, May 28, 2019, *https://oreil.ly/nrJGK*.

26 Sarah Hooker, Aaron Courville, Gregory Clark, Yann Dauphin, and Andrea Frome, "What Do Compressed Deep Neural Networks Forget?" *arXiv*, November 13, 2019, *https://oreil.ly/bgfFX*.

27 Sara Hooker, Nyalleng Moorosi, Gregory Clark, Samy Bengio, and Emily Denton, "Characterising Bias in Compressed Models," *arXiv*, October 6, 2020, *https://oreil.ly/ZTI72*.

28 Hooker et al., "Characterising Bias in Compressed Models."

Within a year, cracks start showing up and it appears that the building might topple. The city decides that this building poses a safety risk and requests for it to be demolished. The contractor's decision to save cost and the owner's decision to save time in the beginning now end up costing the owner much more money and time.

You might encounter this narrative often in ML systems. Companies might decide to bypass ethical issues in ML models to save cost and time, only to discover risks in the future when they end up costing a lot more, such as the preceding case studies of Ofqual and Strava.

The earlier in the development cycle of an ML system that you can start thinking about how this system will affect the life of users and what biases your system might have, the cheaper it will be to address these biases. A study by NASA shows that for software development, the cost of errors goes up by an order of magnitude at every stage of your project lifecycle.[29]

Create model cards

Model cards are short documents accompanying trained ML models that provide information on how these models were trained and evaluated. Model cards also disclose the context in which models are intended to be used, as well as their limitations.[30] According to the authors of the model card paper, "The goal of model cards is to standardize ethical practice and reporting by allowing stakeholders to compare candidate models for deployment across not only traditional evaluation metrics but also along the axes of ethical, inclusive, and fair considerations."

The following list has been adapted from content in the paper "Model Cards for Model Reporting" to show the information you might want to report for your models:[31]

- *Model details*: Basic information about the model.
  - — Person or organization developing model
  - — Model date
  - — Model version
  - — Model type

29 Jonette M. Stecklein, Jim Dabney, Brandon Dick, Bill Haskins, Randy Lovell, and Gregory Moroney, "Error Cost Escalation Through the Project Life Cycle," NASA Technical Reports Server (NTRS), *https://oreil.ly/edzaB*.

30 Margaret Mitchell, Simone Wu, Andrew Zaldivar, Parker Barnes, Lucy Vasserman, Ben Hutchinson, Elena Spitzer, Inioluwa Deborah Raji, and Timnit Gebru, "Model Cards for Model Reporting," *arXiv*, October 5, 2018, *https://oreil.ly/COpah*.

31 Mitchell et al., "Model Cards for Model Reporting."

— Information about training algorithms, parameters, fairness constraints or other applied approaches, and features

— Paper or other resource for more information

— Citation details

— License

— Where to send questions or comments about the model

- *Intended use*: Use cases that were envisioned during development.

  — Primary intended uses

  — Primary intended users

  — Out-of-scope use cases

- *Factors*: Factors could include demographic or phenotypic groups, environmental conditions, technical attributes, or others.

  — Relevant factors

  — Evaluation factors

- *Metrics*: Metrics should be chosen to reflect potential real-world impacts of the model.

  — Model performance measures

  — Decision thresholds

  — Variation approaches

- *Evaluation data*: Details on the dataset(s) used for the quantitative analyses in the card.

  — Datasets

  — Motivation

  — Preprocessing

- *Training data*: May not be possible to provide in practice. When possible, this section should mirror Evaluation Data. If such detail is not possible, minimal allowable information should be provided here, such as details of the distribution over various factors in the training datasets.

- *Quantitative analyses*

  — Unitary results

  — Intersectional results

- *Ethical considerations*

- *Caveats and recommendations*

Model cards are a step toward increasing transparency into the development of ML models. They are especially important in cases where people who use a model aren't the same people who developed this model.

Note that model cards will need to be updated whenever a model is updated. For models that update frequently, this can create quite an overhead for data scientists if model cards are created manually. Therefore, it's important to have tools to automatically generate model cards, either by leveraging the model card generation feature of tools like TensorFlow (*https://oreil.ly/iQtrS*), Metaflow (*https://oreil.ly/nucaZ*), and scikit-learn (*https://oreil.ly/Yk16x*) or by building this feature in-house. Because the information that should be tracked in a model's card overlaps with the information that should be tracked by a model store, I wouldn't be surprised if in the near future, model stores evolve to automatically generate model cards.

Establish processes for mitigating biases

Building responsible AI is a complex process, and the more ad hoc the process is, the more room there is for errors. It's important for businesses to establish systematic processes for making their ML systems responsible.

You might want to create a portfolio of internal tools easily accessible by different stakeholders. Big corporations have tool sets that you can reference. For example, Google has published recommended best practices for responsible AI (*https://oreil.ly/0C30s*) and IBM has open-sourced AI Fairness 360 (*https://aif360.mybluemix.net*), which contains a set of metrics, explanations, and algorithms to mitigate bias in datasets and models. You might also consider using third-party audits.

Stay up-to-date on responsible AI

AI is a fast-moving field. New sources of biases in AI are constantly being discovered, and new challenges for responsible AI constantly emerge. Novel techniques to combat these biases and challenges are actively being developed. It's important to stay up-to-date with the latest research in responsible AI. You might want to follow the ACM FAccT Conference (*https://oreil.ly/dkEeG*), the Partnership on AI (*https://partnershiponai.org*), the Alan Turing Institute's Fairness, Transparency, Privacy group (*https://oreil.ly/5aiQh*), and the AI Now Institute (*https://ainowinstitute.org*).

Summary

Despite the technical nature of ML solutions, designing ML systems can't be confined in the technical domain. They are developed by humans, used by humans, and leave their marks in society. In this chapter, we deviated from the technical theme of the last eight chapters to focus on the human side of ML.

We first focused on how the probabilistic, mostly correct, and high-latency nature of ML systems can affect user experience in various ways. The probabilistic nature can lead to inconsistency in user experience, which can cause frustration—"Hey, I just saw this option right here, and now I can't find it anywhere." The mostly correct nature of an ML system might render it useless if users can't easily fix these predictions to be correct. To counter this, you might want to show users multiple "most correct" predictions for the same input, in the hope that at least one of them will be correct.

Building an ML system often requires multiple skill sets, and an organization might wonder how to distribute these required skill sets: to involve different teams with different skill sets or to expect the same team (e.g., data scientists) to have all the skills. We explored the pros and cons of both approaches. The main cons of the first approach is overhead in communication. The main cons of the second approach is that it's difficult to hire data scientists who can own the process of developing an ML system end-to-end. Even if they can, they might not be happy doing it. However, the second approach might be possible if these end-to-end data scientists are provided with sufficient tools and infrastructure, which was the focus of Chapter 10.

We ended the chapter with what I believe to be the most important topic of this book: responsible AI. Responsible AI is no longer just an abstraction, but an essential practice in today's ML industry that merits urgent actions. Incorporating ethics principles into your modeling and organizational practices will not only help you distinguish yourself as a professional and cutting-edge data scientist and ML engineer but also help your organization gain trust from your customers and users. It will also help your organization obtain a competitive edge in the market as more and more customers and users emphasize their need for responsible AI products and services.

It is important to not treat this responsible AI as merely a checkbox ticking activity that we undertake to meet compliance requirements for our organization. It's true that the framework proposed in this chapter will help you meet the compliance requirements for your organization, but it won't be a replacement for critical thinking on whether a product or service should be built in the first place.

Epilogue

Wow, you made it! You've just finished a pretty technical book of 100,000 words and over 100 illustrations written by a writer who speaks English as her second language. With the help of many colleagues and mentors, I worked really hard on this book, and I'm grateful that you chose to read it out of so many books out there. I hope that the takeaways you can get from this book will make your work a little bit easier.

With the best practices and tooling that we have now, there are already many incredible ML use cases influencing our everyday life. I have no doubt that the number of impactful use cases will grow over time as tooling matures, and you might be among the people who will make this happen. I'm looking forward to seeing what you build!

ML systems have a lot of challenges. Not all of them are fun, but all of them are opportunities for growth and impact. If you want to talk about these challenges and opportunities, don't hesitate to reach out. I can be found on Twitter at @chipro or via email at *chip@claypot.ai*.

Index

C

D

E

F

G

H

I

J

K

L

M

N

O

S

T

U

V

W

X

Z

About the Author

Chip Huyen (*https://huyenchip.com*) is co-founder and CEO of Claypot AI, developing infrastructure for real-time machine learning. Previously, she was at NVIDIA, Snorkel AI, and Netflix, where she helped some of the world's largest organizations develop and deploy machine learning systems.

When a student at Stanford, she created and taught the course TensorFlow for Deep Learning Research. She is currently teaching CS 329S: Machine Learning Systems Design at Stanford. This book is based on the course's lecture notes.

She is also the author of four bestselling Vietnamese books, including the series Xách ba lô lên và Đi (Quảng Văn 2012, 2013). The series was among FAHASA's Top 10 Readers Choice Books in 2014.

Chip's expertise is in the intersection of software engineering and machine learning. LinkedIn included her among the 10 Top Voices in Software Development in 2019, and Top Voices in Data Science & AI in 2020.

Colophon

The animal on the cover of *Designing Machine Learning Systems* is a red-legged partridge (*Alectoris rufa*), also known as a French partridge.

Bred for centuries as a gamebird, this economically important, largely nonmigratory member of the pheasant family is native to western continental Europe, though populations have been introduced elsewhere, including England, Ireland, and New Zealand.

Relatively small but stout bodied, the red-legged partridge boasts ornate coloration and feather patterning, with light brown to gray plumage along its back, a light pink belly, a cream-colored throat, a brilliant red bill, and rufous or black barring on its flanks.

Feeding primarily on seeds, leaves, grasses, and roots, but also on insects, red-legged partridges breed each year in dry lowland areas, such as farmland, laying their eggs in ground nests. Though they continue to be bred in large numbers, these birds are now considered near threatened due to steep population declines attributed, in part, to overhunting and disappearance of habitat. Like all animals on O'Reilly covers, they're vitally important to our world.

The cover illustration is by Karen Montgomery, based on an antique line engraving from *The Riverside Natural History*. The cover fonts are Gilroy Semibold and Guardian Sans. The text font is Adobe Minion Pro; the heading font is Adobe Myriad Condensed; and the code font is Dalton Maag's Ubuntu Mono.

O’Reilly Media, Inc.介绍

O’Reilly以“分享创新知识、改变世界”为己任。40多年来我们一直向企业、个人提供成功必需之技能及思想，激励他们创新并做得更好。

O’Reilly业务的核心是独特的专家及创新者网络，他们通过我们分享知识。我们的在线学习（Online Learning）平台提供独家的直播培训、图书及视频，使客户更容易获取业务成功所需的专业知识。几十年来O’Reilly图书一直被视为学习开创未来之技术的权威资料。我们全年举办的诸多会议是活跃的技术聚会场所，来自各领域的专业人士在此建立联系，讨论最佳实践并发现可能影响技术行业未来的新趋势。

我们的客户渴望作出推动世界前进的创新，我们能祝您一臂之力。

业界评论

“O’Reilly Radar博客有口皆碑。”

——Wired

“O’Reilly凭借一系列（真希望当初我也想到了）非凡想法建立了数百万美元的业务。”

——Business 2.0

“O’Reilly Conference是聚集关键思想领袖的绝对典范。”

——CRN

“一本O’Reilly的书就代表一个有用、有前途、需要学习的主题。”

——Irish Times

“Tim是位特立独行的商人，他不光放眼于最长远、最广阔的视野并且切实地按照Yogi Berra的建议去做了：‘如果你在路上遇到岔路口，走小路（岔路）。’回顾过去Tim似乎每一次都选择了小路，而且有几次都是一闪即逝的机会，尽管大路也不错。”

——Linux Journal